GPI Membrane Anchors

edited by

M. Lucia Cardoso de Almeida
Escola Paulista de Medicina
São Paulo, Brazil

To Wayne Masterson, a superb biochemist and sweet friend who knew how to mingle an earthly love for these anchors with a firm belief in the secrets of heaven.

ACADEMIC PRESS
Harcourt Brace Jovanovich, Publishers
London San Diego New York Boston
Sydney Tokyo Toronto

ACADEMIC PRESS LIMITED
24/28 Oval Road
London NW1 7DX

United States Edition published by
ACADEMIC PRESS INC.
San Diego, CA 92101

A catalogue record for this book is available from the British Library

ISBN 0-12-159390-8

GPI Membrane Anchors

This publication has been reprinted from a special issue of the *Cell Biology International Reports*, Volume **15**, Number 9, September 1991 and from Volume **15**, Number 11, November 1991

Printed in Great Britain by The University Press, Cambridge

Contributors

N.W. Andrews *Department of Cell Biology and Internal Medicine, Yale University School of Medicine, New Haven, Connecticut 06510, USA*

Michael A. Baldwin *Department of Neurology, University of California, San Francisco, California 94143, USA*

C. Beghdadi-Rais *Institute of Biochemistry, University of Lausanne, CH-1066 Epalines, Switzerland*

M. Bernard *Institute of Biochemistry, University of Lausanne, CH-1066 Epalines, Switzerland*

C. Bron *Institute of Biochemistry, University of Lausanne, CH-1066 Epalines, Switzerland*

Ingrid W. Caras *Genetech Inc., San Francisco, California, USA*

M. Lucia Cardoso de Almeida *Escola Paulista de Medicina, São Paulo, Brazil*

Mark Carrington *Department of Biochemistry, Tennis Court Road, Cambridge, CB2 1QW, UK*

Andreas Conzelmann *Institut de Biochimie, Perolles, 1700 Fribourg, Switzerland*

Aline M. da Silva *Departamento de Bioquímica, Instituto de Química, Universidade de São Paulo, Caixa Postal 20780 São Paulo, 01498 Brazil*

Michael A. Davitz *Division of Immunology, Department of Pathology, New York University School of Medicine, New York, USA*

N. Deglon *Institute of Biochemistry, University of Lausanne, CH-1066 Epalines, Switzerland*

Chantal Desponds *Institute of Biochemistry, University of Lausanne, CH-1066 Epalinges, Switzerland*

Margaret dos Santos Medeiros *Membrane Peptidase Research Group, Department of Biochemistry and Molecular Biology, University of Leeds, Leeds, LS2 9JT, UK*

P. Doherty *Department of Experimental Pathology, UMDS, Guy's Hospital, London Bridge, London, SE1 9RT, UK*

N. Fasel *Institute of Biochemistry, University of Lausanne, CH-1066 Epalines, Switzerland*

Michael A.J. Ferguson *Department of Biochemistry, University of Dundee, Dundee, DD1 4HN, Scotland*

Christoph Frankhauser *Institute of Biochemistry, University of Lausanne, CH-1066 Epalinges, Switzerland*

Steve W. Homans *Department of Biochemistry, University of Dundee, Dundee, DD1 4HN, Scotland*

N.M. Hooper *Membrane Peptidase Research Group, Department of Biochemistry and Molecular Biology, University of Leeds, Leeds, LS2 9JT, UK*

H. Ikezawa *Faculty of Pharmaceutical Sciences, Nagoya City University, 3–1 Tanabe-dori, Mizuho-ku, Nagoya, Aichi 467, Japan*

K. Joiner *Department of Cell Biology and Internal Medicine, Yale University School of Medicine, New Haven, Connecticut 06510, USA*

Claudette Klein *Departamento de Bioquímica, Instituto de Química, Universidade de São Paulo, Caixa Postal 20780 São Paulo, 01498 Brazil*

Krishna Kodukula *Roche Institute of Molecular Biology, Roche Research Center, Nutley, NJ 07110, USA*

M.P. Lisanti *Department of Cell Biology and Anatomy, Cornell University Medical College, New York, NY 10021, USA*

Wayne J. Masterson *Department of Biochemistry, University of Dundee, Dundee, DD1 4HN, Scotland*

Malcolm J. McConville *Department of Biochemistry, University of Dundee, Dundee, DD1 4HN, Scotland*

Anant K. Menon *Laboratory of Molecular Parasitology, The Rockefeller University, 1230 York Avenue, New York, NY 10021, USA*

Christine N. Metz *Division of Immunology, Department of Pathology, New York University School of Medicine, New York, USA*

Radmila Micanovic *Roche Institute of Molecular Biology, Roche Research Center, Nutley, NJ 07110, USA*

Stanley B. Prusiner *Departments of Neurology and Biochemistry and Biophysics, University of California, San Francisco, California 94143, USA*

Alessandro Puoti *Institute of Biochemistry, University of Lausanne, CH-1066 Epalinges, Switzerland*

Peter J. Robinson *Transplantation Biology Section, Clinical Research Centre, Watford Road, Harrow, UK*

E. Rodríguez-Boulan *Department of Cell Biology and Anatomy, Cornell University Medical College, New York, NY 10021, USA*

Guillermo Romero *Department of Pharmacology, University of Virginia Health Sciences Center, Charlottesville, VA 22908, USA*

Terrone L. Rosenberry *Department of Pharmacology, Case Western Reserve University, Cleveland, Ohio 44120, USA*

M. Rousseaux *Institute of Biochemistry, University of Lausanne, CH-1066 Epalines, Switzerland*

Sergio Schenkman *Escola Paulista de Medecina, São Paulo, Brazil*

Neil Stahl *Department of Neurology, University of California, San Francisco, California 94143, USA*

A.J. Turner *Membrane Peptidase Research Group, Department of Biochemistry and Molecular Biology, University of Leeds, Leeds, LS2 9JT, UK*

Sidney Udenfriend *Roche Institute of Molecular Biology, Roche Research Center, Nutley, NJ 07110, USA*

F.S. Walsh *Department of Experimental Pathology, UMDS, Guy's Hospital, London Bridge, London, SE1 9RT, UK*

Dawn Walters *Department of Biochemistry, Tennis Court Road, Cambridge, CB2 1QW, UK*

Helena Webb *Department of Biochemistry, Tennis Court Road, Cambridge, CB2 1QW, UK*

P. Webster *Department of Cell Biology and Internal Medicine, Yale University School of Medicine, New Haven, Connecticut 06510, USA*

Alan F. Williams *MRC Cellular Immunology Research Unit, Sir William Dunn School of Pathology, University of Oxford, Oxford, OX1 3RE, UK*

Preface

The articles gathered in this Special Issue on "Glycosylphosphatidyl-inositol (GPI) membrane anchors" might, pictorially, be represented as a young oak tree growing by a lake. Its roots dwell in the fertile ground of relevance to Cell Biology. Its solid stem feeds several branches initiated in its first decade or so of life. Its rapid growth has not only marvelled cell biologists by producing promising new "sprouts", but also disconcerted many by the amount of information arising from a variety of systems and leading in multiple directions.

During this decade we have witnessed the dissection of GPI structure and recognized its originality as an anchoring device, appreciated its ubiquitous distribution, learned about the logical precision of the main steps of its biosynthetic pathway, and understood the elegant means by which a protein signals for GPI attachment. There are "sproutings" of evidence for the uniqueness of GPI as a tag in intracellular traffic and also as a mediator of transmembrane signalling. In the years to come many of the present mysteries will probably be unveiled and hopefully the ability of GPI to interact with other membrane components, its suitability for phospholipase processing, and the structural heterogeneity and concurrent conformational features of GPI may turn out to be associated with the performance of some of its biological functions.

If the contributions gathered together here can help people to find their way through this emerging field, or to benefit from simple interpretations formulated from an intimidating collection of complex data, or else stimulate a promenade down speculative pathways seeking new directions, then it has been worth allowing trees to be felled to make the paper used in the capture of this snapshot of a young oak growing by a lake.

M Lucia Cardoso de Almeida

Contents

STRUCTURAL REQUIREMENTS OF A NASCENT PROTEIN FOR PROCESSING TO A PI-G ANCHORED FORM: STUDIES IN INTACT CELLS AND CELL-FREE SYSTEMS

***Sidney Udenfriend, Radmila Micanovic, Krishna Kodukula**

Roche Institute of Molecular Biology, Roche Research Center, Nutley, NJ 07110, USA.

* Corresponding author.

Over 70 distinct proteins have now been identified as being bound to the plasma membrane by a phosphatidylinositol-glycan (PI-G) anchor. Based on the observed structures of several isolated PI-G tailed proteins and the sequence deduced from the corresponding cDNAs (Ferguson and Williams, 1988; Cross, 1990) it is apparent that the PI-G moiety is added to an internal residue of the nascent protein with the concomitant removal of a COOH-terminal signal peptide of c.a. 17-30, mainly hydrophobic, residues. COOH-terminal processing presumably occurs on the lumenal side of the rough endoplasmic reticulum and the mature PI-G tailed protein is then transported to the cell surface (Fig. 1).

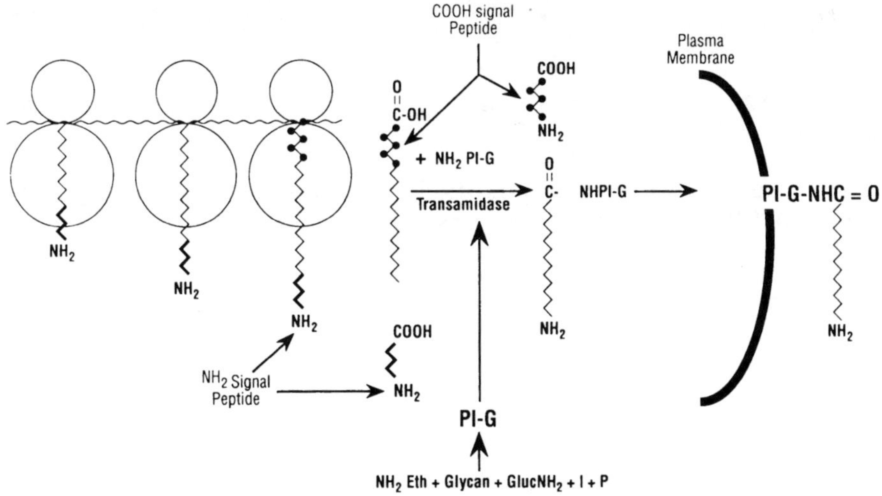

Figure 1.
Biosynthesis of phosphatidylinositol-glycan (PI-G) anchored membrane proteins. NH_2Eth = ethanolamine; GlucNH_2 = glucosamine; I = inositol; P = phosphatide.

Recent studies by Ferguson et al (1988) and Masterson et al (1989) have elucidated the structure and intermediate steps involved in the formation of the PI-G moiety. The latter has also been produced in cell culture and has been partially purified (Gaulton et al, 1988). From Figure 1 it is apparent that there exist two independent pathways, one to generate the PI-G moiety, the other to appropriately modify the nascent protein and transport it to the site where both can be linked with the concomitant cleavage of the COOH-terminal signal peptide.

In our laboratory we have concerned ourselves with the protein component of PI-G tailing process and initially used placental alkaline phosphatase (PLAP) as a model PI-G acceptor. We selected human PLAP as a model for PI-G tailing studies for a number of reasons. PLAP had been widely studied and much was known of its properties. It is also the most stable of the animal alkaline phosphatases (AP), withstanding temperatures of 65°C for 5-10 min. Most importantly the placental isozyme is found only in humans and higher primates. It is therefore possible to transfect other mammalian cells with PLAP cDNA and detect its activity without interference by the endogenous AP of the cell used for transfection. This specificity can be further enhanced by heating the preparation prior to assay to 65°C and also by adding an inhibitor of non-specific AP such as L-homoarginine, which does not inhibit PLAP. Finally, PLAP, like other APs, has a high turnover rate and can hydrolyze phosphate ester bonds in any type of compound. This makes it possible to devise simple and highly sensitive assays for the enzyme in cell and tissue extracts.

PREPARATION OF ANTIBODIES DIRECTED TO SPECIFIC SITES IN PLAP AND PREPROPLAP

Site directed and polyclonal antibodies were important tools in these studies. Based on the cDNA deduced sequence of nascent PLAP (Millan, 1986) three peptides were synthesized and used to produce site-directed antibodies (Garattini et al, 1985; Bailey et al, 1988). One of the peptides comprising the nine COOH-terminal residues of nascent PLAP (residues 505 to 513) yielded antibodies directed to the COOH terminus of the nascent protein (exo-Ab). A second peptide, corresponding to residues 474 to 484, yielded antibodies that reacted with both nascent and mature PLAP in a region expected to be close to the PI-G tail (endo-Ab). In fact, Asp^{484} was later shown to be critical for interaction with this antibody (unpublished data). A third peptide (corresponding to residues 1 to 9 of the mature enzyme) yielded antibodies directed to the amino terminus of mature PLAP (amino-Ab). Amino-Ab does not react with the nascent protein and requires removal of the NH_2 terminal signal peptide to become reactive. A schematic presentation of the location of the three antibody sites in nascent PLAP is shown in Fig. 2.

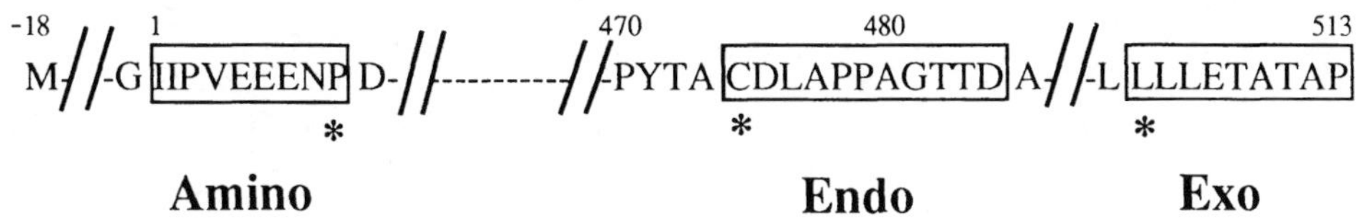

Figure 2.
Amino acid sequences of the three anti-peptide antibodies. The numbers above the sequence indicate the location of each residue in preproPLAP. Sites of conjugation of the peptides to carrier protein prior to immunization are indicated by an asterisk.

DETERMINATION OF THE SITE OF ATTACHMENT OF THE PI-G MOIETY IN PLAP

Endo-Ab to PLAP (Fig. 2) was used in an RIA for monitoring isolation of the carboxyl terminal PI-G tailed peptide from a chymotryptic digest of the mature protein. Commercial PLAP was partially purified then reduced and carboxy-methylated. The modified protein was treated with chymotrypsin and the resulting peptides were resolved by several stages of HPLC. RIA with endo-Ab was used to monitor each stage of the purification. The chymotryptic peptide, which reacted with the endo-Ab, when purified to homogeneity and sequenced, was found to terminate in an Asp representing Asp^{484} of nascent PLAP. The presence of stoichiometric amounts of ethanolamine, glucosamine and hexose in the isolated product identified it as the carboxyl-terminal PI-G tailed peptide (Micanovic et al, 1988). Subsequently Ogata et al (1988) isolated a shorter PI-G tailed peptide from a papain hydrolysate of PLAP and verified the presence of Asp^{484} at the carboxyl terminus of mature PLAP. Thus a 29 residue peptide is cleaved from the carboxyl terminus nascent PLAP-513 during processing to the mature PI-G tailed form.

Although over 70 PI-G tailed proteins have been identified, the site of attachment of the PI-G moiety has been ascertained on only a few. These are shown in Table I to permit comparison with PLAP. It is readily apparent that all the COOH-terminal signal peptides (cleavable residue) contain a core hydrophobic sequence of about 15 to 20 residues. However, their primary sequences show no obvious homology and vary considerably in size. An interesting point is that the PI-G moiety is attached to one of the small amino acids in each of the proteins. This was found to be highly significant as discussed below under Mutations at the Site of PI-G Attachment.

Table 1. COOH-termini of PI-G tailed proteins that have been fully characterized. The amino acid to which the PI-G moiety is attached is shown in bold type and the adjacent small amino acid domains are underlined.

Protein	Sequence
Acetylcholinesterase (Torpedo)[a]	F L P K L L N A T A **C** D G E L S S S G T S S S K G I I F Y V L F S I L Y L I F Y
Alkaline Phosphatase (placenta)[b]	C D L A P P A G T T **D** A A H P G R S V V P A L L P L L A G T L L L L E T A T P
Carcinoembryonic Antigen[c]	N S I V K S I T V S **A** S G T S P G L S A G A T V G I M I G V L V G V
Decay Accelerating Factor[d]	T T P N K G S G T T **S** G T T R L L S G H T C F T L T G L L G T L V T M G L L T
Dictyostelium PsA[e]	P T P S Q T S T T T **G** S A S T V V A S L S L I F S M I L S L C
Prion Protein (hamster)[f]	E S Q A Y Y D G R R **S** S A V L F S S P P V I L L I S F L I F L M V G
MRC OX-45 antigen[g]	Y F I P P C T L A R **S** S G V H W I A A W L V V T L S I I P S I L L A
5' Nucleotidase[h]	Y P A V E G R I K F **S** A A S H Y Q G S F P L I I L S F W A V I L V L Y Q
RT 6.2 (rat)[i]	R K K S N Y N C L Y **S** S A G A R E S C V S L F L V Q L L C L A E P
Thy 1(rat)[j]	I N V I R D K L V K **C** G G I S L L V Q N T S W L L L L L L S L S F L
T.brucei VSG MIT 117a[k]	N C K W E N N A C K **D** S S I L V T K K F A L T V V S A A F V A L L F
T.brucei VSG MIT 118a[l]	T D E P D K E K C R **N** G S F L T S K Q F A F S V V S A A F V A L L F
T.brucei VSG MIT 221[k]	K T G N T N T T G S **S** N S F V I S K T P L W L A V L L F
T.brucei PARP[m]	E P E P E P E P E P **G** A A T L K S V A L P F A I A A A A L V A A F
T.Congolense VSG YNat1.1[n]	E K S G E S S H L P **S** G S S H G T K A I R S I L H V A L L M

References: a) Gibney et al, 1988; Sikorov et al, 1988; b) Micanovic et al, 1988; Ogata et al, 1988; c) Hefta et al, 1988 d) Moran et al, 1991; e) Early et al, 1988; f) Stahl et al, 1990; g) Killeen et al, 1988; h) Misumi et al, 1990; Ogata et al, 1990; i) Koch et al, 1990; j) Seki et al 1985; k) Boothroyd et al, 1981; Holder and Cross, 1981; l) Holder, 1985; Majumdar et al, 1981; m) Mowatt and Clayton, 1989; n) Strikler et al, 1987.

COOH-TERMINAL REQUIREMENTS FOR THE CORRECT PROCESSING OF PI-G ANCHORED PROTEINS

The only similarity among the characterized PI-G-tailed nascent proteins is that they are mainly hydrophilic with the exception of their amino and carboxyl termini. The latter end in a hydrophobic sequence varying from about 15 to 20 residues in length. To determine structural requirements for PI-G tailing we produced mutant forms of PLAP cDNA and expressed them in COS cells (Berger et al, 1987). Wild-type PLAP cDNA was cloned into the eukaryotic expression vector. pBC12BI (Cullen, 1987) and transient transfections were carried out using the DEAE-dextran and chloroquine method. Several types of deletion and substitution mutants were produced (Fig. 3) (Berger et al,

```
                480         490         500         510         520         530         540
        513 -- A G T T D A A H P G R S V V P A L L P L L A G T L L L L E T A T A P
        507 -- A G T T D A A H P G R S V V P A L L P L L A G T L L L L
        502 -- A G T T D A A H P G R S V V P A L L P L L A G
        497 -- A G T T D A A H P G R S V V P A L L
        493 -- A G T T D A A H P G R S V V
        489 -- A G T T D A A H P G
502+(SI4)   -- A G T T D A A H P G R S V V P A L L P L L A G S I I I I
502+(SQ4)   -- A G T T D A A H P G R S V V P A L L P L L A G S Q Q Q Q
        518 -- A G T T D A A H P G R S V V P A L L P L L A G Y C C C C E L L L L E T A T A P
       489G -- A G T T D A A H P G S K L S I A S F F F I I G L I I G F L V L R V G I H L C I K L K H T K K R Q I Y T D I E M N R L G K
```

Figure 3.
The carboxyl termini of deletion and substitution mutants of PLAP-513. Asp[484], the normal site of PI-G attachment is highlighted. PLAP-489G represents a chimer produced by addition of a portion of vesicular stomatis virus through its membrane spanning domain (underlined) to PLAP-489.

1988). To determine the minimal length of the COOH-terminal hydrophobic sequence necessary for successful PI-G tailing we constructed a number of deletion mutants. Deletion of the six COOH-terminal residues from PLAP-513 produced PLAP-507. The latter was expressed on the cell surface in an enzymatically active form and was also PI-G tailed. This was ascertained by demonstrating release by phospholipase C, incorporation of the radioactive precursors of PI-G and cell surface detection by immunohistochemistry. Although further deletions of the COOH-terminus (PLAP 502 to PLAP 489) yielded enzymatically active forms of AP, none of these truncated mutants were attached to the cell membrane but were, instead, secreted into the medium in forms that did not contain the elements of PI-G. Interestingly, the secreted forms were produced in much larger amounts than were the PI-G tailed forms. To investigate the affects of other variables on PI-G tailing we utilized PLAP-502 (secreted) and extended it by cassette mutagenesis to yield PLAP-507 mutants. The latter in its native form is PI-G tailed. With Ser-Ile-Ile-Ile-Ile instead of Thr-Leu-Leu-Leu-Leu at the COOH-terminus of PLAP-507, there was no difference in expression, enzyme activity, PI-G tailing or membrane insertion. However, replacing Thr-Leu-Leu-Leu-Leu with the hydrophilic sequence Ser-Gln-Gln-Gln-Gln yielded enzymatically active PLAP which, like PLAP-502, was secreted and not PI-G tailed. Extension of PLAP-502 beyond 513 (PLAP-518) with mainly hydrophobic residues yielded a PI-G tailed membrane-bound protein. Thus, there is apparently a minimal size of hydrophobic sequence at the COOH-terminus required for PI-G tailing but the exact length can vary considerably. These findings are in accord with the variations in the COOH-termini of the few characterized nascent proteins destined to be PI-G tailed. When a hybrid protein was produced composed of PLAP-489 (which is normally secreted) coupled to the COOH terminus of vesicular stomatitis virus glycoprotein (Fig. 3, PLAP-489G) the mutant was not released by PI-PL-C but was anchored to the plasmalemma by the membrane-spanning peptide sequence of the viral protein (Berger et al, 1989). Although the COOH-terminus of the viral protein provided a hydrophobic sequence as long as and in register through residue 489 of PLAP-513 the cytoplasmic domain of the viral protein signalled the cell to process PLAP-489G to a form bound to the membrane by a peptide-spanning sequence. Hybrids of the opposite type have been reported where the COOH terminus of a PI-G tailed protein is attached to a normally secreted protein (Stroynowski et al, 1987; Caras et al, 1987). The resulting hybrids are in such cases PI-G tailed and membrane-bound. Apparently for proteins containing no other internal membrane-spanning sequences the COOH-terminal sequence provides the signal as to whether the protein is to be secreted or anchored by either a PI-G tail or by a membrane-spanning peptide sequence.

MUTATIONS AT ASP^{484}, THE SITE OF PI-G ATTACHMENT

The COOH-terminal residue to which the PI-G moiety is attached has been established in about twenty proteins (Table I). Thus far only Ser, Asp, Asn, Gly, Cys and Ala have been found to serve this purpose. To determine whether other amino acids can serve as sites for PI-G tailing essentially saturation mutagenesis was carried out at position 484 of PLAP-513, the normal site of PI-G tailing (Micanovic et al, 1990a). Mutants were prepared by oligonucleotide-directed, site-specific mutagenesis. The amino acids substituted for Asp^{484} were Thr, Pro, Met, Val, Leu, Trp, Tyr, Lys, Glu and Gln (category II) and Ser, Ala, Asn, Gly and Cys (category I). It was found that all mutants were expressed to the same extent, i.e. they yielded approximately the same amount of immunoprecipitable PLAP protein for the same amount of cDNA used in transfection.All were glycosylated to about the same extent, indicating that they were all processed in the ER. However, the measured AP enzyme activities of the different 484 mutants varied considerably. Among the category I amino acids Ser and Asn were most active, even more active than wild-type Asp^{484}. Gly and Ala were somewhat less active and Cys was least active. With respect to category II amino acids the Pro mutant yielded no detectable enzyme activity. Most other category II mutants yielded very low but detectable activity (about 1-2 percent of the Ser mutant) except for the Val and Leu mutants, which yielded as much as 5 percent of the activity of the Ser mutant.

Enzymatically active PLAP produced by the mutants, was releasable from the cells by both PI-PLC and bromelain, indicating that active enzyme (even that produced by category II mutants) was on the cell surface and PI-G tailed. Since all site-directed mutants had been expressed to the same extent, such variations in enzyme activity and in releasability by PI-PLC- and bromelain can only be explained by differences in the extent of PI-G tailing. Thus, it can be concluded that all mutants, except Pro, were PI-G tailed to some extent and that PI-G tailing is a requisite for externalization, membrane attachment, protein maturation and enzyme activity. It also suggests that, at least in transfected cells, all mutant proteins, including wild type ASp^{484}, do not yield fully PI-G tailed PLAP. It is likely that incomplete PI-G tailing is an artifact of transfection in COS cells, particularly because they are overproducing PLAP.

The finding that only the six amino acids of category I can serve as efficient cleavage/PI-G acceptor sites indicates a high degree of selectivity. Similar findings were recently reported with cleavage/attachment site directed mutants of decay accelerating factor (Moran et al, 1991). There is, thus, some similarity in specificities between the NH_2-terminal signal peptidase and COOH-terminal processing enzymes(s). Both require a hydrophobic chain of comparable size (c.a.15-20 residues) in one case preceding and in the other following the cleavage site. In each case small amino acids with a monosubstituted ß carbon atom are preferred at the COOH side of the site of cleavage as well as at the adjacent sites. Apparently NH_2-terminal cleavage sites are limited to glycine, alanine, serine and occasionally cysteine (Wickner and Lodish, 1985; Pollitt et al, 1986). These are also preferred for COOH-terminal processing. Aspartic acid and asparagine have not been reported to serve as NH_2-terminal cleavage sites. In making such comparisons it is important to remember that NH_2-terminal signal processing is catalyzed by a peptidase, whereas COOH-terminal processing presumably requires a transamidase (Ferguson and Williams, 1988). The latter will be discussed in the section on cell-free processing.

CELL-FREE COOH-TERMINAL PROCESSING OF NASCENT PROTEIN

Details concerning mechanisms of processing cannot be elucidated from studies with intact cells. To address this issue we developed a cell-free system that can process preproPLAP to mature PLAP (Bailey et al, 1989; Micanovic et al, 1990b). PreproPLAP is generated from PLAP mRNA in an in vitro translation system containing ^{35}S-Methionine; rough microsomal membranes (RM) from CHO cells are used as the source of processing enzyme(s). Evidence for correct processing of nascent PLAP was based largely on the specificities of site-directed antibodies (Fig. 2). Utilizing immunoassays and SDS-PAGE it was possible to demonstrate the cleavage of nascent PLAP-513 at Asp^{484} with the loss of the 29-residue COOH terminus. Three PLAP-related products appeared after incubation with microsomes (Fig. 4).

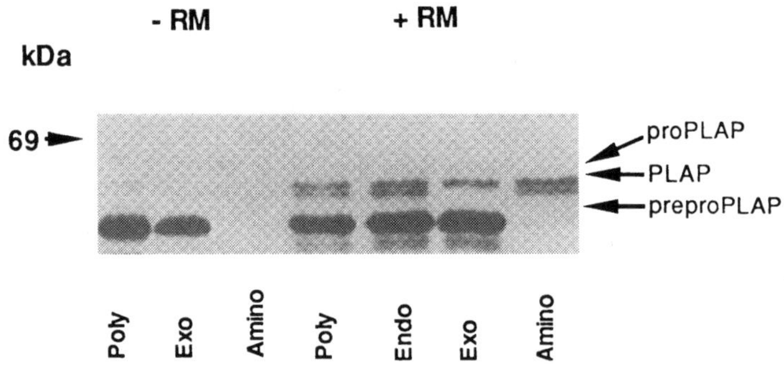

Figure 4.
Processing of wild type preproPLAP-513 (produced in rabbit reticulocyte lysate) by CHO cell RM. After incubation in the presence of protease inhibitors both with and without RM, aliquots were taken for immunoprecipitation with the specific antibodies (poly, antibody to mature PLAP; exo, exo-Ab; endo, endo-Ab; amino, antibody to NH_2 terminus of mature PLAP). Immunoprecipitates were subjected to $NaDoSO_4$/PAGE and fluorography.

The smallest represents unprocessed preproPLAP, which, from its immunological reactivity, contains both the NH_2- and COOH-terminal signal peptides. The two slower migrating protein bands are glycosylated, processed forms of PLAP. The upper one is pro-PLAP, which has lost its NH_2-terminal signal peptide but retains the nascent COOH terminus. The lower band represents mature PLAP that has lost both NH_2- and COOH-terminal signal peptides. The latter still contains the endo-Ab epitope indicating that cleavage occurs after residue 484.

Comparisons were made between the abilities of category I and category II 484-site-directed mutants of preproPLAP to act as substrates in the cell-free system (Micanovic et al, 1990). As shown in Fig. 5, category I mutants containing serine and alanine at residue 484 (also glycine and cysteine; data not shown) yielded two processed products, pro-PLAP and mature PLAP. By contrast, the category II mutants, tryptophan, valine, proline, and leucine, yielded only one N-glycosylated product, pro-PLAP. Category II mutants did not yield the mature form of PLAP, as would be indicated by loss of the exo epitope and a corresponding decrease in size. Thus, in the cell-free system, as in transfected COS cells, only small amino acids with a monosubstituted ß carbon atom can serve as the site of specific cleavage of the COOH-terminal signal peptide.

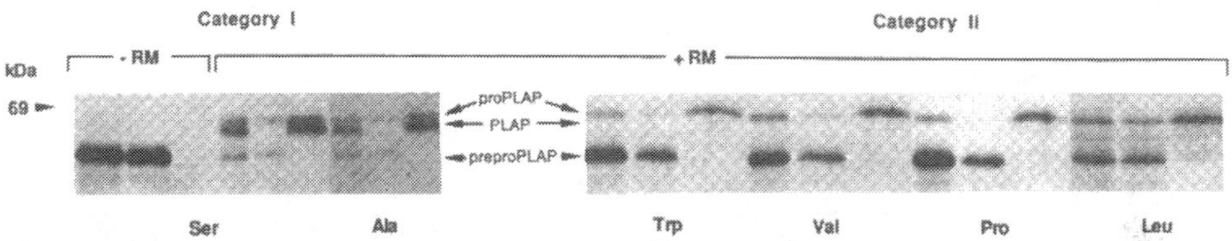

Figure 5.
Processing of residue 484 mutant forms of preproPLAP in the presence of CHO cell RM. The appropriate mRNA was added to the complete system and, after incubation, aliquots were taken from each mutant sample for precipitation with the corresponding antibodies. This was followed by $NaDoSO_4$/PAGE and fluorography. The first three bands were obtained using PLAP-513/Ser-484 mRNA in the absence of RM; the second three bands were in the presence of RM. Each subsequent set of three band represents (from left to right) immunoprecipitates with polyclonal, exo, and amino antibodies.

STUDIES WITH AN ENGINEERED PROTEIN SUBSTRATE

Although the observations on the cell-free processing of PLAP-513 and its site-directed mutants are interesting and significant, preproPLAP is not an ideal substrate for studying NH_2-terminal and COOH-terminal processing. A better substrate would be smaller so that products differing by 2 to 3 kDa, the molecular masses of the NH_2- and COOH-terminal signal peptides, could be readily resolved on SDS-PAGE. PLAP-513 mRNA yields a translation product of 59 kDa and contains N-glycosylation sites which, on processing, yield products larger than the nascent protein (Fig. 4). In fact, most PI-G anchored proteins are glycosylated so that changes between the nascent (prepro) proteins and the mature products are not due simply to the loss of signal peptides. These properties make it difficult to monitor processing of the prepro forms of most PI-G tailed proteins in cell-free systems. To overcome these problems we designed a smaller and simpler substrate for PI-G "transamidase" derived by deletion of approximately 60% of the internal sequence of preproPLAP-513. This engineered protein, prepromini PLAP-208, retains the NH_2-and COOH-terminal signal peptides of PLAP-513 as well as all the epitopes for site-directed antibodies of the latter, but is devoid of glycosylation sites, the active site, and most of the cysteine residues (Fig. 6) (21).

prepromini PLAP

			-17						-11										-1
			Met	Leu	Leu	Leu	Leu	Leu	Leu	Leu	Gly	Leu	Arg	Leu	Gln	Leu	Ser	Leu	Gly
1									10										20
Ile	Ile	Pro	Val	Glu	Glu	Glu	Asn	Pro	Asp	Phe	Trp	Asn	Arg	Glu	Ala	Ala	Glu	Ala	Leu
21									30										40
Gly	Ala	Ala	Lys	Lys	Leu	Gln	Pro	Ala	Gln	Thr	Ala	Ala	Lys	Asn	Leu	Ile	Ile	Phe	Leu
41									50										60
Gly	Asp	Gly	Met	Gly	Val	Ser	Thr	Val	Thr	Ala	Ala	Arg	Ile	Leu	Lys	Gly	Gln	Lys	Lys
61									70										80
Asp	Lys	Leu	Gly	Pro	Glu	Ile	Pro	Leu	Ala	Met	Ala	Met	Met	Met	Arg	Ser	Met	Met	Met
81									90										100
Lys	Ala	Tyr	Thr	Val	Leu	Leu	Tyr	Gly	Asn	Gly	Pro	Gly	Tyr	Val	Leu	Lys	Asp	Gly	Ala
101									110										120
Arg	Pro	Asp	Val	Thr	Glu	Ser	Glu	Ser	Gly	Ser	Pro	Glu	Tyr	Arg	Gln	Gln	Ser	Ala	Val
121									130										140
Pro	Leu	Asp	Glu	Glu	Thr	His	Ala	Gly	Glu	Asp	Val	Ala	Val	Phe	Ala	Arg	Gly	Pro	Gln
141									150										160
Ala	His	Leu	Val	His	Gly	Val	Gln	Glu	Gln	Thr	Phe	Ile	Ala	His	Val	Met	Ala	Phe	Ala
161									170										180
Ala	Cys	Leu	Glu	Pro	Tyr	Thr	Ala	Cys	Asp	Leu	Ala	Pro	Pro	Ala	Gly	Thr	Thr	Asp	Ala
181									190										200
Ala	His	Pro	Gly	Arg	Ser	Val	Val	Pro	Ala	Leu	Leu	Pro	Leu	Leu	Ala	Gly	Thr	Leu	Leu
201																			
Leu	Leu	Glu	Thr	Ala	Thr	Ala	Pro												

Figure 6.
The deduced amino acid sequence of prepromini PLAP. The NH_2- and COOH-terminal signal peptides are underlined; epitopes of the site-specific antibodies are boxed in and the Met-rich linker sequence is in a double box.

Prepromini PLAP is generated from the corresponding mRNA and is processed to mini PLAP by CHO cell RM. Assay of the putative PI-G transamidase activity in this system is specific and highly reproducible. With prepromini PLAP, as a substrate, one can demonstrate its stepwise conversion to the pro form and then to the mature form.

When miniPLAP mRNA was translated in the presence of RM, three products were generated that were readily separated on SDS-PAGE (Fig. 7) and further identified by their interactions with the specific site-directed antibodies (Fig. 8) (Kodukula et al, 1991).

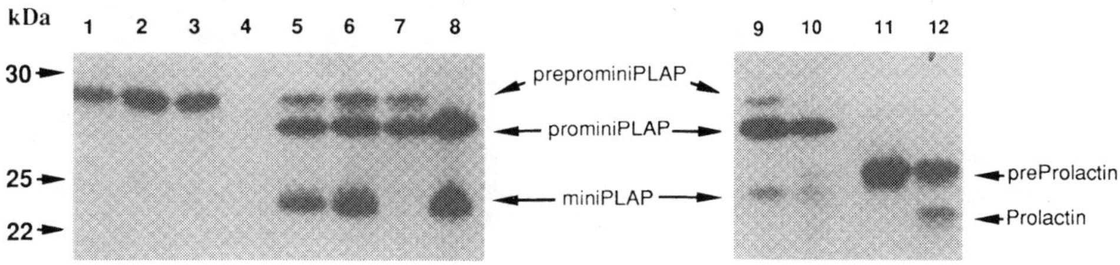

Figure 7.
Processing of preprominiPLAP both in vitro and in vivo. Following incubation, samples were immunoprecipitated with the antibodies and subjected to SDS-PAGE and fluorography. Lanes 1-4 are in vitro translation products in the absence of CHO RM and lanes 5-8 are processed products in the presence of CHO RM. Lanes 9 and 10 are lysates of COS cells transfected with preprominiPLAP cDNA. Lanes 11 and 12 are preprolactin translated in vitro in the absence and presence of CHO RM. Lanes 1,5 and 9 = polyclonal Ab; lanes 2 and 6 = endo-Ab; lanes 3 and 7 = exo-Ab; lanes 4,8 and 10 = amino-Ab.

When the corresponding mRNAs were translated in the presence of CHO RM, only those coding for category I amino acids (Ser, Gly, Cys, Ala) at residue 179 yielded mature miniPLAP. As with the Asp^{179} form, three bands were obtained during processing with each category I mutant after immunoprecipitation and SDS-PAGE. The three were identified by their size as well as by their reactivity with the specific antibodies. Each of the category I mutants is therefore a substrate of the COOH-terminal cleavage/attachment enzyme. The relative extents of processing of each mutant are shown in Table II. Here again, the Ser mutant was the most efficient substrate. When Trp, Val, Pro, and Leu were substituted at residue 179 of preprominiPLAP, translation of the corresponding mRNAs in the presence of RM followed by immunoprecipitation and SDS-PAGE yielded only two bands, identified as the prepro and the pro forms of miniPLAP by size and immunoreactivity. With miniPLAP as with PLAP-513 the putative transamidase exhibits great specificity for the amino acid at the COOH-terminal cleavage/attachment site.

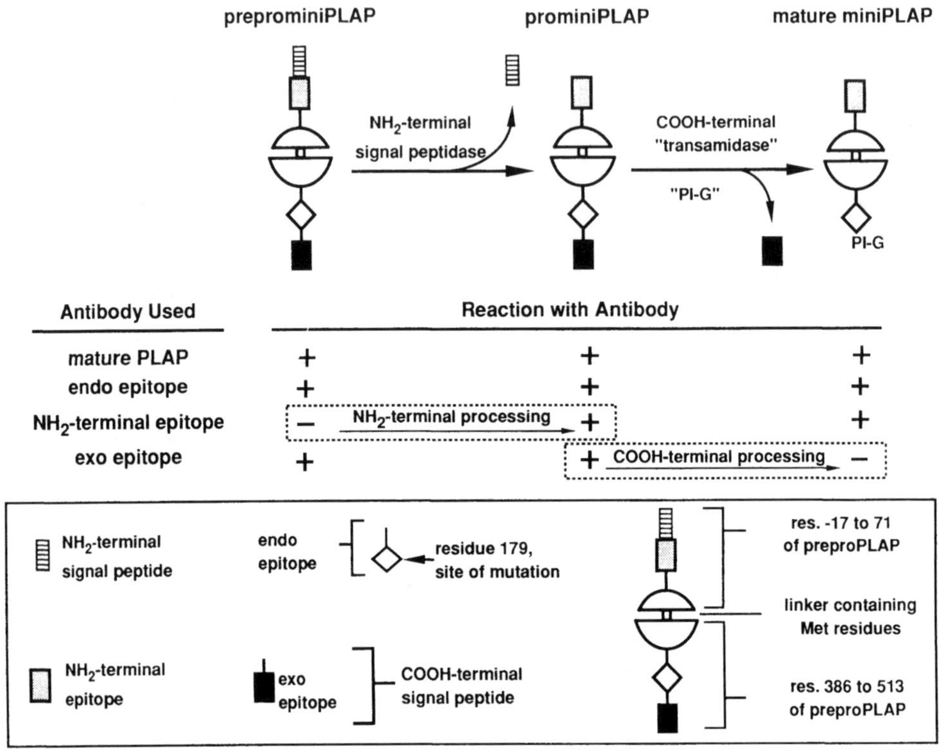

Figure 8.
Comprehensive scheme of processing of preprominiPALP. The structural relationship of preprominiPLAP to wild-type preproPLAP and the immunologic determinants that were used to elucidate the steps in processing to the pro and mature forms.

KINETICS OF PROCESSING OF PREPROMINIPLAP

The time course of production and processing of preprominiPLAP Ser^{179} in the cell free system is shown in Fig. 9 (Kodukula et al, 1991). It is apparent that processing by NH_2-terminal signal peptidase by the RM was so rapid that little preprominiPLAP accumulated. The amount of prominiPLAP increased for about 30 min before conversion to mature miniPLAP became apparent. Mature miniPLAP then increased rapidly at the expense of prominiPLAP. The changes in amounts of preprominiPLAP and the mature form are typical of a precursor-product relationship. Since little preprominiPLAP accumulated during these incubations, conversion from prominiPLAP to mature miniPLAP could be calculated directly from the corresponding densitometric measurements. After 90 min over 40% of the total translated material had been converted to miniPLAP. It has also been demonstrated that

COOH-terminal processing is a function of enzyme concentration by varying the amounts of RM used to process prepromini PLAP Ser^{179} (data not shown).

Relative Proportions of Products of Translation and Processing (% of total)

Forms of miniPLAP	category I					category II			
	Asp	Ser	Gly	Cys	Ala	Trp	Val	Pro	Leu
prepro	16.5	12.9	12.3	8.6	8.2	19.3	31.1	17.7	19.9
pro	52.3	45.2	65.9	74.2	68.4	80.7	68.9	82.3	80.1
mature	31.3	42.0	21.7	17.2	23.4	0.0	0.0	0.0	0.0

Table 2.
Relative proportions of product of translation and processing of miniPLAP mRNA. Incubations were carried out as described under methods and samples were immunoprecipitated with polyclonal-Ab and subjected to SDS-PAGE and fluorography. Values represent the average of absorbance obtained in three experiments and are expressed as the fraction of the total of prepro, pro and mature forms for each mutant.

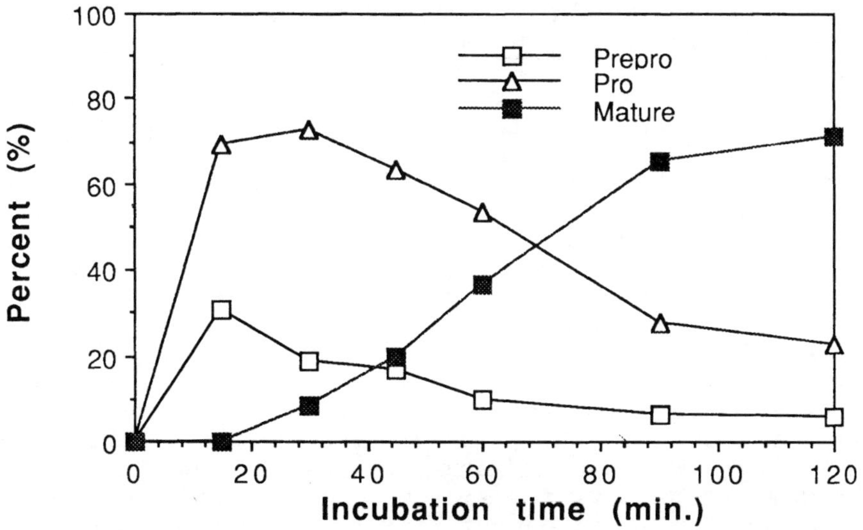

Figure 9.
Rate of processing of prepromini PLAP. The absorbance of each band obtained on processing the various site directed mutants was measured and plotted as a function of time of incubation.

DISCUSSION

PLAP-513 and miniPLAP-208 and the various mutant forms of these two proteins, along with site-specific antibodies have proven to be useful tools to investigate details of the PI-G tailing process. They have made it possible: 1) to determine the type of signals that target a nascent protein for PI-G tailing; 2) to determine the signal that designates a specific internal amino acid residue in the nascent protein to act as the receptor for the PI-G moiety; and 3) to investigate COOH-terminal processing in cell-free systems.

By comparing the different PI-G tailed proteins that have been fully characterized one can make some generalizations about the sizes and compositions of COOH-terminal signal peptides that satisfy requirements for PI-G tailing. We have been able to confirm these generalizations experimentally by using two proteins with established sites of PI-G tailing and asking specific questions that could be answered by appropriate mutational modifications. We showed experimentally that for successful PI-G tailing the hydrophobic sequence at the COOH-terminus could be shortened slightly but could be lengthened appreciably. We also demonstrated experimentally that hydrophobicity, and not primary structure, is important. Using the same protein we showed that addition of a hydrophobic cytoplasmic sequence following the COOH-terminal hydrophobic sequence no longer yielded a PI-G tailed protein. Instead the hydrophobic sequence acted as a typical transmembrane anchor.

Again, by examination of the completely characterized proteins one can see that only a few amino acids can act as receptor for the PI-G moiety. By using one protein and a modified variant we were able to show convincingly that only six amino acids, Asp, Asn, Ser, Gly, Aly and Cys can serve effectively as PI-G acceptors. We also showed that Leu and Val had slight activity and might conceivably be found as a PI-G site in some proteins. Currently, additional information is being gathered about the structural requirements at the cleavage/attachment site. From a few proteins that have been fully characterized (Table I) it is apparent that the PI-G moiety is almost always attached to the first of a sequence of three small amino acids. We have referred to this as the "small amino acid domain" (Micanovic et al, 1990a). We have recently mutated the amino acid next to the PI-G tailed site (COOH-terminal) and shown experimentally that small amino acids are also required in that position to permit correct processing (to be published). The "small amino acid domain" apparently required for COOH-terminal processing brings to mind the -3, -1 rule processing of NH_2-terminal signal peptides (Nilsson and von Heijne, 1991). In the latter case, recognition by NH_2-terminal signal peptidase requires two small amino acids, not consecutively, but at the -3 and -1 positions.

With respect to PI-G, tailing many questions still remain. For example, examination of the sequences in Table I indicates that some proteins have more than one small amino acid domain within a reasonable range for PI-G tailing. Why is one domain selected over another? In addition, some small amino acid domains are larger than three residues and it is not apparent why a specific residue within a sequence of four or more small amino acids is selected to accept the PI-G anchor.

Obviously, elucidation of the mechanism of PI-G tailing will require purified enzymes and synthetic substrates. As a start in this direction we demonstrated that, at least, part of the PI-G tailing mechanism, cleavage, resides in the RM. Our published studies were all carried out with RM obtained from CHO cells. However, we have now been able to show that the specific cleavage of prominiPLAP to miniPLAP is a property of RM produced from most eukaryotic cells (to be published). That the cleavage obtained in the cell-free system is related to the PI-G tailing process is based on several observations: (a) the changes in immunoreactivities during processing from the prepro form to the mature form; (b) the relative mobilities of the miniPLAP precursors and products on PAGE; (c) the finding that only those nascent proteins with either Asp, Asn, Ala, Gly, Ser or Cys at the cleavage/substitution site are processed to the mature form; and (d) mutants with Ser at the cleavage/ substitution site are most active both in transfected cells and in the cell-free system and with both PLAP and miniPLAP. In addition we have recently found that mutants that are partially truncated at the COOH-terminus and are, therefore, secreted and not PI-G tailed in intact cells (Fig. 6, miniPLAP 197 and 184), are also not cleaved at Asp^{179}, in the cell-free system (to be published).

We are convinced that the cleavage observed in the cell-free system is carried out by the PI-G tailing enzyme(s). We have not as yet shown condensation with PI-G but believe that we have not achieved the right experimental conditions. At present we can generate about 10-50 fmol of mature miniPLAP per experiment so that incorporation of one equivalent of a ^{14}C or ^{3}H labelled component of PI-G cannot be detected. However, even when we introduced ^{32}P and ^{32}ATP during incubation or into intact cells prior to preparing RM we failed to detect incorporation of label into mature PLAP. Since we had not made attempts to generate labelled PI-G or to remove a possible pool of unlabelled PI-G, failure to observe incorporation of exogenous PI-G is not surprising. If our observation is correct, that cleavage can occur without condensation, are we dealing with a peptidase which is then followed by a condensing enzyme? There is general agreement in the field that PI-G tailing is catalyzed by a transamidase or transpeptidase (Ferguson and Williams, 1988; Cross, 1990). For example, the condensation of D-alanine to a glycine residue of the bacterial cell wall protein is carried out by a transpeptidase (Tipper and Strominger, 1965). This is the enzyme that is inactivated by penicillin. In such reactions the energy of the peptide bond is maintained in the enzyme-substrate complex so that an external energy source is not required (Fig. 10).

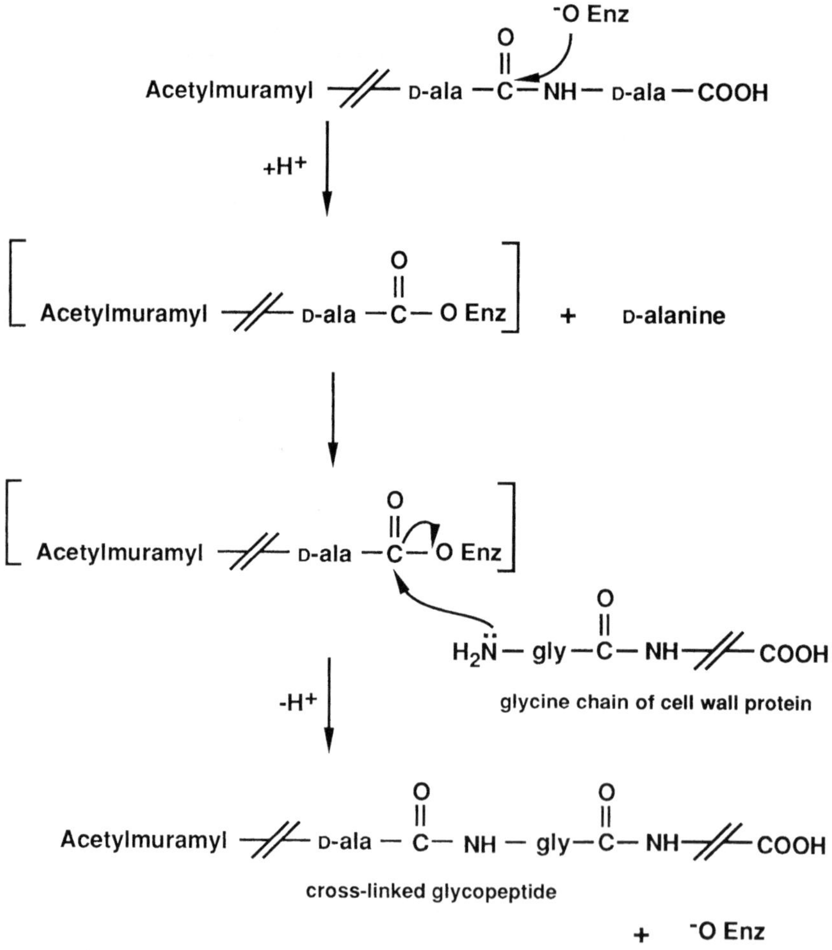

Figure 10.
Transpeptidase mechanism proposed for the crosslinking of the bacterial cell wall. Penicillin reacts irreversibly with the enzyme-acetylmuramyl-D-ala complex. (After Tipper and Strominger).

The mechanism of addition of the PI-G moiety to Asp[484] of PLAP-513 is presented in a similar manner in Fig. 11. As demonstrated by Tipper and Strominger (1965) the activated carbonyl group of the complex condenses with the amino group of the acceptor which apparently is also bound to the enzyme. However, in the absence of the specific, enzyme bound acceptor (Fig. 11) (which in our case is PI-G), the activated carbonyl group would react (more slowly) with water resulting in hydrolysis and an apparent peptidase action.

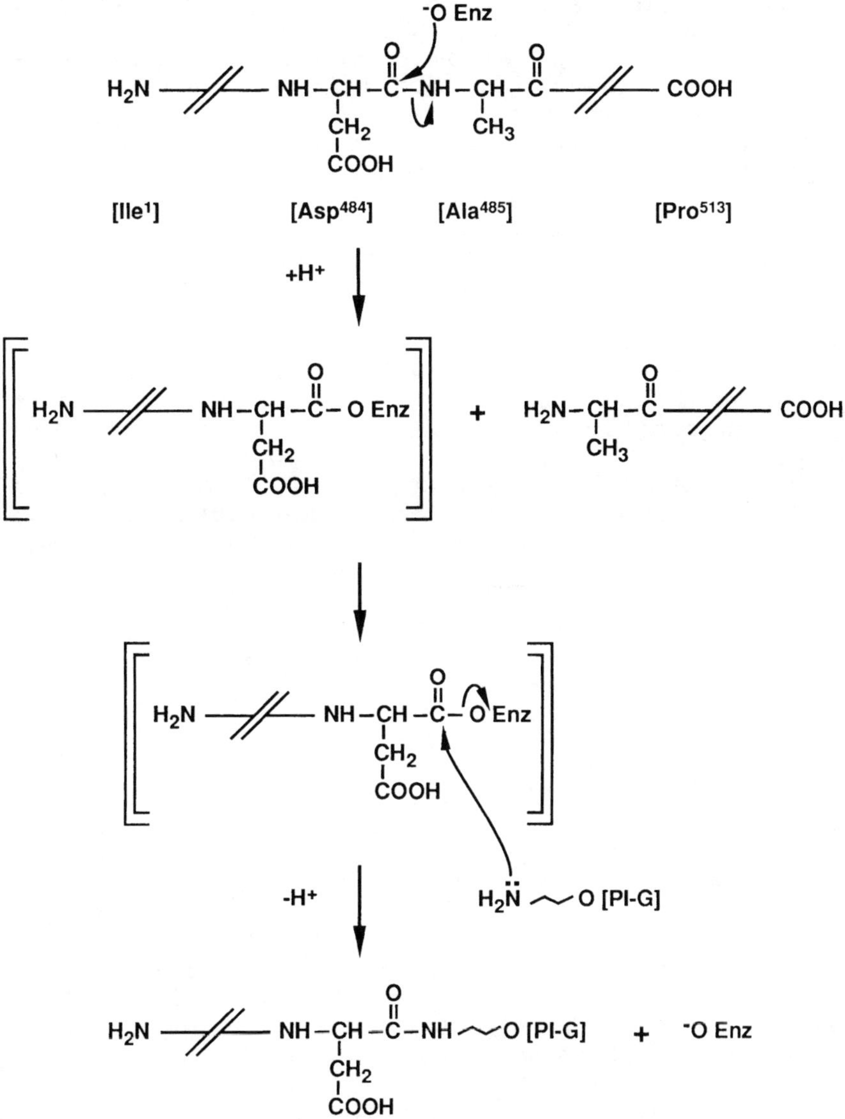

Figure 11.
Proposed mechanism for PI-G tailing of PLAP and other nascent proteins.

Nucleophiles such as hydrazine and hydroxylamine would also be expected to react with a transamidase-substrate complex. Solubilization and purification of the putative transamidase and the availability of the appropriate highly labelled PI-G acceptor are needed to unequivocally determine the mechanism of PI-G tailing.

REFERENCES

Bailey, C. A., Gerber, L., Howard, A. D . and Udenfriend, S. (1989). Processing at the carboxyl terminus of nascent placental alkaline phosphatase in a cell-free system: Evidence for specific cleavage of a signal peptide. Proc. Natl. Acad. Sci. USA **86:** 22-26.

Bailey, C. A., Howard, A., Micanovic, R., Berger, J., Heimer, E., Felix, A., Gerber, L. and Udenfriend, S. (1988). Site-directed antibodies for probing the structure and biogenesis of phosphatidylinositol glycan-linked membrane proteins: application to placental alkaline phosphatase. Analyt. Biochem. **170:** 532-541.

Berger, J., Howard, A. D., Brink, L., Gerber, L., Hauber, J., Cullen, B. R. and Udenfriend, S. (1988). COOH-terminal requirements for the correct processing of a phosphatidyl-inositol-glycan anchored membrane protein. J. Biol. Chem. **263:** 10016-10021.

Berger, J., Howard, A. D., Brink, L., Gerber, L., Hauber, J., Cullen, B. R. and Udenfriend, S. (1987). Expression of active membrane-bound human placental alkaline phosphatase by transfected simian cells. Proc. Natl. Acad. Sci. USA **84:** 4885-4889.

Berger, J., Micanovic, R., Greenspan, R. J. and Udenfriend, S. (1989). Conversion of placental alkaline phosphatase from a phosphatidylinositol-glycan-anchored protein to an integral transmembrane protein. Proc. Natl. Acad. Sci. USA **86:** 1457-1460.

Boothroyd, J. C., Paynter, C. A., Cross, G. A. M., Bernards, A. and Borst, P. (1981). Variant Surface Glycoproteins of *Trypanosoma brucei* are synthesised with cleavable hydrophobic sequences at the carboxy and amino termini. Nuleic Acids Res. **9:** 4735-4743.

Caras, I. W., Weddell, G. N., Davitz, M. A., Nussenzweig, V. and Martin, D. W. Hr. (1987). Signal for attachment of a phospholipid membrane anchor in delay-accelerating factor. Science **238:** 1280-1283.

Cross, G. A. M. (1990). Glycolipid anchoring of plasma membrane proteins. Ann.Rev. Cell Biol. **6:** 1-39.

Cullen, B. R. (1987). Use of eukaryotic expression technology in the functional analysis of cloned genes. Meths.Enzymol. **152:** 684-704.

Early, A. E., Williams, J. G., Meyer, H. E., Por, S. B. Smith, E. et al. (1988). Structural characterisation of *Dictyostelium discoideum* prespore-specific gene D19 and of its product, cell surface glycoprotein PsA. Mol. Cell Biol. **8:** 3458-3466.

Ferguson, M. A. J. and Williams, A. F. (1988). Cell-surface anchoring of proteins via glycosyl-phosphatidyl-inositol structures. Ann. Rev. Biochem. **57:** 285-320.

Ferguson, M. A. J., Homans, S. W., Dwek, R.A. and Rademacher, T. W. (1988). Glycosyl-phosphatidyl-inositol moiety that anchors *Trypanosoma brucei* variant surface glycoprotein to the membrane. Science **239:** 753-759.

Garattini, E., Margolis, J., Heimer, E., Felix, A. and Udenfriend, S. (1985). Human placental alkaline phosphatase in liver and intestine. Proc. Natl. Acad. Sci.USA **82:** 6080-6084.

Gaulton, G. N., Kelly, K. L., Pawlowski, J., Mato, J. M. and Jarett, L. (1988). Regulation and function of an insulin-sensitive glycosyl-phosphatidylinositol during T-lymphocyte inactivation. Cell **53:** 963-970.

Gibney, A., MacPhee-Qiugley, K., Thompson, B., Vedvick, T. Low, M.G. et al (1988). Divergence in primary structure between the molecular forms of acetylcholinesterase. J. Biol. Chem. **263:** 1140-1145.

Hefta, S. A., Hefta, L. J. F., Lee, T. D., Paxton, R. J. and Shively, J. E. (1988). Carcinoembryonic antigen is anchored to membranes by covalent attachment to a glycosophosphatidylinositol moiety; identification of ethanolamine linkage site. Proc. Natl. Acad. Sci. USA **85:** 4648-4652.

Holder, A. A. (1985). Glycosylation of the variant surface antigens of *Trypanosoma brucei*. Curr. Tops. Microbiol. Immunol. **117:** 57-74.

Holder, A. A. and Cross, G. A. M. (1981). Glycopeptides from variant surface glycoproteins of *Trypanosoma brucei*: C-terminal location of antigenically cross-reacting carbohydrate moieties. Mol. Biochem. Parasitol. **2:** 135-150.

Killeen, N., Moessner, R., Arvieux, J., Willis, A. and Williams, A. F. (1988). The MRC OX-45 antigen of rat leucocytes and endothelium is in a subset of the immunoglobulin superfamily with CD2, LFA-3 and carcino-embryonic antigens. EMBO J. **7:** 3087-3091.

Koch, F., Haag, F., Kashan, A. and Thiele, H. G.(1990). Primary structure of the rat RT6.2, a nonglycosylated phosphatidylinositol-linked surface marker of postthymic cells. Proc. Natl. Acad. Sci. USA **87:** 964-967.

Kodukula, K., Micanovic, R., Gerber, L., Tamburrini, M., Brink, L. and Udenfriend. S. (1991). Biosynthesis of Phosphatidylinositol Glycan-anchored Membrane proteins. Design of a simple protein substrate to characterise the enzyme that cleaves the COOH-terminal signal peptide. J. Biol. Chem. **266:** 4464-4470.

Majumdar, H. K., Boothroyd, J. C. and Weber, H. (1981). Homologous 3′-terminal regions of mRNAs for surface antigens of different antigen variants of *Trypanosoma brucei*. Nucl. Acids Res. **9:** 4745-4753.

Masterson, W., Doering, T. L., Hart, G. W., and Englund, P. T. (1989). A novel pathway for glycan assembly: biosynthesis of the glycosyl-phosphatidylinositol anchor of the trypanosome variant surface glycoprotein. Cell **56:** 793-800.

Micanovic, R., Bailey, C. A., Brink, L., Gerber, L., Pan, Y.-C. E., Hulmes, J. D. and Udenfriend, S. (1988). Aspartic acid-484 of nascent placental alkaline phosphatase condenses with a phosphatidylinositol glycan to become the carboxyl terminus of the mature enzyme. Proc. Natl. Acad. Sci. USA **85: 1398-1402.**

Micanovic, R., Gerber, L. D., Berger, J., Kodukula, K. and Udenfriend, S. (1990a). Selectivity of the cleavage/attachment site of phosphatidylinositol-glycan-anchored membrane proteins determined by site-specific mutagenesis at Asp-484 of placental alkaline phosphatase. Proc. Natl. Acad. Sci. USA **87:** 157-161.

Micanovic, R., Kodukula, K., Gerber, L. D. and Udenfriend. S. (1990b). Selectivity at the cleavage/attachment site of phosphatidylinositol-glycan anchored membrane proteins is enzymatically determined. Proc. Natl. Acad. Sci. USA **87:** 7939-7943.

Millan, J. L. (1986). Molecular cloning and sequence analysis of human placental alkaline phosphatase. J. Biol. Chem. **261:** 3112-3115.

Misumi, Y., Ogata, S., Hirose, S. and Ikehara, Y. (1990). Primary structure of rat liver 5′-nucleotidase deduced from the cDNA: presence of COOH-terminal hydrophobic domain for possible post-translational modification by glycophospholipid. J. Biol. Chem. **265:** 2178-2183.

Moran, P., Raab, H., Kohr, W.J. and Caras, I. W. (1991). Glycophospholipid membrane anchor attachment: Molecular analysis of cleavage/ attachment site. J. Biol. Chem. **266:** 1250-1257.

Mowatt, M. R. and Clayton, C. E. (1989). Glycophospholipid membrane anchor attachment: Molecular analysis of cleavage/ attachment site. J. Biol. Chem. **264:** 15088-15093.

Nilsson, I. M. and von Heijne, G. (1991). A de novo Designed signal Peptide Cleaver Cassette functions in vivo. J. Biol. Chem. **266:** 3408-3410.

Ogata, S., Hayashi, Y., Takami, N. and Ikehara, Y. (1988). Chemical characterisation of the membrane anchoring domain of human placental alkaline phosphatase. J. Biol. Chem. **263:** 10849-10494.

Ogata, S., Hayashi, Y., Misumi, Y. and Ikahara, Y. (1990). Membrane anchoring domain of rat liver 5'-nucleotidase: Identification of COOH-terminal Serine 523 covalently attached with a glycolipid. Biochemistry **29:** 7923-7927.

Pollitt, S., Inouye, S. and Inouye, M. (1986). Effect of amino acid substitutions at the signal peptide cleavage site of the *Escherichia coli* outer membrane lipoprotein. J. Biol. Chem. **261:** 1835-1837.

Seki, T., Moriuchi, T., Chang, H.-C., Denome, R. and Silver, J. (1985). Structural organisation of the rat Thy-1 Gene. Nature **313:** 485-487.

Stahl, N., Borchett, D. R. and Prusiner, S. B. (1990). Differential release of cellular and scrapie prion proteins from cellular membranes by phosphatidylinositol-specific phospholipase C. Biochemistry **29: 5405-5412.**

Strickler, J. E., Binder, D.A., L'Italien, J. J., Shimamoto, G. T., Wait, S. W. et al. (1989). *Trypanosoma conglense*: structure and molecular organisation of the surface glycoproteins of two early blood stream variants. Biochemistry **26:** 796-805.

Stroynowski, I., Soloski, M., Low, M.G. and Hood, L. (1987). A single gene encodes soluble and membrane-bound forms of the major histocompatability Qa-2 antigen: anchoring of the product by a phospholipid tail. Cell **50:** 759-768.

Tipper, D. J. and Strominger, J. L. (1965). Mechanism of action of penicillins: a proposal based on their structural similarity to acyl-D-alanyl-D-alanine. Proc. Natl. Acad. Sci. USA **54:** 1133-1141.

Wickner, T. W. and Lodish, F. H. (1985). Multiple mechanisms of protein insertion into and across membranes. Science **230:** 400-407.

SIGNAL TRANSDUCTION BY GPI-ANCHORED MEMBRANE PROTEINS

***Peter J Robinson**

Transplantation Biology Section, Clinical Research Centre, Watford Road, Harrow, UK.

* Corresponding author.

INTRODUCTION

Antibodies specific for glycosyl-phosphatidylinositol (GPI)-anchored membrane proteins can cause mitogenesis, lymphokine production and other intracellular changes in haematopoietic cells. It has therefore been suggested that transmembrane signalling may be an intrinsic and functionally important property of these molecules. Lipid anchors do not span the cell membrane and therefore have no direct contact with the cytoplasm of the cell. How these molecules are able to transduce activation signals is not known. There is some evidence for the involvement of other, integral membrane proteins in GPI-mediated signal transduction. In this article, I shall discuss the molecular basis of signalling in relation to other antibody-induced changes, such as redistribution and internalisation of membrane proteins.

FUNCTIONS OF GPI-ANCHORS

GPI is now well established as a highly efficient alternative means of protein anchoring in membranes. Protein anchoring is, in fact, the only clear physiological function which can be attributed to these structures. Other properties have been described and reviewed elsewhere (Ferguson and Williams, 1988; Low, 1989). GPI-anchored molecules have, for example, the potential for attaining a higher packing density in the plane of the membrane, which may be important in maintaining an impermeable surface coat in parasites. The potentially greater lateral mobility of GPI-anchored proteins compared to their transmembrane counterparts may be of advantage for rapid binding to cellular or soluble ligands. Finally, the exquisite sensitivity of these anchors to cleavage by phospholipases C and D may facilitate dissociation of cell-cell conjugates.

One characteristic feature of GPI-anchored proteins is their high level of structural and functional diversity (Low, 1989). It is notable, however, that many of these molecules are specific for extracellular ligands or substrates and a number of them (e.g. NCAM, Thy-1) are members of the immunoglobulin superfamily of interaction molecules. Thus, GPI-linked membrane molecules may serve two functions, firstly in interacting with other cells and secondly in initiating biological responses. A good example of this is lymphocyte activation.

ACTIVATION OF LYMPHOCYTES

Most of the available information on GPI-mediated signalling phenomena has been obtained using mammalian T lymphocytes. T cells can respond to a range of mitogenic stimuli including specific antigen, lectins, antibodies and lymphokines. Mitogenic effects mediated by GPI-anchored molecules have been observed in man, mice, rats and guinea pigs. Proliferative responses and lymphokine production occur in resting T cells and thymocytes as well as in T cell lines and hybridomas. Other physiological changes involving GPI have been reported in B lymphocytes, monocytes and granulocytes. Requirements for T lymphocyte signalling have been reviewed recently (Robinson, 1991).

The best studied GPI-linked protein with respect to T cell activation is Thy-1, a differentiation antigen present on rodent thymocytes and in the central nervous system of most mammals. The six other GPI-linked molecules implicated in T cell activation (see Robinson, 1991) to date have little structural similarity, if any, to Thy-1 at the protein level. Nevertheless, the parameters and requirements for cell activation are qualitatively similar for all of these molecules suggesting that the anchor is essential for mitogenesis. This was confirmed using T cells from transgenic mice expressing GPI-linked or transmembrane forms of the histocompatibility antigen Qa-2 and H-2D^b (Robinson et al, 1989).

Binding of antibodies to GPI-anchored proteins initiates early intracellular changes such as cytoplasmic free [Ca^{2+}] and phosphoinositide turnover. These effects are strongly enhanced by agents which promote clustering of cell surface immune complexes, such as anti-immunoglobulin antibodies or accessory cells bearing receptors for immunoglobulin (Fc receptors). Interestingly, immobilisation of antibodies or anti-immunoglobulin, which strongly promotes signalling via the T cell antigen-receptor (TCR) pathway, inhibits GPI-mediated signalling, suggesting that lateral mobility is important for coupling of GPI-anchored molecules to the signalling machinery.

Early changes in cytoplasmic free calcium and inositol phosphate turnover rates also occur when the antigen receptor is activated, suggesting that the GPI and TCR pathways may be coupled. The dependence of GPI-mediated proliferative responses on co-expression of a functional antigen receptor complex complicates the task of unravelling the origins of the various second messengers. To distinguish these two pathways, even earlier signalling events, such as tyrosine kinase activation, are of importance. Anti-Thy-1 antibodies have been shown to induce tyrosine phosphorylation of the T cell receptor zeta chain polypeptide (Hsi et al, 1989) and other proteins. We have found that the mitogenic effects of Qa-2 specific antibodies are abrogated by low concentrations of Genistein, a tyrosine kinase inhibitor, whilst other routes of activation are unaffected. Also, mobilisation of a tyrosine kinase by anti-Thy-1 antibodies appears not to involve the antigen receptor directly because cell lines deficient in the entire receptor complex show specific phosphorylation of certain protein substrates (Hsi et al, 1989).

In resting T cells, proliferative and lymphokine responses require an additional mitogenic signal which can be delivered by activators of protein kinase C (PKC), such as phorbol myristate acetate (PMA). Immortalised cells, such as T cell hybridomas, do not require this second signal for lymphokine production, suggesting that clustering of GPI-linked proteins alone delivers a mitogenic signal in resting cells, which requires the co-operation of other pathways to cause proliferation.

MOLECULAR ASSOCIATIONS ON THE CELL SURFACE

Several independent lines of evidence indicate physical interactions between GPI-linked molecules and other less mobile membrane components. Addition of antibodies to Thy-1 or Qa-2 on mouse lymphocytes results in rapid redistribution and capping, implying coupling to cytoskeletal components. Consistent with this, photobleaching recovery techniques show that only about half of the Thy-1 molecules have rapid translational diffusion rates expected for lipid-linked structures (Ishihara and Jacobsen, 1987). Even before they were known to be GPI-anchored, several cell surface molecules were found to be difficult to solubilise in nonionic detergents such as NP40 and Triton X-100 (Hoessli showed that Thy-1 isolated from a T-lymphoma cell line using Triton X-100 detergent was mainly in the form of large complexes of molecular weight 15 x 10^6 daltons. Bourguignon and co-workers (1986) obtained similar results and detected other proteins physically associated with Thy-1. By surface labelling with ^{125}I, Hoessli detected a protein of molecular wight 55 kilodaltons and small amounts of GPI-linked Qa-2. In addition, several other unlabelled proteins of molecular weights between 30 and 70 kilodaltons were found in this complex. Bourguignon and co-workers (1986) demonstrated surface-labelled polypeptides with molecular weights of 85 and 180 kilodaltons in association with Thy-1. Of particular interest was the direct association between Thy-1 and an intracellular phosphoprotein of molecular weight 68 kilodaltons, which was cross-reactive with antibodies against the erythrocyte component 4.1. This protein was found to undergo reversible association with Thy-1 in detergent/urea mixtures. The 4.1-like molecule was also found to associate with the cytoskeletal component fodrin and is phosphorylated on serine residues in response to phorbol ester treatment of cells. Thus, under conditions optimal for GPI-mediated lymphocyte proliferation, the 4.1-like protein is probably phosphorylated. This raises the interesting possibility that cytoskeletal interactions may be involved in lymphocyte signalling.

In another report (Volarevic et al, 1990), Thy-1 was found to associate with the leukocyte common antigen, CD45. This abundant cell surface molecule has tyrosine phosphatase activity and has been implicated in lymphocyte proliferative responses (Pingel and Thomas, 1989). However, it has also been shown by Bourguignon and co-workers (1985) that CD45 associates with fodrin, suggesting that Thy-1 and CD45 may not interact directly but may be linked through another molecule, possibly the 4.1-like phosphoprotein.

SURFACE MOBILITY AND SIGNALLING

Redistribution, capping and endocytosis of GPI-linked molecules, induced by antibodies, appear to be linked in some way to the process of transmembrane signalling. Under conditions which prevent molecular clustering, e.g. antibody immobilisation, proliferative responses are inhibited. This may occur in one of two ways. Firstly, as was proposed by Bamezai et al (1989), immobilisation of stimulatory antibodies may prevent internalisation of immune complexes, thus abrogating an essential signal. A second possibility is that restriction of the surface mobility of these molecules may render impossible the task of bringing together all necessary components of the signalling complex on both the outer and inner faces of the cell membrane.

It is now well-established that GPI-linked membrane proteins do not use the classical route of receptor-mediated endocytosis involving clathrin coated pits (Rothberg et al, 1990; Van den Bosch et al, 1988). Instead, another route involving endosomes is used. Since inclusion in coated pits is, at least partly, a passive process, it appears that active exclusion of GPI-linked molecules occurs. Active exclusion from coated pits has been found for a transmembrane protein, CD4, (M.Marsh, personal communication) and requires the participation of a cytoplasmic component, $p56^{lck}$ tyrosine kinase. It seems likely, therefore, that cytoplasmic proteins, associating with GPI-anchors via an intermediate membrane-spanning molecule, may determine the intracellular fate of endocytosed GPI-linked proteins.

CONCLUSIONS

The stimulatory properties of antibodies specific for GPI-anchored membrane proteins suggest that their natural ligands or substrates may also affect the activation state of the cells. Evaluation of this hypothesis may require the construction of an artificial test system, since there are few known ligands for GPI-linked proteins which are associated with suitably responsible cells. At present, T lymphocytes are the most suitable responders, but secretion of cytokines by cells of the macrophage lineage is also of interest.

Biochemical data reviewed here predict the existence of one or more integral membrane proteins which associate with GPI-anchors. These associations may pre-exist on unstimulated cells or may be formed when antibody is present. I favour the view that GPI-anchors are associated with several other molecules in resting T cells and that additional polypeptides may be sequestered as a result of antibody binding. It is now important to characterise these molecules further.

Some components of the proposed signalling complex are likely to exist in different tissues. Use of the conserved anchor to couple different proteins to this complex enables signalling to operate in a variety of cell types. The specificity of the stimulus is therefore determined by interaction of an appropriate GPI-linked receptor with its ligand. Tissue specific differences in the composition of the signalling complex may resolve the hitherto unexplained phenomenon of anchor resistance to phospholipase C treatment (Low et al, 1988). The versatility of this signalling pathway may help to explain the pleiotropic responses which occur during inflammatory processes.

REFERENCES

Bamezai, A., Goldmacher, V., Reiser, H. and Rock, K. L. (1989). Internalization of phosphatidylinositol-anchored lymphocyte proteins I. Documentation and potential significance for T-cell stimulation. J. Immunol. **143:** 3107-3116.

Bourguignon, L. Y. W., Suchard, S. J., Nagpal, M. L. and Glenney, J. R., Jr. (1985). A T-lymphoma transmembrane glycoprotein (gp 180) is linked to the cytoskeletal protein, fodrin. J. Cell Biol. **101:** 477-487.

Bourguignon, L. Y. W., Suchard, S. J. and Kalomiris, E. L. (1986). Lymphoma Thy-1 glycoprotein is linked to the cytoskeleton via a 4.1-like protein. J. Cell Biol. **103:** 2529-2540.

Ferguson, M. A. J. and Williams, A. F. (1988). Cell surface anchoring of proteins via glycosyl-phosphatidyl inositol structures. Ann. Rev. Biochem. **57:** 285-320.

Hoessli, D. and Rungger-Brändle, E. (1985). Association of specific cell surface glycoproteins with a Triton X-100 resistant complex of plasma membrane proteins isolated from T-lymphoma cells (P1798) Exp. Cell Res. **156:**. 239-250.

Hooper, N. M. and Turner, A. J. (1988). Ectoenzymes of the kidney microvillar membrane. Biochem. J. **250:** 865-869.

Hsi, E. D., Siegel, J. N., Minami, Y., Luong, E. T., Klausner, R. D. and Samelson, L. E. (1989). T cell activation induces rapid tyrosine phosphorylation of a limited number of cellular substrates. J. Biol. Chem. **264:** 10836-10842.

Ishihara, A., Hou, Y. and Jacobson, K. (1987). The Thy-1 antigen exhibits rapid lateral diffusion in the plasma membrane of rodent lymphoid cells and fibroblasts. Proc. Natl. Acad. Sci. USA **84:** 1290-1293.

Low, M. G. (1989). The glycosyl-phosphatidylinositol anchor of membrane proteins. Biochim. Biophys. Acta. **988:** 427-454.

Low, M. G., Stiernberg, J., Waneck, G. L., Flavell, R. A. and Kincade, P. W. (1988). Cell-specific heterogeneity in sensitivity of phosphatidyl-anchored membrane antigens to release by phospholipase C. J. Immunol. Methods. **113:** 101-111.

Pingel, J. T. and Thomas, M. L. (1989). Evidence that the leukocyte-common antigen is required for antigen-induced T lymphocyte proliferation. Cell **58:** 1055-1065.

Robinson, P. J., Millrain, M., Antoniou, J., Simpson, E. and Mellor, A. L. (1989). A glycophospholipid anchor is required for Qa-2-mediated T cell activation. Nature **342:** 85-87.

Robinson, P. J. (1991). Phosphatidylinositol membrane anchors and T-cell activation. Immunol. Today **12:** 35-41.

Rothberg, K.G., Ying,Y., Kolhouse,J.F., Kamen,B.A. and Anderson, R. G. W. (1990). The glycophospgolipid linked folate receptor internalises folate without entering the clathrin-coated pit endocytic pathway. J. Cell Biol. **110:** 637-649.

van den Bosch, R. A., du Maine, A. P. M., Geuze, H. J., van der Ende, A. and Strous, G. J. (1988). Recycling of 5′-nucleotidase in a rat hepatome line. EMBO J. **7:** 3345-3351.

Volarevic, S., Burns, C. M., Sussman, J. J. and Ashwell, J. D. (1990). Intimate association of Thy-1 and the T cell antigen receptor with the CD45 tyrosine phosphatase. Proc. Natl. Acad. Sci. USA **87:** 7085-7089.

EMERGENCE OF THE LY-6 SUPERFAMILY OF GPI-ANCHORED MOLECULES

***Alan F Williams**

**MRC Cellular Immunology Research Unit,
Sir William Dunn School of Pathology,
University of Oxford, England, OX1 3RE.**

* Corresponding author.

INVERTEBRATE NEURAL MOLECULES OF THE IMMUNOGLOBULIN SUPERFAMILY

In the early 1980s structural studies had led to the conclusions that the Thy-1 cell surface molecule was related to a single immunoglobulin (Ig) domain and was attached to the lipid bilayer by a non-protein anchor attached at the COOH-terminus (Williams and Gagnon, 1982). At that time the HLA MHC Class-1 molecule and the immunoglobulins themselves were the only molecules known to be in the Ig-superfamily. Taken together the data for Ig, MHC and Thy-1 suggested that Ig-superfamily molecules would be commonly found at cell surfaces and that many Ig-superfamily molecules would be involved in cell-cell recognition events that were not related to immunity (Williams, 1982).

In an attempt to prove that Ig-superfamily molecules had first evolved to mediate non-immune cell-cell interactions, a homologue of Thy-1 was sought in neural tissue of an invertebrate species. The optic nervous tissue of the squid provided a suitable source of material for biochemical studies and glycoproteins were purified by lentil lectin affinity chromatography and fractionated by gel filtration. A major glycoprotein fraction that had broadly similar properties to Thy-1 was obtained and the sequence of a peptide from this supported the idea of a relationship to Thy-1 (Williams and Gagnon, 1982). Further analysis revealed that the glycoprotein fraction contained two molecules that were named Sgp1 and Sgp2 and both were sequenced (Williams, Tse and Gagnon, 1988). Both molecules had short peptide chains (84 residues for Sgp1 and 92 for Sgp2) and both had glycosyl phosphatidylinositol (GPI) anchors at their COOH-termini. The peptide with similarities to Thy-1 was from Sgp1, but analysis of the complete sequence did not support the preliminary data. There was no convincing argument for a relationship by divergent evolution between Sgp1 or Sgp2 and Thy-1, or other Ig-superfamily molecules. Subsequently the discovery of Ig-superfamily molecules in invertebrate nervous tissue has been made by others (Harrelson and Goodman, 1988).

Lymphoid Ly-6 Antigens and Sgp2.

The mouse Ly-6 antigens were discovered by the classical immunogenetic approach of searching for polymorphic determinants by immunising between inbred strains of mice and also by immunisations between species in which monoclonal antibodies (MAbs) that were mitogenic for mouse T cells were sought (reviewed Shevach and Kirty, 1989). The antigens were found to be small glycoproteins with GPI anchors. It is a major enigma that GPI-anchored molecules are a potent target for mitogenic effects that are seen when cells are cultured in the presence of phorbol esters and the cell surface molecules are cross-linked with MAbs plus a second anti-Ig antibody (reviewed Robinson, 1991). When MAbs have been screened by this procedure the antigens that have most commonly been found are Thy-1 and antigens of the Ly-6 group. There are numerous genes in the Ly-6 group but for present purposes there are two main categories, namely the Ly-6A(E) (LeClair et al, 1986) and Ly-6C (Palfree et al, 1988) prototypes and these are 50% identical in their amino acid sequences. Rat Ly-6 antigens have also been sequenced and those identified thus far are most similar to the mouse Ly-6A(E) type (Friedman et al, 1990).

The squid sequence SGp2 initially showed no similarities to any other sequence but publication of the Ly-6E.1 and Ly-6C.2 sequences indicated the possibility that Sgp2 was an invertebrate homologue of these molecules. The proteins were all of similar size and all were attached to GPI anchors. In each case the sequences contained 10 cysteine residues and these could be aligned along the sequences beginning at the amino terminus and ending at the likely point of attachment of the GPI anchor. Sequence alignments are shown in Figure 1. The position of GPI attachment is known for Sgp2 but not for the Ly-6 antigens whose sequences come from cDNA sequencing. Thus the Ly-6 sequences include the signal sequence for GPI attachment and this has been left out in the alignments in Figure 1. The suggested points of attachment for the Ly-6 antigens are compatible with the loose consensus for GPI linkage (Ferguson and Williams, 1988) and also with sequence identities between Sgp2 and the Ly-6 antigens.

Other Sequences in the Ly-6 Family and Analysis of the Relationships.

The new sequences have recently been added to this superfamily namely that of the human CD59 cell surface antigen (Davies, et al, 1989) and that of the urokinase

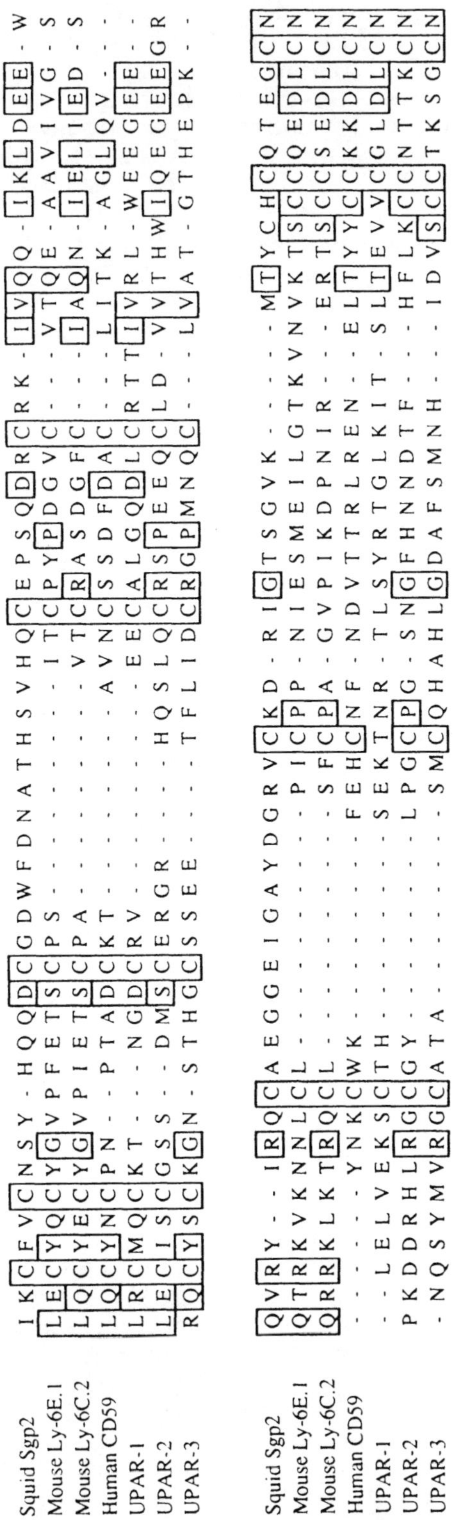

Figure 1. Alignment of sequences in the Ly-6 superfamily.

The sequences are from the references given in the text and they have been aligned by eye. Boxes are placed around residues that are identical in any three sequences. For Sgp2, the full mature sequence of 92 residues is shown and it has established that the GPI anchor is attached to the carboxyl group of Asn 92. The Ly-6E.1 and Ly-6C.2 sequences were determined by sequencing of CDNA and following the sequences shown (79 and 76 residues) there are a further 29 amino acids that are believed to constitute the signal sequence for GPI attachment. The same situation holds for CD59 where 33 amino acids are predicted from cDNA following the shown sequence of 70 amino acids. For the UPAR sequences the domains shown include residues 1-77 UPAR1), 93-177 (UPAR2), and 192-272 (UPAR3). On the basis of sequence patterns the site for GPI attachment in the UPAR sequence may be at Ser 282 or Gly 283. The predicted signal sequence continues to residue 313.

plasminogen activator receptor (UPAR) (Roldan et al, 1990; Plough et al, 1991; Palfree, 1991). CD59 is a single domain structure like the Ly-6 antigens and Sgp2 whilst the UPAR contains three Ly-6 type domains. Both CD59 and UPAR have GPI anchors. Alignments of these sequences with the Ly-6 antigens and Sgp2 is shown in Figure 1. All the sequences except UPAR domain 1 contain 10 cysteine residues and there are clear patches of strong identity at the beginning and ends of the domains. The absence of the seventh and eighth cysteines in UPAR domain 1 suggests that these cysteines may form a disulphide bond.

Comparison of Sequences in the LY6-superfamily by Analysis Using the ALIGN Programme - Scores in SD Units

	Sgp2	Ly6E	Ly6C	CD59	UPAR1	UPAR2	UPAR3
Sgp2		3.7	3.8	4.0	5.0	4.6	4.6
Ly6E	3.7		19.0	5.2	5.0	5.4	5.6
Ly6C	3.8	19.0		5.7	4.7	6.1	5.5
CD59	4.0	5.2	5.7		4.2	4.8	6.7
UPAR1	5.0	5.0	4.7	4.2		3.8	3.5
UPAR2	4.6	5.4	6.1	4.8	3.8		6.7
UPAR3	4.6	5.6	5.5	6.7	3.5	6.7	

Figure 2. Analysis of relationships in the Ly-6 superfamily using the ALIGN programme.

The sequences as shown in figure 1 were compared using the ALIGN programme of Dayhoff, et al, (1983) using the 250 PAMS mutation data matrix plus a bias of 6 and with a gap penalty of 6. The number of random comparisons used to determine the random mean scores and random standard deviation was 100 for each comparison. The scores shown are the number of standard deviations (SD) by which the real comparisons exceeded the mean random scores. Scores greater than 3SD units are considered to be of interest (see Williams, (1987) for data on significance of ALIGN scores).

The sequence similarities were analyzed using the ALIGN program of Dayhoff et al, (1983). In this analysis a score is calculated based on sequence identities and conservative substitutions and this score is expressed as the number of standard deviations away from a mean score resulting from a large number of comparisons using the same sequences that have been randomly scrambled. Figure 2 shows comparisons between all the domains as in Figure 1 and it can be seen that the scores between the diverse sequences are roughly similar. Thus there is a good case for arguing that all these sequences are in a superfamily of molecules related by divergent evolution. In two aspects the Sgp2 sequence can be distinguished from all the others. Firstly it has 13 amino acids between cysteines 6 and 7 compared with 3-5 amino acids in this region for the other sequences. Secondly, in all the sequences except Sgp2, the Cys 8 and Cys 9 residues are adjacent whereas in Sgp2 they are interrupted by one residue. This latter difference would lead to a different orientation of disulphide bonds, but it is not obvious that this would alter the pairing of cysteine residues in disulphide bond formation. If the disulphide bonds in this group of sequences all showed the same connectivity this would strongly support the argument of a superfamily relationship.

Gene Structure.

The arrangement for the Ly-6C gene has been determined (Bothwell et al, 1988). In this gene, exon 1 is non-coding and exon 2 encodes all but 3 residues of the leader sequence. Exon 3 encodes the final 3 amino acids of the leader sequence plus the protein sequence until residue 36. Exon 4 encodes the rest of the protein sequence plus the signal sequence for GPI attachment followed by a considerable amount of non-coding sequence. Data is also available for some rat Ly-6 genes (Freidman et al, 1990) and this suggests a conserved intron/exon pattern for the Ly-6 family.

Given the arrangement of the Ly-6C gene it seems likely that the Ly-6 family of leukocyte antigens has evolved by duplications of full length genes. The absence of an intron between the end of the extracellular sequence and the GPI signal sequence would make it difficult for the extracellular motif to be duplicated and diversified in combination with exons encoding motifs of other superfamilies. If this gene organisation is the general case for the Ly-6 superfamily, it may account for the fact that thus far all but one molecule in this group encodes only a single domain and that mixing with other superfamilies has not yet been seen.

Functions in the Ly-6 Superfamily.

The function of the Sgp2 molecule is unknown and that of the Ly-6 antigens remains mysterious. The Ly-6 antigens can be targets for mitogenic effects at the cell surface but what is the meaning of this? On mouse T cells there are at least four categories of GPI-linked molecules that can be potent targets for mitogenesis including Thy-1, Ly-6, Qa-2 and the BCM1 (OX45/Blast 1) antigen (Robinson, 1991; Wong et al, 1990; and Wong Y-W and Williams, A.F., unpublished data). These molecules are very diverse in their protein sequences and thus the mitogenic effect would seem to be somehow linked to the GPI anchor. However, it is not clear that these mitogenic effects reflect the events of normal T cell activation and thus far no ligands are known for any of these lymphoid antigens.

For the cD59 and UPAR molecules the situation is much clearer. CD59 is a molecule that opposes the final stages of the assembly of the complement attack complex by binding to complement component 8 in the attack complex (Meri et al, 1990). This protection occurs in a species specific manner such that human complement is down-regulated but that of other species acting on human cells is not. CD59 can thus be considered to be a regulatory binding motif at the cell surface. The UPAR molecule might also be considered within this functional framework. UPAR binds plasminogen activator and thus presumably concentrates this protease at a cell surface where it is required. It is thought that plasminogen activator may be involved in the penetration of extracellular matrices and that its function may be important in metastasis as well as in non-pathogenic functions (Roldan et al, 1990).

Protein Superfamilies Based on Stable Protein Motifs that Diversify in Evolution by Duplication and Divergence - Evolution From Cell Surface Coat Proteins of Primitive Eukaryotes.

There are now 10 known sequence motifs that are found in sequences of cell surface molecules and these are often, but not always, encoded by single exons (Baron et al, 1991). With the motifs in single exons, gene duplication of the exon leads to replication and then divergence of the whole exon, as a basis for evolution of new proteins. Furthermore, in some cases the placing of intron: exon boundaries is such that the motifs can be mixed to form a new structure by insertion of single exons into an intron region of another gene (Pathy, 1987). The motifs concerned, with references in Baron et al, (1991) unless stated otherwise are: the Ig domain, the EGF domain, the complement control protein (CCP) repeat, the fibronectin type 3 repeat, the LDL receptor repeat, the nerve growth factor receptor repeat (Mallett et al, 1990), the link protein domain, the lectin domain, the CD5 antigen motif (Freeman et al, 1990) and the Ly-6 domain. Of these only the Ig and Ly-6 domains have been found as single domains. For the Ig superfamily single domains have been found as cell surface molecules with either GPI or protein anchors, whilst the single Ly-6 domains have thus far been found only as GPI anchored molecules.

Concepts for the Ig superfamily may provide a prototype for thinking about all of these domains. In this case it has been argued that the domain evolved as a cell surface structure that was resistant to proteolysis that became involved in cell-cell interactions (Williams, 1982, 1987). Increasingly sophisticated molecular signalling was required as multi-cellular organisms developed and the Ig superfamily molecules appear to be the largest group of control proteins. Equally the other superfamilies that are involved in cell-cell interactions may also have first evolved as coat proteins.

It appears to be the case that most molecules at the surface of single cell eukaryotes have GPI anchors. Most of the data comes from work on mammalian parasites but the results should reflect the case for free-living single cell eukaryotes provided that the parasites can be fairly regarded as diverse organisms that have become parasites by uncorrelated mechanisms. The fact that the coat protein of free-living *Paramecium aurelia* is also GPI-anchored supports this view. Single cell eukaryotes have the problem of living in hostile environments without the benefit of a cell wall as seen in bacteria and plants and thus may have had a particular problem in producing a cell surface that could stand to hostile environments. The GPI anchor may have been the solution to this problem in that the surface coat molecules could be anchored directly into the bilayer to provide a tightly packed surface. In contrast protein anchored molecules may perhaps always lead to some disruption of the bilayer. If this thinking is correct, it seems likely that the coat proteins that first functioned to mediate the formation of multi-cellular organisms leading to animal species would have been attached to the cell surface by GPI anchors.

REFERENCES

Baron, M., Norman, D. G. and Campbell, I. D. (1991). Protein molecules. TIBS **16:** 13-17.

Bothwell, A., Pace, P. E. and LeClair, K. P. (1988). Isolation and expression of an IFN-responsive Ly-6C chromosomal gene. J.Immunol. **140:** 2815-2820.

Davies, A., Simmons, D. L., Hale, G., Harrison, R. A., Tighe, H., Lachman, P. J. and Waldmann, H. (1989). CD59, an Ly-6-like protein expressed in human lymphoid cells, regulates the action of the complement membrane complex on homologous cells. J. Exp. Med. **170:** 637-654.

Dayhoff, M. O., Barker, W. C. and Hunt, L. T. (1983). Establishing homologies in protein sequences. Methods in Enzymol. **91:** 524-545.

Freeman, M., Ashkenas, J., Rees, D. J. G., Kingsley, D. M., Copeland, N. G., Jenkins, N. A. and Krieger, M. (1990). An ancient highly conserved family of cysteine rich protein domains revealed by cloning type 1 and type 2 murine macrophage scavenger receptors. Proc. Natl. Acad. Sci. USA **87:** 8810-8814.

Freidman, S., Palfree, R. G. E., Sirlin, S. and Hammerling, U. (1990). Analysis of three distinct Ly6-A-related cDNA sequences isolated from rat kidney. Immunogenetics **31:** 104-111.

Harrelson, A. L. and Goodman, C. S. (1988). Growth cone guidance in insects: Fasciclin 2 is a member of the immunoglobulin superfamily. Science **242:** 700-708.

LeClair, K. P., Palfree, R. G. E., Flood, P. M., Hammerling, U. and Bothwell, A. (1986). Isolation of a murine Ly6 cDNA reveals a new multigene family. EMBO J. **5:** 3227-3234.

Mallett, S., Fossum, S. and Barclay, A. N. (1990). Characterisation of the MRC OX40 antigen of activated CD4 positive T lymphocytes- a molecule related to nerve growth factor receptor. EMBO J. **4:** 1063-1068.

Meri, S., Morgan, B. P., Davies, A., Daniels, R. H., Olavesen, M. G., Waldmann, H. and Lachman, P. J. (1990). Human pronectin (CD59), an 18,000-20,000 MW complement-lysis restricting factor, inhibits C5b-8 catalysed insertion of C9 into bilayers. Immunology **71:** 1-9.

Palfree, R. G. E. (1991). The Urokinase-type plasminogen activator receptor is a member of the Ly6 superfamily. Immunol. Today. In press.

Palfree, R. G. E., Sirlin, S., Dumont, F. J. and Hammerling, U. (1988). N-terminal cDNA characterisation of murine lymphocyte antigen Ly6C.2. J.Immunol. **140:** 305-310.

Pathy, L. (1987). Intron-dependent evolution: preferred types of exons and introns. FEBS Lett. **214:** 1-7.

Ploug, M., Ronne, E., Behrendt, N., Jensen, A. L., Blasi, F. and Dano, K. (1991). Cellular receptor for urokinase plasminogen activator. J. Biol. Chem. **266:** 1926-1933.

Robinson, P. J. (1991). Phosphatidylinositol membrane anchors and T-cell activation. Immunol. Today. **12:** 34-41.

Roldan, A. L., Cubellis, M. V., Masucci, M. T., Behrendt, N., Lund, L. R., Dano, K., Appella, E. and Blasi, F. (1990). Cloning and expression of the receptor for human urokinase plasminogen activator,a central molecule in cell surface,plasmin dependent proteolysis. EMBO J. **9:** 467-474.

Schevach, E. M. and Korty, P. E. (1989). Ly6: a Multigene family in Search of a Function. Immunol. Today. **10:** 195-200.

Williams, A. F. (1982). Surface Molecules and Cell Interactions. J. Theoret. Biol. **98:** 221-234.

Williams, A. F. (1987). A year in the life of the immunoglobulin superfamily. Immunol. Today. **8:** 298-303.

Williams, A. F. and Gagnon, J. (1982). Neuronal Cell Thy-1 glycoprotein: Homology With Immunoglobulin. Science **216:** 696-703.

Williams, A. F. Tse, A. G-D. and Gagnon, J. (1988). Squid Glycoproteins with Structural Similarities to Thy-1 and Ly-6 Antigens. Immunogenetics **27:** 265-272.

Wong, Y. W., Williams, A. F., Kingsmore, S. F. and Seldin, M. F. (1990). Structure, expression and genetic linkage of the mouse BCMI (OX45 or Blas1) antigen. J. Exp. Med. **171:** 2115-2130.

GLYCOSYLATED-PHOSPHATIDYLINOSITOLS AS VIRULENCE FACTORS IN *LEISHMANIA*

***Malcolm J McConville**

Department of Biochemistry, University of Dundee, Dundee, DD1 4HN, Scotland.

* Corresponding author.

INTRODUCTION

Protozoan parasites of the genus *Leishmania* cause a number of important diseases in humans and undergo a complex life cycle, alternating between an insect (sandfly) vector and their mammalian hosts. These organisms occur as extra-celluar promastigotes in the digestive tract of the sandfly and as obligate intracellular amastigotes in the phagolysosomes of mammalian macrophages. There is evidence that surface molecules play a key role in mediating these host-parasite interactions and in protecting the parasite in these hydrolytic environments. The surface membrane of both stages is coated by a glycocalyx which is unusual in being rich in glycosylated-phosphatidylinositol (GPI) glycolipids. Some of these glycolipids act as membrane anchors for the major cell surface proteins (Bordier 1985, Murray et al 1990, Lohman et al 1990). This type of membrane anchor is partic-ularly common in the protozoas, but has also been found in all other eukaryotes which have been investigated (reviewed by Ferguson and Williams 1988, Low 1989, Thomas et al 1990). Recent studies on the structure of the GPI anchor from the promastigote surface protease (also referred to as gp63) of *L. major* (Schneider et al 1990)indicate that the protein anchors produced by these parasites are structurally similar to those produced by other protozoa (Ferguson et al 1988) and by higher eukaryotes (Homans et al 1989), Roberts et al 1989) *Leishmania* also produce two other classes of GPI that are not linked to protein and which are the major cellular glycoconjugates. These are the heterogeneous lipophosphoglycans (LPGs) and a family of low molecular weight glycoinositolphospholipids (GIPLs). In this chapter, I will review recent work on the structure and function of these novel GPIs.

LIPOPHOSPHOGLYCAN

2.1 Structure of promastigote LPG

LPGs are produced by all species of *Leishmania* and are the major macromolecule on the promastigote cell surface. They were initially detected as prominent antigens which could be metabolically labelled with (^{3}H)-sugars, but not amino acids, and had a heterogeneous molecular weight by SDS-polyacrylamide gel electrophoresis and gel filtration (Handman et al 1984, Turco et al 1984, 1989,) Studies on the promastigote LPGs from *L. donovani* (Orlandi and Turco 1987, Turco et al 1987, 1989, Thomas et al 1991), *L. major* (McConville et al 1987, 1990a) and *L. mexicana* (Ilg et al 1991) suggest that these molecules have the generic structure shown in Fig 1.

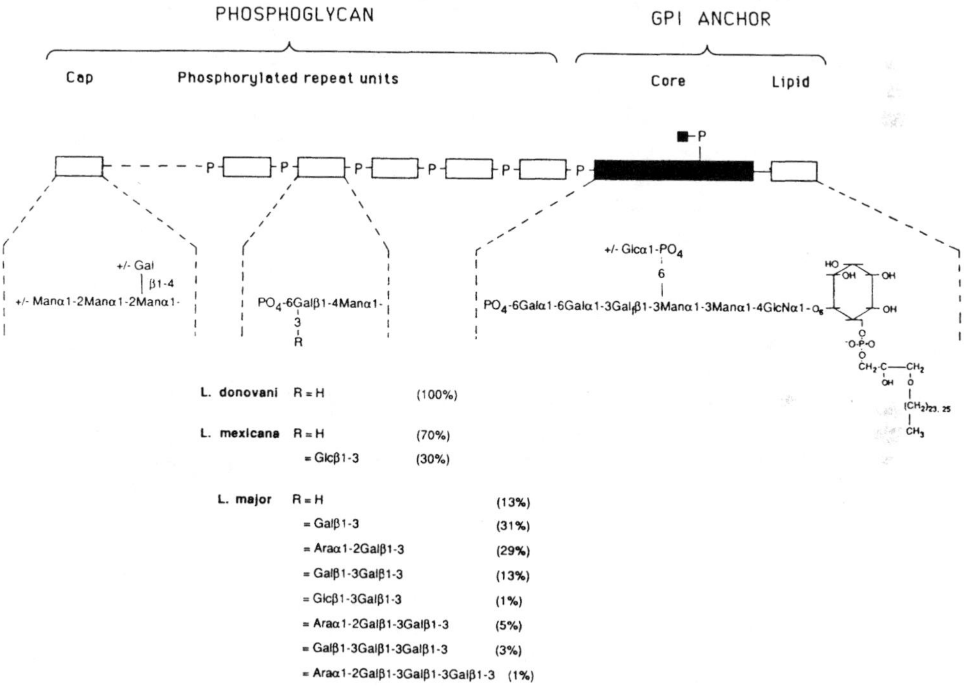

Figure 1. Generic structure of *Leishmania* LPG showing the species-specific and common structural features. These structures are based on the following studies; *L.donovani* LPG (Orlandi and Turco 1987, Turco, et al. 1987, 1990, Thomas, et al. 1991), *L.mexicana* LPG (Ilg, et al. 1991) and *L.major* LPG (McConville, et al. 1987, 1990a, McConville and Homans, unpublished data). The major species-specific differences in LPG structure occur in the nature of the oligosaccharide side-chains on the phosphorylated disaccharide repeat units and the frequency of substitution. Differences also occur in the degree to which the second mannose residue of the core glycan is substituted with glucose-1-phosphate (100% in *L.donovani* and *L.mexicana*, LPG, 60% in *L.major* LPG) and the relative amounts of each cap structure.

They all contain a heterogeneous phosphoglycan moiety (M_r4000-40,000) which is linked to the membrane by a novel GPI anchor. The phosphoglycans are made up of repeating units of PO_4-6Galß1-4Manα1-, where the 3-position of the Gal residues can either be unsubstituted (as in *L. donovani* LPG), partially substituted with ßGlc residues (as in *L. mexicana* LPG) or highly (>90%)substituted with a variety of side chains containing ßGal, ßGlc or αArap (as in *L. major* LPG) (Fig 1). Some of these structures, including the backbone disaccharide sequence and the terminal α-D-arabinopyranose residues of the *L. major* LPG, have not been identified previously in any glycoconjugate. The presence of a common backbone sequence which is variably elaborated with species-specific side chains is consistent with the serological studies which indicate the presence of both conserved and species-specific epitopes (Tolsen et al 1989). The ends of the phosphoglycan chains, distal to the membrane, are capped by a number of neutral oligosaccharides which all contain the sequence Manα1-2Manα1- (McConville et al 1990, Thomas et al 1991, Ilg et al 1991). This sequence may constitute another epitope common to all LPGs. The phosphoglycan is linked to the membrane by a GPI anchor which contains a phosphorylated hexasaccharide head group and an unusual *lyso*alkyl-PI lipid moiety, with predominantly C24:0 or C26:0 alkyl chains (Fig 1). The glycan moiety contains the core sequence Manα1-4GlcNH$_2$α1-6*myo*inositol, which is common to all GPIs, but diverges from the protein-linked GPIs beyond this sequence. In particular, the LPG cores are unusual in containing an internal galactofuranose residue. This residue was previously identified as αGal$_f$ (Turco et al 1989) but has now been shown to be in the ß-configuration (McConville et al 1990a, Thomas et al 1991). The terminal galactose residue of the core is always phosphorylated on the 6-position and appears to be the exclusive site of attachment for the phosphorylated repeat units (McConville et al 1990a, Thomas et al 1991, McConville and Homans unpublished date). Recent studies have shown that the second mannose residue of the core may also be substituted, not with repeat units, but with glucose-1-phosphate (Thomas et al 1991, McConville and Homans unpublished date).

In addition to the species-specific differences in LPG structure, there is evidence that the LPGs are modified during metacyclogenesis, when promastigotes make the transition from log growth phase to stationary phase. This developmental modification of promastigote LPG may be important in the increased infectivity of metacyclic promastigotes (see 2.5). Sacks and co-workers showed that metacyclic LPG from *L. major* had a higher apparent molecular weight on SDS-polyacrylamide gel electrophoresis, no longer bound the lectin peanut agglutinin (PNA) and contained a novel carbohydrate epitope (Sacks and da Silva 1987). Recent studies show that there is an approximate doubling of the number of repeat units in LPG during metacyclogenesis (Turco and Sacks 1990) and that the lack of PNA binding is due to capping of the galactose-containing side chains with α-arabinopyranose residues (McConville, Ferguson, Sacks and Turco unpublished results). Loss of PNA binding also occurs during metacyclogenesis in *L. donovani* (Cooper et al 1988), even though its LPG does not contain arabinose. It is possible that PNA binds to the neutral cap structures in log phase cells (see Fig 1) which become cryptic in metacyclic LPG.

2.2 Three dimensional structure of LPG

A dramatic and compelling image of how *Leishmania* LPG might appear was provided by a molecular modelling study on the LPG from *L. donovani* (Homans 1990). In this structure the oligosaccharide repeat units form an extended chain with a helical pitch. Several stable conformers are possible due to flexibility around the phosphodiester bonds (Fig 2) which may allow the phosphoglycan chain to expand or contract, reminiscent of a slinky spring (Homans, S.W., Melhert, A. and Turco, S.J. unpublished data). A feature of the different conformers is that the mannose and phosphate residues are buried within the helix while the galactose residues exist on the outside of the helix (Fig 2). The peripheral kisposition of the galactose residues is important as it suggests that the side chains of the *L. major* and *L. mexicana* LPGs can be accommodated without major conformational changes in the repeat backbone. These chains are likely to be arranged perpendicular to the axis of the main chain and thus greatly increase the cross-sectional area of the LPG molecule. The core glycan also exists in an extended configuration and has regions of mobility around the Galα1-6Gal linkage and within the ß-galactofuranose residue (Homans 1990). These studies predict that the LPG will extend away from the plasma membrane for some distance and that the LPGs from *L. major* and *L. mexicana,* will have a bottlebrush appearance. It has been estimated that there are 1-5 X 10^6 molecules of LPG on the cell surface (Orlandi and Turco 1987, McConville and Bacic 1990, McConville and Blackwell 1991), which would effectively coat the majority of the promastigote surface. This coat, which can be visualized by electron microscopy (Pimenta et al 1989), may form a macromolecular diffusion barrier as antibodies are unable to bind LPG epitopes near the plasma membrane (Tolsen et al 1989).

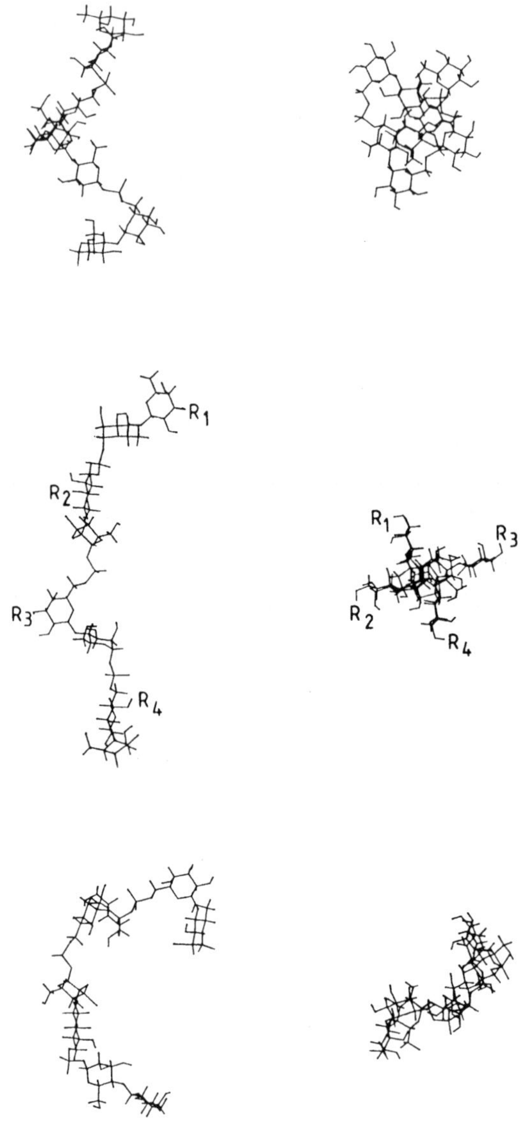

Figure 2. Predicted three dimensional structure of the *L. donovani* LPG phosphoglycan. Shown are the three major conformers (1-3) of a sequence containing four repeat units of Galβ1-4Manα1-PO_4-6 (from Homans, S.W., Melhert, A. and Turco, S.W. unpublished data). These structures were deduced by computational chemistry and by direct experimental determination of spin coupling constants by homo- and heteronuclear NMR. Each conformer is shown as viewed longitudinally (on the left) or down the long axis of the repeats (on the right). R denotes the 3-position of the repeat unit galactose which may be substituted with oligosaccharide side-chains in *L. major* and *L. mexicana* LPGs (see **Fig 1**).

2.3 Relationship between cellular LPG and 'excreted factor'

LPG labelled on the surface of promastigotes is rapidly released into the culture supernatant (Handman et al 1984, King et al 1987). A significant fraction of this material still retains its lipid moiety (Ilg et al 1991) and may interact specifically with serum proteins that contain a hydrophobic pocket (King et al 1987). It is likely that this LPG is being passively released from the plasma membrane as its attachment to the bilayer by a single alkyl chain will be weak. This is supported by chemical analysis of the cellular and extracelluar forms of the *L. mexicana* LPG which indicates that there is preferential loss of LPG molecules from the plasma membrane with longer carbohydrate chains and shorter alkyl groups (Ilg et al 1991). Phosphoglycan chains which are indistinguishable from cellular LPG, except that they lack any part of the GPI anchor are also released into the culture supernatant (Greis, K., Turco, S.J., Homans, S.W., McConville, M.J., Thomas, J. and Ferguson, M.A.J. unpublished results). It is unknown whether these soluble chains are released by a membrane-bound endoglycosidase or phosphodiesterase or whether they represent separate biosynthetic products. Together, the released intact LPG and the soluble phosphoglycan (PG) appear to account for a major fraction of the excreted factor observed in *Leishmania* cultures.

2.4 Amastigote LPG

There appears to be species-specific differences in the expression of LPG in the amastigotes stage which infects mammalian macrophages. LPG was not detected in *L. donovani* amastigotes by direct chemical analysis or by specific radiochemical labelling procedures (McConville and Blackwell 1991). In contrast, an LPG-like molecule has been detected in *L. major* amastigotes although at much lower levels than in promastigotes (Handman et al 1989), Glaser et al 1991, Turco and Sacks 1991). Preliminary characterization suggests that the repeat units of this amastigote LPG are distinct (antigenically and structurally) from promastigote LPG (Glaser et al 1991, Turco and Sacks 1991).

2.5 Function of LPG

After being introduced into the bloodstream of the mammalian host, promastigotes must evade being lysed by the serum complement system and then bind to and be phagocytosed by host macrophages. There is evidence that LPG plays a key role in these different steps (Table 1). The mechanism by which LPG increases the complement resistance of metacyclic promastigotes has been studied in *L. major* (Puentes et al 1991). Paradoxically, LPG appears to be the major activator of complement on the

Table 1 Proposed functions of *Leishmania* promastigote LPG

Function	Reference
In the sandfly:	
Attachment to epithelial cells in the sandfly midgut	Davies et al 1990
In the mammalian host:	
Activation of complement via alternative or classical pathways	Puentes et al 1988
Resistance to complement-mediated lysis (in metacyclic promastigotes)	Puentes et al 1990
Attachment to macrophages and phagocytosis	Handman and Goding 1985, Handman et al 1986, Russell and Wright 1988, Da Silva et al 1989, Talamas-Rohana et al 1990
Inhibition of the oxidative burst and scavenger of oxygen radicals	McNeely et al 1989, McNeely and Turco 1990, Chan et al 1989
Chelation of divalent cations	Eilam et al 1985
Inhibition of lysosomal enzymes	El-On et al 1980
Inhibition of selected macrophage signal transduction pathways	Descoteaux et al 1991

surface of both log phase and metacyclic promastigotes (Puentes et al 1988). On log phase promastigotes complement is deposited primarily as the C3b, which leads to the formation of the C5b-9 complex and membrane lysis (Puentes et al 1988, 1990). While C3b is also deposited on the LPG of metacyclic promastigotes, the C5b-9 complex that subsequently forms is released from the cell surface as a soluble complex before it can insert into the parasite membrane (Puentes et al 1990). A likely mechanism to account for these differences is that activation of complement occurs further from the plasma membrane on metacyclic promastigotes due to the presence of longer LPG chains preventing the stable insertion of the membrane attack complex into the membrane. In contrast, deposition of C3b on the shorter LPG chains of log phase promastigotes leads to the formation of a C5b-9 complex which is close enough to the plasma membrane to stably insert into the bilayer and cause cell lysis (Puentes et al 1990). Strains of *Salmonella* bacteria which produce long lipopolysaccharide chains may evade complement-mediated lysis in a similar way (reviewed Joiner 1989). The mechanism by which LPG is able to activate the complement pathway will be of interest as both the alternative and classical pathways are thought to be involved (Puentes et al 1988). One possibility is that the terminal mannose caps of the LPG are recognized by the mannose-binding protein in mammalian serum (McConville and Feizi, unpublished date), which is able to activate complement via both pathways (Ikeda et al 1987, Schweinle et al 1989). Activation of complement at the terminus of LPG, distal to the plasma membrane, would increase the steric hindrance to C5b-9 insertion as well as increase the accessibility of these complement components to macrophage receptors (see below).

Binding of complement components, in particular C3b, to surface LPG may be required for parasite uptake by macrophages via the complement receptors CR1 or CR3. This is suggested by the finding that binding of metacyclic promastigotes of *L. major* to macrophages in vitro can be inhibited by down-regulating surface CR1 receptors (da Silva et al 1989). Uptake via the CR1 receptor may be advantageous for the parasite as this receptor does not activate the microbicidal oxidative burst. However, experiments performed in the absence of serum components indicate that LPG may also bind directly to macrophage receptors. This was initially shown by Handman and Goding (1985) who were able to inhibit the binding of *L. major* promastigotes to macrophages by preincubating the parasites with the anti-LPG monoclonal antibody WIC-79.3. Interestingly, variant strains of *L. major* (Handman et al 1986, Elhay et al 1990) or *L. donovani* (McNeely and Turco 1990) which are deficient in surface LPG, are also phagocytosed by macrophages but are rapidly killed in the phagolysosome, suggesting that they may be taken up via other receptors which activate various microbicidal processes (eg oxidative burst, nitrous oxide production). After passive transfer of LPG into the membranes of these variant strains, they had similar kinetics of uptake as the wild-type strains and were able to survive for longer in the phagolysosome of macrophages (Handman et al 1986, McNeely and Turco 1990). LPG-coated beads have been used to identify some of the receptors on the macrophage which can recognize *L. major* LPG (Russell and Wright 1988, Talamus-Rohana et al 1990). Down-regulation of several integrins of the CD18 complex, notably CR3 and p150,95, diminished the binding of the beads to the macrophages suggesting that LPG may be recognised by these proteins via their lectin-like domains. Recent studies suggest that some of these receptors may bind to specific side chains of the *L. major* LPG. In particular, attachment of promastigotes to macrophages could be inhibited in the presence of purified repeat units containing Galβ1-or Galβ1-3Galβ1-side chains, but not by repeat units with Araα1-2Galβ1-side chains (Kelleher, Handman and Bacic, unpublished results). Other macrophage receptors, such as the mannose-fructose receptor may also bind to LPG (McConville and Feizi, unpublished results) and mediate promastigote phagocytosis. This interaction may be particularly important in facilitating the attachment of *L. donovani* promastigotes to macrophages (Channon, et al., 1983, Blackwell, et al., 1985, Wilson and Pearson,

1988). Taken together, the available information suggests that LPG plays a key role in mediating phagocytosis of *Leishmani* promastigotes, potentially interacting with a variety of different macrophage receptors via a number of distinct mechanisms.

Once inside the phagolysosome, LPG may protect the parasite surface from hydrolytic enzymes (El-On et al 1980) or attenuate the effects of the oxidative burst by scavenging oxygen radicals (Chan et al 1989), or chelating divalent cations in the phagolysosome which are necessary to sustain the burst (Eilam et al 1985). LPG is also a potent inhibitor of protein kinase C (McNeely et al 1989) and it has been proposed that released LPG may cross the membrane of the parasitophorous vacuole and gain access to cytosolic protein kinase C species which are involved in activating the oxidative burst (McNeely and Turco 1990) and host cell signal transduction (Descoteaux et al 1991).

There is comparatively little information on the function of promastigote LPG in the insect vector. It is likely that LPG is involved in mediating the attachment of promastigotes to the gut wall as the secreted LPG appears to bind specifically to the epithelial cells of the gut wall (Davies et al 1990). Furthermore, Davies et al (1990) showed that as the promastigotes detached from the gut wall and migrated to the pharynx of the sandfly, they expressed the unique epitopes of metacyclic LPG, suggesting that developmental modification of LPG may be important in regulating the forward migration of promastigotes along the sandfly gut.

There have been no studies on the function of amastigote LPG. However, the absence of detectable LPG in *L. donovani* amastigotes indicates that, at least in some species, other surface molecules may perform some of the above functions or that amastigotes invade macrophages and evade the microbicidal processes in the phagolysosome by different mechanisms.

GIPLs

3.1 Structures

The GIPLs are the major class of glycolipid in these parasites. They are present in very high copy numbers, ca.10^7 per cell, in both the promastigote and amastigote stages and may account for as much as 20% of the cellular carbohydrate (McConville and Blackwell 1991). Structural analysis of the GIPLs from *L. donovani* (McConville and Blackwell 1991) and *L. major* (McConville and Bacic 1989, McConville et al 1990b) indicate that there are species-specific differences as well as stage-specific differences in these structures. The GIPLs from *L. donovani* contain linear or branched glycan chains with 2-4 mannose residues linked via non-N-acetylated glucosamine to an 1-alkyl-2-acyl-PI lipid moiety (see Fig 3, McConville and Blackwell 1991). The glycan structures of the promastigote GIPLs resemble the protein GPI anchors and the LPG anchor in containing mannose branches on both the 6- and 3-position of the first mannose residue, respectively. By contrast

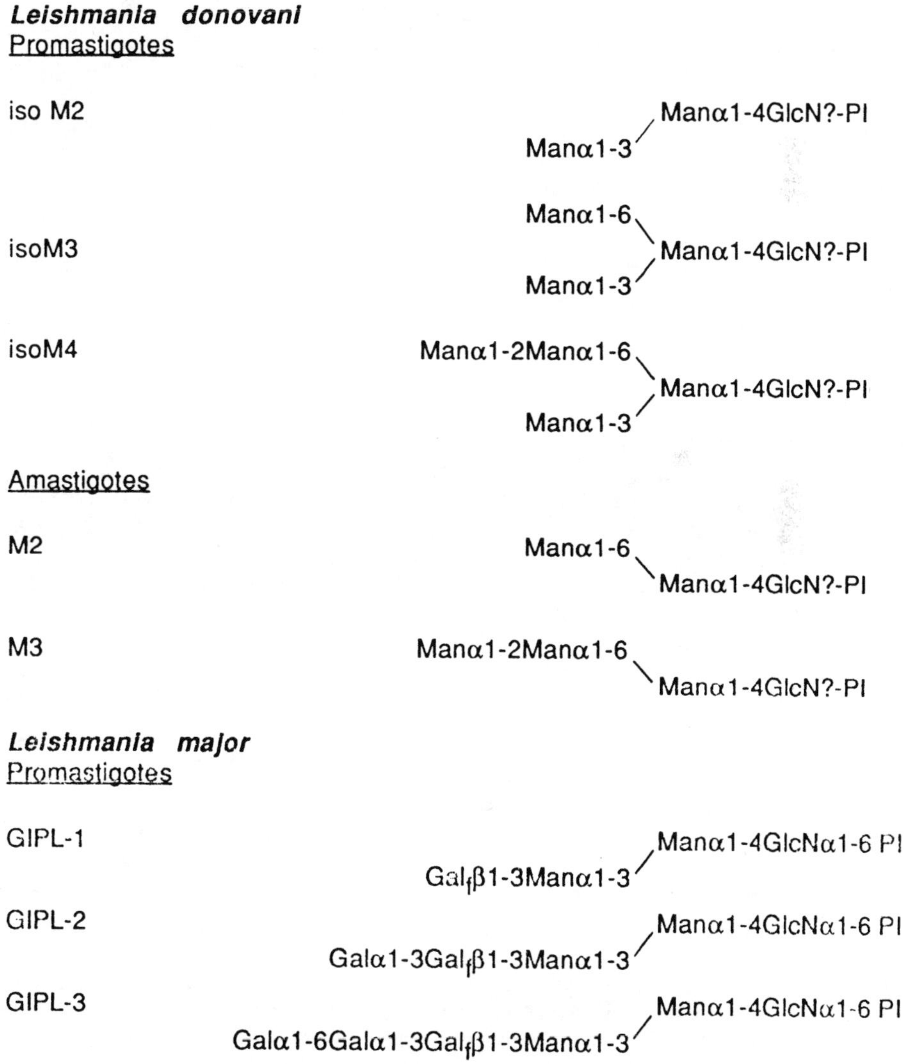

Figure 3. Structures of the major GIPLs from *L. donovani* and *L. major* (from McConville, et al. 1990a,b, and McConville and Blackwell 1991).

the amastigote GIPLs lack the mannose branch on the 3-position and have the same glycan structure as the pg63 GPI anchor (Fig 3, McConville and Blackwell 1991, Schneider et al 1990). *L. major* promastigotes synthesize a third series of GIPLs which are structurally homologous to the LPG anchor in containing the core sequence Galfß1-3Manα1-3Manα1-4GlcNα1-6 *myo*inositol (Fig 3, McConville et al 1990a,b). Significantly different structures were reported from another strain of *L. major* (Rosen et al 1989) although reanalysis has revealed that they are in fact the same as described above (Rosen et al 1990). An interesting aspect of the *L. major* GIPLs is the finding that there is an increase in the chain length of the alkyl with elongation of the glycan head group (McConville and Bacic 1989, McConville et al 1990b). Thus the smaller GIPLs have predominantly C18:0 alkyl chains while the larger GIPLs have predominantly C24:0 or C26:0 alkyl chains. Whether this indicates that some of the glycosyl transferases have a specificity for precursors with longer alkyl chains or that a novel type of lipid remodelling process is occurring is unknown. Lipid remodelling steps do occur in *Leishmania* as some of the GIPLs are selectively deacylated by an endogenous phospholipase A_2 (McConville and Bacic 1989, McConville et al 1990).

3.2 Function of the GIPLs

Immunofluorescence and surface labelling experiments indicate that a significant proportion of the cellular GIPLs are expressed at the cell surface (McConville and Bacic 1989, 1990). It has been estimated that they may cover the same proportion of the promastigote membrane as LPG and most of the surface of the smaller amastigotes (McConville and Bacic 1990, McConville and Blackwell 1991). While the function of these surface glycolipids is unknown, it is likely that they will influence the physico-chemical properties of the plasma membrane. In this regard it is of interest that all these glycolipids contain ether-linked fatty groups which may protect the membranes from chemical or enzymatic degradation in the hydrolytic environments of the sandfly gut or the phagolysosome. The species-specific differences in glycan head group structure are striking and may indicate a requirement for specific structures on the cell surface which may be involved in specific host-parasite interactions. Alternatively, these differences may not be functionally significant but simply reflect the utilization of different enzymes systems which are already involved in the synthesis of either the protein or LPG anchors. From estimates of surface expression, it is clear that a significant proportion of the GIPLs must have an intracellular location (McConville and Blackwell 1991). Some of these GIPLs are structurally homologous to the GPI anchors of surface proteins or LPG and may function as precursors to these anchors.

GPIs AND THE HOST IMMUNE RESPONSE

4.1 Humoral responses

Patients with leishmaniasis and experimentally infected mice produce high titres of antibodies directed against LPG or GIPL epitopes (Schnur 1982, Rosen et al 1988, Avila et al 1989, McConville et al 1990). Many of these antibodies belong to the IgG isotypes, IgG2a, IgG2b and IgG3 (Handman and Hocking 1982, McConville and Bacic 1989), indicating that some of the B-cells may be stimulated by T-cells (see below). Of interest is the finding that many patients with leishmaniasis have elevated levels of IgG antibodies with an apparent specificity for terminal Galα1-3Gal epitopes (Towbin et al 1987, Avila et al 1988a,b). Although the fine specificity of these antibodies has not been rigorously determined they appear to differ from the anti-Gal antibodies which account for up to 1% of the circulating IgG in normal human serum (Galili et al 1984) and may be directed against the terminal Galα1-3Gal_f epitopes of some of the GIPLs (eg. GIPL-2) (McConville et al 1991). There is no evidence that these antibodies are involved in host resistance, although it is possible that they may act as opsonins on the parasite surface and stimulate complement activation.

4.2 T-cell responses

Early experiments showed that purified *L. major* LPG (Handman and Mitchell 1985, McConville et al 1987), could vaccinate highly susceptible strains of mice against infection with *L. major*. As resistance to leishmaniasis appears to be mediated by T-cells (reviewed by Muller et al 1989) these results indicate that LPG causes the specific activation of host-protective T-cells. Further evidence supporting this notion has been recently obtained (Moll et al 1989, Kemp et al 1991). Studies on the T-cell responses in mice showed that vaccination with purified LPG (together with *Corynebacterium parvum* as adjuvant) induced the production of *L. major* reactive T-cells (Moll et al 1989). The activated lymphocytes were of the $CD4^+$ (T-helper) subclass (Moll et al 1989) which are involved in mediating host immunity. Indirect support for the activation of T-cells in vivo was provided by the ability of LPG to elicit a specific delayed-type hyposensitivity response in resistant mouse strains and the demonstration that the anti-LPG antibodies were of the IgG subtype (Moll et al 1989). Interestingly, the presence of the *lyso*alkyl-PI moiety appears to be essential for induction of protective T-cells as immunization with the phosphoglycan moiety results in exacerbation of lesion development (Mitchell and Handman

1986). Human T-cells also appear to recognize LPG as peripheral blood mononuclear cells from cured kala azar patients can be specifically activated by purified *L. major* LPG (Kemp et al 1991). These results support the intriguing notion that in some circumstances, non-peptide antigens can stimulate T-cells, although whether this occurs in the same way as peptide antigens (i.e. in association with MHC glycoproteins) is unknown. In this regard it is significant that LPG epitopes have been detected on the surface of infected macrophages (Handman and Goding 1985, Handman 1990, Tolsen et al 1990). While some of these epitopes may be associated with macrophage surface proteins (Handman 1990) it is likely that the passive transfer of LPG from the promastigote to the macrophage membrane during phagocytosis may account for most of these epitopes (Tolsen et al 1990).

CONCLUSIONS

In addition to utilizing GPI glycolipids as anchors for surface proteins, *Leishmania* parasites produce a number of novel GPIs which are not linked to protein. These molecules form a glycocalyx over the surface membrane and are major antigens to be targeted by the host immune response. Structural studies on the LPGs from different species have revealed common architectural elements as well as species-specific and developmentally-regulated modifications. The absence of similar molecules in closely related kinetoplastid parasites indicates that the LPGs may have appeared relatively recently in evolutionary terms, after these organisms had developed their specialized parasitic life-style. This is consistent with the finding that these molecules play key roles in parasite infectivity and survival in both the insect vector and the mammalian host. In this regard, information on LPG structure has provided insights into the mechanisms of action of these multifunctional molecules and the ways in which *Leishmania* parasites interact with host cells. Considerable information is also available on the structures of the GIPLs although little is known about their function. Possible pathways of metabolism of the LPGs, the GIPLs and the protein linked GPI anchors in the different developmental stages of the parasite are indicated in Fig 4. In view of the abundance of GPI glycoconjugates in these organisms, they are likely to provide an interesting system for isolating and studying the enzymes involved in GPI biosynthesis.

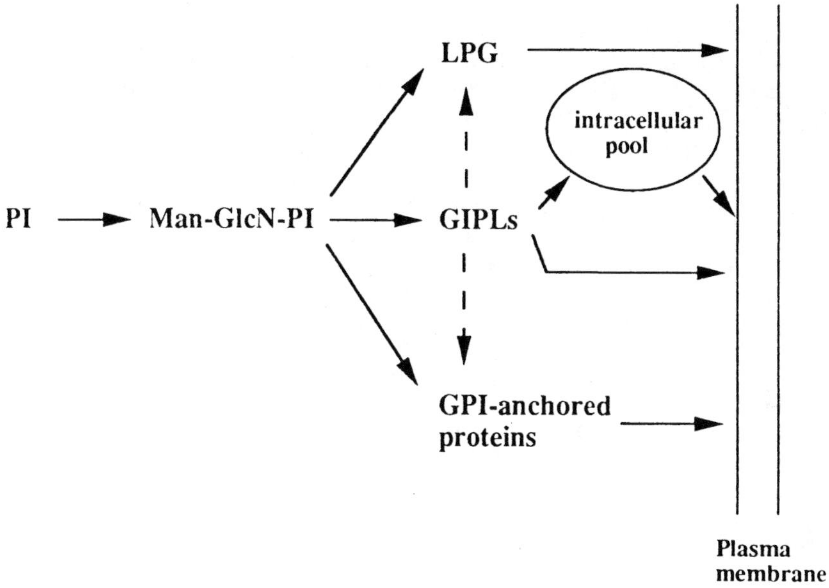

Figure 4. Possible pathways of GPI metabolism in the different developmental stages of *Leishmania*. All the GPI glycoconjugates of *Leishmania* contain the core sequence Manα1-4GlcNα1-6*myo*inositol and probably share the same initial biosynthetic steps. Some of the GIPLs are structurally similar to the GPI anchors of the surface proteins and LPG and may act as biosynthetic precursors to these anchors (dotted lines). However, most of the GIPLs appear to be abundant metabolic end products which are expressed either on the cell surface or in, as yet unidentified, intracellular pools. The importance of these different pathways may vary in the two major developmental stages of the parasite. In particular, there is evidence that LPG (McConville and Blackwell 1991) and the levels of some GPI-anchored proteins (Medina-Acosta, et al. 1989) are down-regulated in the amastigote stage, while the celluar level of the GIPLs are comparable in both stages (McConville and Blackwell 1991).

ACKNOWLEDGEMENT

I would like to thank my colleagues for making data available to me prior to publication, in particular, Dr Steve Homans for providing the figure on LPG conformation. I also thank Dr Mike Ferguson for valuable discussions. This work was supported by an Australian National Health and Medical Research Council C.J. Martin Fellowship and the Wellcome Trust.

REFERENCES

Avila, J.L., Rojas, M., and Galili, U. (1989). Immunogenic Galα1-3Gal carbohydrate epitopes are present on pathogenic American *Trypanasoma* and *Leishmania*. J. Immunol. **142:** 2828-2834.

Avila, J.L., Rojas, M. and Garcia, L. (1988). Persistence of elevated levels of galactosyl-α(1-3)galactose antibodies in sera from patients cured of visceral leishmaniasis. J. Clin. Microbiol., **26:** 1842-1847.

Blackwell, J.M., Ezekowitz, A.B., Roberts, M.B., Channon, J.Y., Sim, R.B. and Gordon, S. (1985). Macrophage complement and lectin-like receptors bind *Leishmania* in the absence of serum. J. Exp. Med. **162:** 324-331.

Chan, J., Fujiwara, T., Brennan, P., McNeil, M., Turco, S.J., Sibille, J.C., Snapper, M., Aisen, P. and Bloom, B.R. (1989). Microbial glycolipids: Possible virulence factors that scavenge oxygen radicals. Proc. Natl. Acad. Sci. USA. **86:** 2453-2457.

Cooper, A., Rosen, H. and Blackwell, J.M. (1988). Monoclonal antibodies that recognise distinct epitopes of the macrophage type three complement receptor differ in their ability to inhibit binding of *Leishmania* promastigotes harvested at different phases of their growth. Immunology. **65:** 511-514.

Davies, C.R., Cooper, A.M., Peacock, C., Lane, R.P and Blackwell, J.M. (1990). Expression of LPG and GP63 by different developmental stages of *Leishmania major* in the sandfly *Phlebotomus papatasi*. Parasitology. **101:** 337-343.

Descoteaux, A., Turco, S.J., Sacks, D.L. and Matlashewski, G. (1991). *Leishmania donovani* lipophosphoglycan selectively inhibits signal transduction in macrophages. J. Immunol. **145:** in press.

Eilam, Y., El-On, J. and Spira, D.T. (1985). *Leishmania major*: excreted factor, calcium ions and the survival of amastigotes. Exp. Parasitol. **59:** 161-168.

El-On, J., Bradley, D.J. and Freeman, J.C. (1980). *Leishmania donovani*: Action of excreted factor on hydrolytic enzyme activity of macrophages from mice with genetically different resistance to infection. Experimental Parasitology. **49:** 167-174.

Elhay, M.J., Kelleher, M., Bacic, A., McConville, M.J., Tolsen, D.L., Pearson, T.W. and Handman, E. (1989). Lipophosphoglycan expression and virulence in variants of *Leishmania major*. Mol. Biochem. Parasitol. **40:** 167-174.

Ferguson, M.A.J., Homans, S.W., Dwek, R.A. and Rademacher, T.W. (1988). Glycosylphosphatidylinositol moiety that anchors *Trypanosama brucei* variant surface glycoprotein to the membrane. Science. **239:** 753-759.

Galili, U., Shohet, S.B., Kobrin, E., Stults, C.L.M. and Macher, B.A. (1988). Man, apes and old world monkeys differ from other mammals in the expression of α-Galactosyl epitopes on nucleated cells. J. Biol. Chem. **263:** 17755-17762.

Glaser, T.A., Moody, S.F., Handman, E., Bacic, A. and Spitall, T.W. (1991). Mol. Biochem. Parasitol **45:** 337-344.

Handman, E. (1990). Study of *Leishmania major* infected macrophages by use of LPG-specific monoclonal antibodies. Infect. Immun. **58:** 2297-2302.

Handman, E. and Hocking, R.E. (1982). Stage-specific, strain-specific, and cross reactive antigens of *Leishmania* species identified by monoclonal antibodies. Infect. Immun. **37:** 28-33.

Handman, E., Greenblatt, C.L. and Goding, J.W. (1984). An amphipathic sulphated glycoconjugate of *Leishmania*: characterisation with monoclonal antibodies. EMBO. J. **3:** 2301-2306.

Handman, E. and Goding, J.W. (1985). The *Leishmania* receptor for macrophages is a lipid-containing glycoconjugate. EMBO. J. **4:** 329-336.

Handman, E., Schnur, L.F., Spithill, T.W. and Mitchell, G.F. (1986). Passive transfer of *Leishmania* lipopolysaccharide confers parasite survival in macrophages. J. Immunol. **137:** 3608-3613.

Handman, E., Mitchell, G.F. and Goding, J.W. (1987). *Leishmania major*: A very sensitive Dot-blot ELISA for detection of parasites in cutaneous lesions. Mol. Biol. Med. **4:** 377-383.

Homans, S.W., Ferguson, M.A.J., Dwek, R.A., Rademacher, T.W., Anand, R. and Willians, A.F. (1988). Complete structure of the glycosylphosphatidyl-inositol membrane anchor of rat brain Thy-1 glycoprotein. Nature. **333:** 269-272.

Homans, S.W. (1990). The architecture of GPI anchor glycans. TIGG. **2:** 144-155.

Ikeda, K., Sannoh, T., Kawasaki, N., Yamishina, I. (1987). Serum lectin of known structure activates complement through the classical pathway. J. Biol. Chem. **262:** 7451-7451.

Ilg, T., Etges, R., Overath, P., Thomas-Oates, M.J., McConville, M.J., Homans, S.W and Ferguson, M.A.J. (1990). Structure of *Leishmania mexicana* lipophosphoglycan. J. Biol. Chem. manuscript submitted.

Joiner, K. (1988). Complement evasion by bacteria and parasites. Ann Rev. Microbiol. **42:** 201-230.

Kemp, M., Theander, T.G., Handman, E., Hey, A.S., Kurtzhals, J.A.L., Hviid, L., Sorensen, A.L., Were, J.O.B., Koech, D.K. and Kharazami, A. (1991). Activation of human T lymphocytes by *Leishmania* lipophosphoglycan. Scand. J. Immunol. **33:**

King, D.L., Chang, Y.D. and Turco, S.J. (1987). Cell surface lipophosphoglycan of *Leishmania donovani*. Mol. Biochem. Parasitol. **24:** 47-53.

Lohman, K.L., Langer, P.J. and McMahon-Pratt, D. (1990). Molecular cloning and characterisation of the immunologically protective surface glycoprotein GP46/M-2 of *Leishmania amazonensis*. Proc. Natl. Acad. Sci. USA. **87:** 8393-8397.

Low, M.G. (1989). The glycosyl-phosphatidylinositol anchor of membrane proteins. Biochim. Biophys. Acta. **988:** 427-454.

McConville, M.J., Bacic, A., Mitchell, G.F. and Handman, E. (1987). Lipophosphoglycan of *Leishmania major* that vaccinates against cutaneous leishmaniasis contains an alkylglycerophosphoinositol lipid anchor. Proc. Natl. Acad. Sci. USA. **84:** 8941-8945.

McConville, M.J. and Bacic, A. (1989). A family of glycoinositol phospholipids from *Leishmania major*: isolation, characterization and antigenicity. J Biol Chem. **264:** 757-766.

McConville, M.J. and Bacic, A. (1990). The glycinositolphospholipid profiles of two *Leishmania major* strains that differ in lipophosphoglycan expression. Mol. Biochem. Parasitol. **38:** 57-68.

McConville, M.J., Thomas-Oates, J.E., Ferguson, M.A.J. and Homans, S.W. (1990a). Structure of the lipophosphoglycan from *Leishmania major*. J. Biol. Chem. **265:** 19611-19623.

McConville, M.J., Homans, S.W., Thomas-Oates, J.E., Dell, A. and Bacic, A. (1990b). Structures of the glycoinositolphospholipids from *Leishmania major* J. Biol Chem. **265:** 7385-7394.

McConville, M.J. and Blackwell, J.M. (1991). Developmental changes in the glycosylated phosphatidylinositols of *L.Donovani:* characterisation of the promastigote an amastigote glycolipids. Unpublished data.

McNeely, T.B., Rosen, G., Londer, M.V. and Turco, S.J. (1989). Inhibitory effects on protein kinase C activity by lipophosphoglycan fragments and glycosylphosphatidylinositol antigens of the protozoan parasite *Leishmania*. Biochem. J. **259:** 601-604.

McNeely, T.B. and Turco, S.J. (1990). Requirement of liophosphoglycan for intracellular survival of *Leishmania donovani* within human monocytes. J. Immunol. 3099-3106.

Medina-Acosta, E., Karess, R.E., Schwartz, H. and Russell, D.G., (1989). Promastigote surface protease (gp63) of *Leishmania* is expressed but differently processed and localised in the amastigote stage. Mol. Biochem. Parasitol. **38:** 263-273.

Moll, H., Mitchell, G.F., McConville, M.J. and Handman, E (1989). Evidence for T-cell recognition in mice of a purified lipophosphoglycan from *Leishmania major*. Infect. Immun. **57:** 3349-3356.

Muller, I., Pedrazzini, T., Farrell, J.P. and Louis, J. (1989). T-cell responses and immunity to experimental infection with *Leishmania major*. Ann. Rev. Immunol. **7:** 561-578.

Murray, P.J., Spithill, T.W. and Handman, E. (1989). The PSA-2 glycoprotein complex of *Leishmania major* is a glycosylphosphatidylinositol-linked promastigote surface antigen. J. Immunol. **143:** 4221-4226.

Orlandi, P.A., Jr. and Turco, S.J. (1987). Structure of the lipid moiety of the *Leishmania donovani* lipophosphoglycan. J. Biol. Chem. **262:** 10384-10391.

Pimenta, P.F.P., Da Silva, R.P., Sacks, D.L. and da Silva, P.P. (1989). Cell surface nanoanatomy of *Leishmania major* as revealed by fracture-flip. A surface meshwork of 44 nm fusiform filaments identifies infective developmental stage promastigotes. Eur. J. Cell. Biol. **48:** 180-190.

Puentes, S.M., Sacks, D.L., Da Silva, R.P. and Joiner, K. (1988). Complement binding by two developmental stages of *Leishmania major* promastigotes varying in expression of a surface lipophosphoglycan. J. Exp. Med. **167:** 887-902.

Puentes, S.M., da Silva, R.P., Sacks, D.L., Hammer, C.H. and Joiner, K.A. (1990). Serum resistance of metacyclic stage *Leishmania major* promastigotes is due to release of C5b-9. J Immunol. **145:** 4311-4316.

Roberts, E.L., Santikarn, S., Reinhold, V.N. and Rosenberry, T.L. (1989b). Structural characterization of the glycoinositol phospholipid membrane anchor of human erythrocyte acetylcholinesterase by fast atom bombardment spectrometry. J. Biol. Chem. **263:** 18776-18784.

Rosen, G., Londer, M.V., Sevlever, D. and Greenblatt, C.L. (1988). *Leishmania major*:glycolipid antigens recognised by immune human sera. Mol. Biochem. Parasitol. **27:** 93-100.

Rosen, G., Pahlsson, P., Londer, M.V., Westerman, M.E. and Nilsson, B. (1989). Structural analysis of glycosyl-phosphatidylinositol antigens of *Leishmania major*. J. Biol. Chem. **264:** 10457-10463.

Rosen, G., Pahlsson, P., Londner, M.V.Westerman, M.E. and Nilsson, B. (1990). Structural analysis of glycosyl-phosphaidylinositol antigens of *Leishmania major*. J. Biol. Chem. **265:** 7708.

Russell, D.G. and Wright, S.D. (1988). Complement receptor type 3(CR3) binds to an Arg-Gly-Asp-containing region of the major surface glycoprotein gp63, of *Leishmania* promastigotes. J. Exp. Med. **168:** 279-292.

Russell, D.G. and Talamas-Rohana, P. (1989). *Leishmania* and the macrophage: a marriage of convenience. Immunology Today. **10:** 328-333.

Sacks, D.L. (1989). Metacyclogenesis in *Leishmania* promastigotes. Exp. Parasitol. **69:** 100-103.

Sacks, D.L., Hieny, S. and Sher, A. (1984). Identification of cell surface carbohydrate and antigenic changes between noninfective and infective developmental stages of *Leishmania major* promastigotes. J. Immunol. **135:** 564-569.

Sacks, D.L. and Da Silva, R.P. (1987). The generation of infective stage *Leishmania major* promastigotes is associated with the cell-surface expression and release of a developmentally regulated glycolipid. J. Immunol. **139:** 3099-3106.

Sacks, D.L., Brodin, T.N. and Turco, S.J. (1990). Developmental modification of the lipophosphoglycan from *Leishmania major* promastigotes during metacyclogenesis. Mol. Biochem. Parasitol. **42:** 225-234.

Schneider, P., Ferguson, M.A.J., McConville, M.J., Mehlert, A., Homans, S.W. and Bordier, C. (1990). Structure of the glycosyl-phosphatidylinositol membrane anchor of the *Leishmania major* promastigote surface protease. J. Biol. Chem. **265:** 16955-16964.

Schweinle, J., Ezekowitz, R.A.B., Tennder, A. and Joiner, K. (1989). Human mannose-binding protein activates the alternative complement pathway and enhances serum bactericidal activity on a mannose-rich isolate of *Salmonella*. (1989) J. Clin. Invest. **84:** 1821-1829.

Da Silva, R.P., Hall, B.F., Joiner, K.A. and Sacks, D.L. (1989). CR1, the C3b receptor, mediates binding of infective *Leishmania major* metacyclic promastigotes to human macrophages. J. Immunol. **143:** 617-622.

Schnur, L.F (1982). The immunological identification and characterization of *Leishmania* stocks and strains, with special reference to excreted factors serotyping. In Biochemical Characterization of *Leishmania*. World Health Organization /TDR/WSG/IMMLEISH, pp 25-45.

Talamas-Rohana, P., Wright, S.D., Lennartz, M.R. and Russell, D.G (1990). Lipophosphoglycan from *Leishmania mexicana* promastigotes binds to members of the CR3, p150,95 and LFA-1 family of leukocyte integrins. J.Immunol. **144:** 4817-4824.

Thomas, J.R., Dwek, R.A. and Rademacher, T.W. (1990a). Structure, biosynthesis, and function of glycosyl-phosphatidylinositols. Biochemistry. **29:** 5413-5422.

Thomas, J.R., Thomas-Oates, J., McConville, M.J., Homans, S.W., Ferguson, M.A.J., Greis, K. and Turco, S.J. (1990b). The refined structure of *Leishmania donovani* lipophosphoglycan. Unpublished data.

Tolsen, D.L., Turco,S.J., Beecroft, R.P. and Pearson, T,W. (1989). The immunochemical structure and surface arrangement of *Leishmania donovani* lipophosphoglycan determined using monoclonal antibodies. Mol. Biochem. Parasitol. **35:** 109-118.

Tolsen, D.L., Turco, S.J. and Pearson, T.W. (1990). Expression of a repeating phosphorylated disaccharide lipophosphoglycan epitope on the surface of macrophages infected with *Leishmania donovani*. Infect. Immun. **58:** 3500-3507.

Towbin, H., Rosenfelder, G., Wieslander, J., Avila, J.L., Rojas, M., Szarfman, A., Esser, K., Nowack, H. and Timpl, R. (1987). Circulating antibodies to mouse laminin in Chagas disease, American cutaneous leishmaniasis, and normal individuals recognize terminal galactosyl-(α1-3)galactose epitopes. J. Exp. Med. **166:** 419-432.

Turco, S.J., Wilkerson, M.A. and Clowson, D.R. (1984). Expression of an unusual acidic glycoconjugate in *Leishmania donovani*. J. Biol. Chem. **259:** 3883-3889.

Turco, S.J., Hull, S.R., Orlandi, P.A., Jr. and Shepherd, S.D. (1987). Structure of the major carbohydrate fragment of the *Leishmania donovani* lipophosphoglycan. Biochemistry. **26:** 6233-6238.

Turco, S.J., Orlandi, P.A., Jr., Homans, S.W., Ferguson, M.A.J., Dwek, R.A. and Rademacher, T.W. (1989). Structure of the phosphosaccharide-inositol core of the *Leishmania donovani* lipophosphoglycan. J. Biol. Chem. **264:** 6711-6715.

Turco, S.J. and Sacks, D.L. (1991). Expression of a stage-specific lipophosphoglycan in *Leishmania major* amastigotes. Mol. Biochem. Parasitol. **45:** 91-95.

Wilson, M.E. and Pearson, R.D. (1988). Roles of CR3 and mannose receptors in the attachment and ingestion of *Leishmania donovani* by human mononuclear phagocytes. Infect. Immun. **56:** 363-369.

WHY DO SO MANY SURFACE PROTEINS OF TRYPANOSOMATIDS HAVE GPI-ANCHORS?

***Webster P., Joiner K. and Andrews N. W.**

Department of Cell Biology and Internal Medicine, Yale University School of Medicine, New Haven, Connecticut 06510, USA.

* Corresponding author.

INTRODUCTION

Glycosyl-phosphatidylinositol (GPI)-anchors are common features of proteins from many different eukaryotic cells (Ferguson and Williams, 1988), including *Trypanosoma brucei*, in which this form of membrane attachment was first described (Ferguson et al, 1988). Recent work on the characterization of surface proteins of trypanosomatids has shown that most of these proteins have GPI-anchors and are present in abundance on cell surface. Why these protozoan parasites should have large amounts of GPI-anchored proteins on their surface is unknown, as are the functions of these proteins. Similarly, the regulatory mechanisms exerted on the distribution of the components in the parasite are not understood.

In contrast, a substantial amount of information now exists on the control of GPI-anchored proteins expression in higher eukaryotes. In this paper, we will first review current information on membrane traffic in higher eukaryotic cells and on control of GPI-anchored protein distribution, which will serve as a basis for discussion and analysis of similar issues in trypanosomatid parasites. Based on differences and similarities when the two systems are compared we will conclude with speculation on the functional role of GPI-anchored proteins in trypanosomatids.

MEMBRANE TRAFFIC AND MEMBRANE DOMAINS IN HIGHER EUKARYOTES

The plasma membrane of eukaryotic cells exhibit functional specializations indicative of a highly organized macromolecular substructure. Proteins inserted into these membranes are differentially distributed throughout the cell forming domains with different functional and structural characteristics. These features are well illustrated in polarized epithelial cells.

The functional polarity of differentiated epithelial cells

is a result of differences in the apical and basolateral surfaces generated by differential targeting mechanisms and maintained by tight junctions (Rodrigues-Boulan, 1983; Simons and Fuller, 1985). The initial targeting signal may be contained within the cytoplasmic domains of the membrane proteins (Mostov et al, 1987; Hunziker and Mellman, 1989). The integrity of other membrane domains in cells are probably maintained by linkage of structural elements to the hydrophobic transmembrane domains and cytoplasmic tails common to most integral membrane proteins. Such mechanisms are believed to play a role in restricting lateral diffusion of proteins on the cell surface (Gumbiner and Louvard, 1985) and directing proteins to specific membrane domains involved with intracellular transport (Farquhar, 1985), intercellular junctions (Stevenson and Paul, 1989) substrate attachment (Burridge et al, 1988) and endocytosis (Anderson, 1990). Coated pits, the best characterized membrane domains involved with endocytosis, are formed by the deposition of clathrin on the cytoplasmic face of the membrane. The clathrin folding presumably causes the associated membrane to form a depression on the cell surface which becomes a clathrin-coated vesicle, enclosing endocytosed material from the extracellular environment. Other molecules, known as "adaptins", which also associate with coated pits, are believed to connect the clathrin lattice to the cytoplasmic tails of selected transmembrane proteins thus providing the coated pits with their characteristic specificity (Morris et al, 1989, Keen, 1990). In the above example, physical linkage between the cytoplasmic tail of membrane proteins and other protein components of the cell dictates cellular localization.

The glycosyl-phophatidylinositol (GPI)-linked proteins, however, are attached to the outer leaflet of the extracellular face of the cell plasma membrane by a glyco-inositol phospholipid located at the COOH-terminus (Ferguson and Williams, 1988). Without a cytoplasmic tail it is unclear how cells monitor and control these proteins. Fluorescence photobleaching and recovery experiments have demonstrated that the GPI-anchor of a class I MHC molecule in mammalian cells enables this protein to pass boundaries which restrict transmembrane proteins (Edidin and Strynowski, 1991) suggesting that GPI-anchored proteins are free to diffuse in the membrane. Studies on the Thy-1 antigen (Lemanski et al, 1990) and a GPI-anchored human Fc receptor (Anderson, 1989) suggest that they may not be internalized by cells, presumably because they cannot interact with the adaptins. In addition, molecules which normally undergo efficient endocytosis, such as the polymeric Ig receptor, the LDL receptor, the EGF receptor and the transferrin receptor are generally excluded from coated pits and are poorly endocytosed when expressed as mutated forms lacking part or all of the cytoplasmic domains (Lisanti et al, 1990).

Lipid domains may play a role in maintaining the distribution of GPI-anchored proteins. Cholesterol in the plasma membrane of mammalian cells has been shown to be responsible for the maintenance of membrane domains containing enriched amounts of a GPI-anchored protein (Rothberg et al, 1990a; 1990b). The folate receptors on the plasma membranes of a monkey kidney cell line cluster in cholesterol-rich domains from where they are eventually internalized by the cells (Rothberg et al, 1990b). For epithelial cells, it has been postulated that GPI-anchored proteins may be selectively incorporated into apically targeted glycosphingolipid-rich domains in the Golgi complex (Simons and van Meer, 1988; Brown et al, 1989).

DISTRIBUTION OF GPI-ANCHORED PROTEINS IN TRYPANOSOMATIDS

In contrast to higher eukaryotes, the surface of trypanosomatids is covered almost exclusively with GPI-anchored proteins. This situation may result either from the structural constraints imposed on plasma membrane proteins in trypanosomatids or from specific targeting signals in the surface proteins. A discussion of this issue follows.

The pellicular microtubular cytoskeleton of the trypanosomtid parasites may act as a barrier to the cytoplasmic control structures, resulting in an accumulation of GPI-anchored proteins on the cell surface. These cells may have eliminated, or never developed, the requirement for specialized regions over most of the plasma membrane, keeping the transmembrane control mechanisms restricted to one specialized region of the cell. In that circumstance, most of the plasma membrane functions such as endocytosis, would be restricted to this specialized membrane domain where the classical transmembrane proteins should be found. Certainly, for *T. brucei* it appears that all the endocytic processes studied so far have occurred only at the site of the flagellar pocket membrane. This membrane, a common morphological feature of all members of the *Kinetoplastida,* encloses an extracellular lumen cut off from the surrounding environment by a close apposition of plasma membrane with the flagellar membrane (Fig. 1). The mechanisms by which molecules enter the flagellar pocket, however, are not clear. Although evidence exists to suggest that the flagellar pocket of *T. brucei* is openly accessible to extracellular substances (Coppens et al, 1987) conflicting evidence demonstrates the inaccessibility of the flagellar pocket, at 0°C, to both soluble and particulate markers (Webster, 1990).

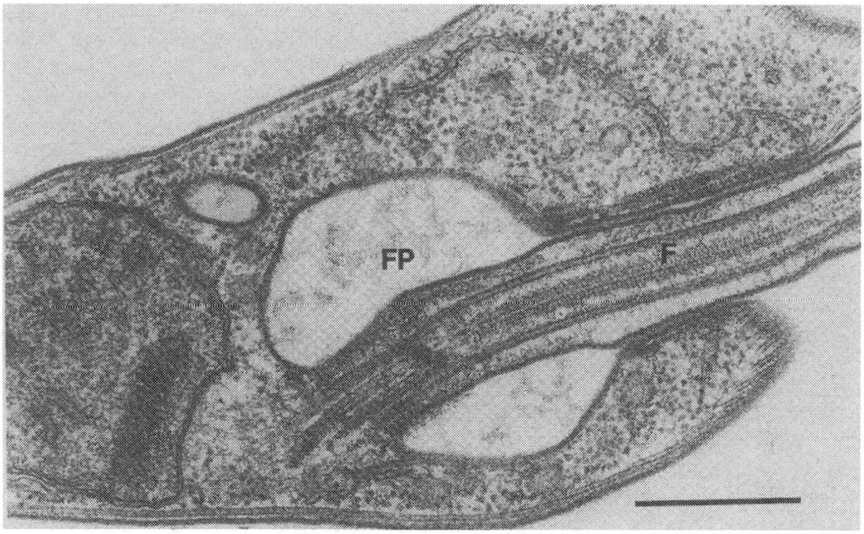

Figure 1:
Trypanosoma brucei: A section through the posterior part of a cell embedded in epoxy resin showing the flagellar pocket (FP). Bar = 0.5μm.

Structures resembling clathrin-coated pits and vesicles which form from the flagellar pocket membrane in *T. brucei* are responsible for the internalization of many various particulate and soluble substances (Langreth and Balber 1975; Coppens et al, 1987; Duszenko et al, 1989; Webster and Grab 1989; Webster 1989; Webster and Fish 1989). Receptor-mediated endocytosis of transferrin and low density lipoproteins (LDLs) seem only to occur from this site (Coppens et al, 1987; Webster and Grab, 1988). A protein similar in structure to the human LDL receptor has been identified in procyclic forms of *T. brucei* and is restricted to the flagellar pocket and intracellular membranes of the parasite (Lee et al, 1990). The LDL-receptor-like protein, called CRAM (Lee et al, 1990), of these cultivated, insect form trypanosomes has a sequence analysis that predicts a classical hydrophobic transmembrane attachment. Most probably, as more receptor proteins are identified in trypanosomatids, they will turn out to have integral membrane domains and cytoplasmic tails and be localized only on the flagellar pocket membrane. A GPI-anchored transferrin-binding protein of *T. brucei* was recently described (Schell et al, 1991), but it remains to be determined if it functions as a receptor for the uptake of transferrin since it allows no structural relationship to the mammalian receptor.

It would probably be incorrect, however, to suggest that there is no cellular control of the GPI-anchored surface proteins in trypanosomatid parasites. Movement of the VSG on the surface of *T. brucei* does occur (Bülow et al, 1988) and directional movement of a single layer of surface bound antibodies has been reported for *Leishmania* (Doyle et al, 1975), *T. cruzi* (Abelha et al, 1981; Teixeira and Santana, 1989), *T. lewisi* (Cherian and Dusanic, 1978), and *T. brucei* (Webster et al, 1990). VSG-bound antibodies on the surface of *T. brucei* are rapidly swept off the cell surface, probably onto and down the flagellum, into the flagellar pocket and the cell (Webster et al, 1990). Although Bülow et al (1988) observed much slower surface movement of VSG, these results require confirmation in the absence of metabolic inhibitors and using trypanosome isolation techniques which minimalize biochemical changes in the parasite (Lonsdale-Eccles and Grab, 1987).

The colocalization of VSG molecules with endocytosed transferrin-gold particles (Webster and Grab, 1988) suggests that the VSG-bound antibodies may also be taken up by endocytic structures. In the case of *T. brucei* even single layers of $F(ab)_1$ fragments of VSG specific antibodies are cleared from the cell surface and internalized in a directional manner (Webster et al, 1990) suggesting that the VSGs are constantly moving across the cell surface. This continuous surface sweeping and internalization of the VSG molecules has been proposed to be a mechanism to avoid complement-mediated lysis of *T. brucei* (Balber et al, 1975; Webster et al, 1990). Perhaps it is also a mechanism by which substances, non-specifically bound to the VSG, are carried to the flagellar pocket where they can bind to specific receptors.

If this controlled movement of GPI-anchored proteins over the cell surface does occur, then cholesterol, shown to be present in elevated amounts in the plasma membrane of *T. brucei* (Voorheis et al 1979) and *T. cruzi* (Franco Da Silveira and Colli, 1981) may play a major role in the mechanisms involved. However, the existence of a lipid-based regulation of GPI-anchored proteins in trypanosomatids must remain speculative, as must the possible mechanisms of such control, until more information is obtained.

GPI-ANCHORS AND TARGETING TO THE FLAGELLAR POCKET

In addition to being the site of endocytosis in trypanosomatids, the flagellar pocket may also be the site at which newly synthesized GPI-anchored proteins leave the cell for the parasite surface. Here we present, for the first time, evidence that Ssp-4, a GPI-anchored protein of *T. cruzi*, appears in the flagellar pocket of trypomastigotes transforming into amastigotes in vitro before it appears on the cell surface (Fig. 2). Perhaps,

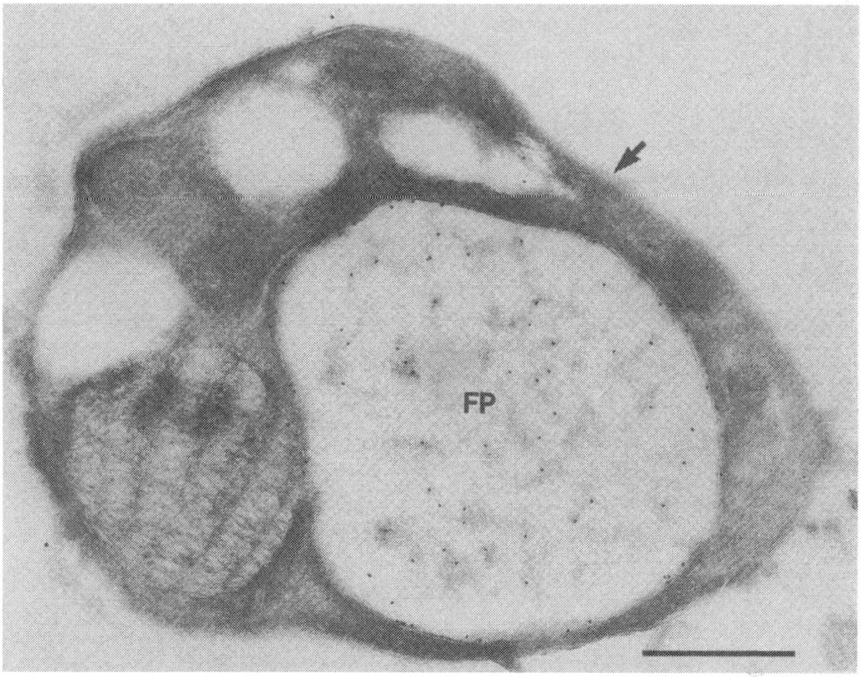

Figure 2:
Trypanosoma cruzi: A thawed cryosection through the flagellar pocket of a cultured trypomastigote form parasite which has been labelled with antibodies to Ssp-4 and protein A-gold. The cell was part of a population transforming to amastigotes, which express Ssp-4 on their surface. The flagellar pocket (FP) of this cell is labelled but not the cell surface (arrow). Bar = 0.5μm.

as suggested for apical targeting in polarized epithelial cells (Lisanti et al, 1990), the GPI-anchor plays a role in directing the intracellular proteins to the flagellar pocket membrane, from where they can diffuse onto the cell surface. However, it is just as likely that the barrier of pellicular microtubules results in all exocytic traffic being directed only to the flagellar pocket membrane.

Morphological evidence indicates that some of the GPI-anchored proteins become detached from the flagellar pocket membrane and move into the lumen. This can be observed with Ssp-4 in *T. cruzi* (Fig. 2), the VSGs in *T. brucei* (Webster and Grab, 1988 and Fig. 3) and the gp63 GPI-anchored surface protease of *Leishmania* (Medina-Acosta et al, 1989). In the case of gp63, the luminal protein is not only detached from the flagellar pocket membrane but also lacks the GPI-anchor (Medina-Acosta et al, 1989). These detached proteins may indicate the activity of specific phospholipases within, or close to the flagellar pocket.

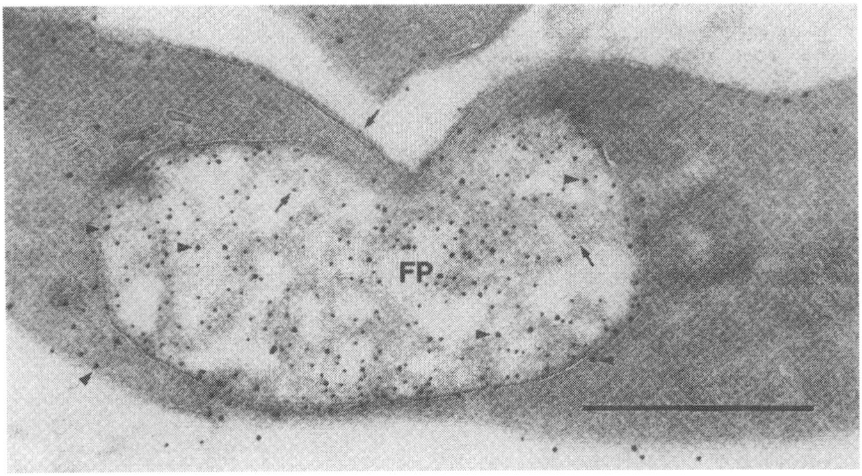

Figure 3:
Trypanosoma brucei: A thawed cryosection through the flagellar pocket of a cell from a cloned population (MITat 1.2) of trypanosomes isolated from infected rat blood. The section was double labelled with specific antibodies and two sizes of protein A-gold. The smaller size gold particles (arrows) visualize antibodies which recognize the whole VSG molecule. The larger gold particles (arrowheads) show binding of anti-cross reacting determinant (CRD) antibodies. In addition to labelling the surface and flagellar pocket membranes both antibodies label the lumen of the flagellar pocket (FP).Bar = 0.5µm.

GPI-SPECIFIC PHOSPOLIPASES

The presence of GPI-specific phospholipases has been reported in *T. cruzi* (Andrews et al, 1988; Schenkman et al, 1988) and *T. brucei* (Cardoso de Almeida and Turma, 1983, Ferguson et al, 1985; Hereld et al, 1986) and both have been implicated in the removal of GPI-anchored surface proteins during transformation of the parasites (Overath et al, 1983; Shapiro, 1986; Andrews et al, 1988). One characteristic common to all GPI-anchored proteins in trypanosomatids is that their expression is coincident with differentiation stages of the parasites. The VSG is only expressed on mammal-infective metacyclic and trypomastigote forms of *T. brucei* and Ssp-4 only expressed on amastigote forms of *T. cruzi* (Fig. 4).

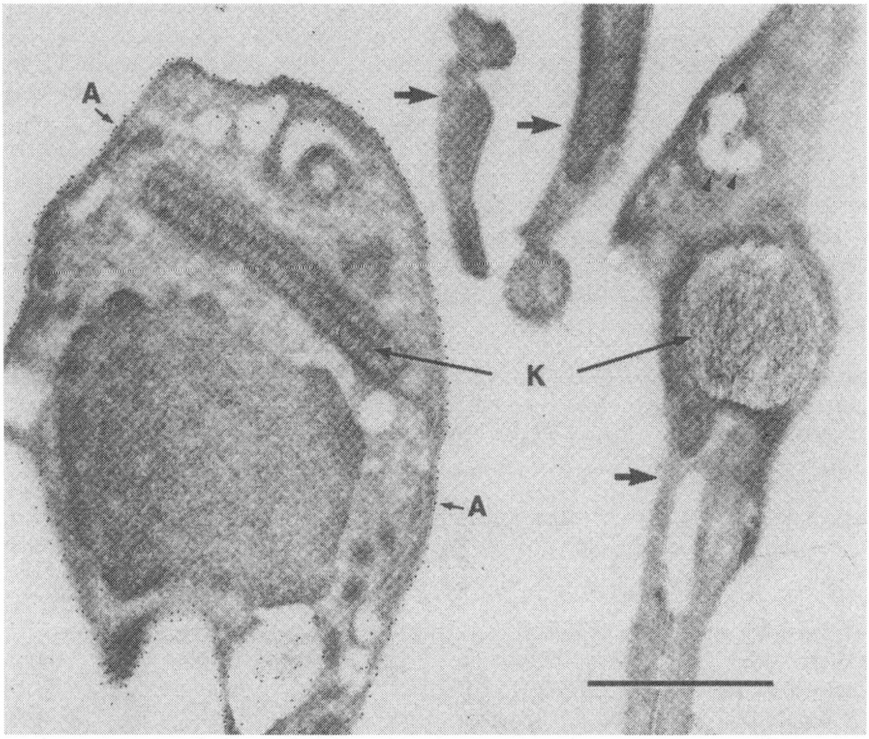

Figure 4:
Trypanosoma cruzi: A thawed cryosection, labelled with Ssp-4 antibodies, through a mixed population of cultivated trypomastigote form parasites transforming into amastigotes. The Ssp-4 antibodies label the surface and many intracellular structures of the amastigote (A) but do not label the trypomastigote surface (arrows). The intracellular labelling of one of the trypomastigote forms (arrowhead) is probably on the flagellar pocket membrane. The kinetoplasts (K) of both trypomastigote and amastigote have the morphological characteristics common to each form. Bar = 1.0μm.

The GPI-specific phospholipase C present in *T. brucei* cleaves the terminal glycosyl-phosphatidylinositol of the VSG (Ferguson et al, 1985) to produce 1, 2 diacylglycerol and glycosyl-inositol phosphate linked to protein. This phospholipase C is localized on the cytoplasmic face of intracellular vesicles (Bülow et al, 1989) and its synthesis is suppressed during a second differentiation step when the bloodstream form trypanosomes are ingested by the insect vector (Bülow and Overath, 1985). During this differentiation step, which can be reconstructed in vitro, the VSG is gradually lost from the surfact (Bülow et al, 1989) and replaced by another protein, called procyclin (Roditi et al, 1990), with a structurally different GPI-anchor to that of the VSG that is also insensitive to the *T. brucei* glycan-specific phospholipase C (Clayton and Mowatt, 1989). One product of the cleavage of GPI-anchors by phospholipase C, diacylglycerol, is an activator of protein kinase C, a central regulatory protein involved in membrane signal transduction (Nishizuka, 1984) and cellular differentiation (Ebling et al, 1985, Melloni et al, 1987). Diacylglycerol can be converted via diacylglycerol kinase into phosphatidic acid, which has growth factor properties (Moolenaar et al, 1986).

Although GPI-anchored proteins released from the surface of parasites have been proposed to generate signals for differentiation, only circumstantial evidence exists to suggest this. Release of Ssp-4 from the surface of intracellular amastigotes of *T. cruzi* occurs prior to transformation to trypomastigote forms and lysis of the host cell plasma membrane (Andrews et al, 1988). When amastigotes were incubated in vitro, most of the surface Ssp-4 was shed from the parasite surface in a soluble form reactive with antibodies to the cross reacting determinant (Barbet and McGuire, 1978), indicating that the release was mediated by a phospholipase C (Andrews et al, 1988). It remains to be determined whether diacylglycerol generated in the membrane as a product of the phospholipase C cleavage reaction, is involved in triggering the differentiation events that follow the release of Ssp-4.

CONCLUSIONS

Evidence indicating that GPI-anchors are the preferred mode of attachment of surface proteins in trypanosomatids is growing. Here we have discussed biological characteristics of these organisms which may provide explanations for this preference. The usually high cholesterol/phospholipid ratios observed in the plasma membrane of some of these protozoans may be the major reason why these cells have high levels of GPI-anchored proteins over the cell surface. Sequestration of plasma membrane functions requiring transmembrane and cytoplasmic domains, such as endocytosis, to the flagellar pocket may be a consequence of these high levels. The presence of the GPI-specific phospholipase C in the parasites suggests that anchor cleavage and diacylglycerol production may be the chosen mechanisms for both differentiation signals and a rapid, efficient compositional change of their surface coats during differentiation. The fluidity expected of a homogeneous GPI-anchored surface protein coat suggests intriguing possibilities concerning the mechanisms used by the parasites to avoid attack by serum components and perhaps in some cases, obtain nutrient molecules bound to the cell surface.

REFERENCES

Abelha, J., Azevedo, M. O. and Teixeira, A. R. L. (1981). *Trypanosoma cruzi*: Antigen-receptor mediated endocytosis of antibody. Mem. Inst Oswaldo Cruz. **76:** 213-225.

Anderson, C. L. (1989). Human IgG Fc Receptors. Clinical Immunol. Immunopath. **53** S63-71.

Anderson, R. G. W. (1990). Molecular motors that shape endocytic membrane, in Steer, C. J. and Hanover, J. (eds) Intracellular trafficking of proteins. London: Cambridge University Press.

Andrews, N. W., Robbins, E. S., Ley, V., Hong, K. S., Nussenzweig, V. (1988). Developmentally regulated, phospholipase C-mediated release of the major surface glycoprotein of amastigotes of *Trypanosoma cruzi*. J. Exp. Med. **167:** 300-314.

Balber, A. E., Bangs, J. D., Jones, S. E. and Proia, R. L. (1979). Inactivation or elimination of potentially trypanolitic complement-activating immune complexes by pathogenic trypanosomes. Infec. Immun. **24:** 617-627.

Bansal, V. S. and Majerus, P. W. (1990). Phosphatidyl-inositol-derived precursors and signals. Annu. Rev. Cell Biol. **6:** 41-67.

Barbet, A. F. and McGuire, T. C. (1978). Crossreacting determinants in variant-specific surface antigens of African trypanosomes. Proc. Natl. Acad. Sci. USA **83** :3207-3211.

Brown, D. A., Crise, B. and Rose, J. (1989). Mechanisms of membrane anchoring affects polarized expression of two proteins in MDCK cells. Nature **245:** 1499-1501.

Bülow, R., Griffiths, G., Webster, P., Stierhof, Y-D., Opperdoes, F. and Overath, P. (1989a). Intracellular localization of the glycosyl-phosphatidylinositol-specific phospholipase C in *Trypanosoma brucei*. J. Cell Sci. **93:** 233-240.

Bülow, R., Nonnengasser, C. and Overath, P. (1989b). Release of the variant surface protein during differentiation of bloodstream to procyclic forms of *Trypanosoma brucei*. Mol. Biochem. Parasitol. **32:** 85-92.

Bülow, R and Overath, P. (1985). Synthesis of a hydrolase for the membrane-form variant surface glycoprotein is repressed during transformation of *Trypanosoma brucei*. FEBS Letters **187:** 105-109.

Bülow, R., Overath, P. and Devoust, J. (1988). Rapid lateral diffusion of the variant surface glycoprotein in the coat of *Trypanosoma brucei*. Biochemistry **27:** 2384-2388.

Burridge,K. Fath, K., Kelly, T., Nuckolls, G. and Turner, C. (1988). Focal adhesions: transmembrane junctions between the extracellular matrix and the cytoskeleton. Annu. Rev. Cell Biol. **4:** 487-525.

Cardoso de Almeida, M-L. and Turner, M.J. (1983). The membrane form of variant surface glycoproteins of *Trypanosoma brucei*. Nature **302:** 349-352.

Cherian, P. V. and Dusanic, D. G. (1978). *Trypanosoma lewisi*: Ultrastructural observations of surface antigen movement induced by antibody. Exp. Parasitol. **44:** 14-25.

Clayton, C. E. and Mowatt, M.R. (1989). The procyclic acidic repetitive proteins of *Trypanosoma brucei*. Purification and post-translational modification. J. Biol. Chem. **264:** 15088-15093.

Coppens, I., Opperdoes, F. R., Courtoy, P. J. and Bauduin, P. (1987). Receptor mediated endocytosis in the bloodstream forms of *Trypanosoma brucei*. J. Protozool. **34:** 465-473.

Cross, G. A. M. (1990). Cellular and immunological aspects of antigenic variation in trypanosomes. Annu. Rev. Immunol. **8:** 83-110.

Doering, T. L., Masterson, W. J., Hart, G. W. and Englund, P. T. (1990). Biosynthesis of glycosyl phosphatidylinositol membrane anchors. J. Biol. Chem. **265:** 611-614.

Doyle, J. J., Behin, R., Mauel, J. and Rowe, D. S. (1975). Antibody-induced movement of protozoan surface membrane antigens. Ann. Rev. N.Y. Acad. Sci. **254:** 315-325.

Duszencko, M., Ivanov, I., Ferguson, M. A. J., Pleskin, H. and Cross G. A. M. (19880. Intracellular transport of a variant surface glycoprotein in *Trypanosoma brucei*. J. Cell Biol. **106:** 77-86.

Ebling, J. G., Vandenbark, G. R., Kuhn, L. J., Ganong, B. R., Bell, R. M. and Niedel, J. E. (1985). Diacylglycerols mimic phorbol diester induction of leukemic cell differentiation. Proc. Nat. Acad. Sci. USA **82:** 815-819.

Edidin, M. and Stroynowski, I. (1991). Differences between the lateral organization of conventional and inositol phospholipid-anchored membrane proteins. A further definition of micrometer-scale membrane domains. J. Cell Biol. **112:** 1143-1150.

Farquhar, M. G. (1985). Progress in unravelling pathways of Golgi traffic. Annu Rev. Cell Biol. **1:** 447-488.

Ferguson, M. A. J., Halder, K. and Cross, G. A. M. (1985). *Trypanosoma brucei* variant surface glycoprotein has a sn-1,2-dimyristyl glycerol membrane anchor at its COOH terminus. J. Biol. Chem. **260:** 4963-4968.

Ferguson, M.A.J., Hommans, S. W., Dwek, R. A. and Rademacher, T. W. (1988). Glycosyl-phosphatidylinositol moiety that anchors *Trypanosoma brucei* variant surface glycoprotein to the cell membrane. Science **239:** 753-759.

Ferguson, M. A. J. and Williams, A. F. (1988). Cell-surface anchoring of proteins via glycosyl-phosphatidylinositol structures. Ann. Rev. Biochem. **57:** 285-320.

Franco Da Silveira, F. and Colli, W. (1981). Chemical composition of the plasma membrane from epimastigote forms of *Trypanosoma cruzi*. Biochem. Biphys Acta **644:** 341-350.

Gumbiner, B. and Louvard, D. (1985). Localized barriers in the plasma membrane: a common way to form domains. Trends in Biol. Sci. **10:** 435-438.

Hereld, D., Krakow, J. L., Bangs, J. D., Hart, G. W. and Englund, P. (1986). A phospholipase C from *T. brucei* which selectively cleaves the glycolipid on the variant surface glycoprotein. J. Biol. Chem. **261:** 13818-13823.

Hunziker, W. and Mellman, I. (1989). Expression of macrophage-lymphocyte Fc receptors in Madin-Darby canine kidney cells: polarity and transcytosis differ for isoforms with or without coated pit localization domains. J. Cell Biol. **109:** 3291-3302.

Jing, S., Spencer, T., Miller, K., Hopkins, C. and Trowbridge, I. S. (1990). Role of the human transferrin receptor cytoplasmic domain in endocytosis: localization of a specific signal sequence for internalization. J. Cell Biol. **110:** 283-294.

Keen, J. H. (1990). Clathrin and associated assembly and disassembly proteins. Annu. Rev. Biochem. **59:** 415-438.

Langreth, S. G. and Balber, A. E. (1975). Protein uptake and digestion in bloodstream and culture forms of *Trypanosoma brucei*. J. Protozool. **22:** 40-53.

Lee, M. G., Bilhain, B. E., Russell, D. G., Deckelbaum, R. J. and Van der Ploeg, L. H. (1990). Characterization of a cDNA encoding a cysteine-rich cell surface protein located in the flagellar pocket of the protozoan *Trypanosoma brucei*. Mol. Cell Biol. **10:** 4506-4517.

Lemanski, P., Fatemi, S. H., Gorican, B., Mayale, S., Rossero, R. and Tartakoff, A. M. (1990). Dynamics and longevity of the glycolipid-anchored membrane protein, Thy-1. J. Cell Biol. **110:** 1525-1531.

Lisanti, M. P., Rodrigues-Boulan, E. and Saltiel, A. R. (1990). Emerging functional role for the glycosyl-phosphatidylinositol membrane protein anchors. J. Membrane Biol. **117:** 1-10.

Lonsdale-Eccles, J. D. and Grab, D. J. (1987). Purification of African trypanosomes can cause biochemical changes in the parasite. J. Protozool. **34: 405-408.**

Medina-Acosta, F., Karess, R. E., Schwartz, H. and Russell, D. G. (1989). The promastigote surface protease (gp63) of *Leishmania* is expressed but differentially processed and localized in the amastogote stage. Mol. Biochem. Parasitol. **37:** 263-273.

Melloni, E., Pontremoli, S., Michetti, M., Sacco, O., Cakirogolu, A. G., Jackson, J. F., Rifkind, R. A. and Marks, P. A. (1987). Protein kinase C activity and hexamethulethylenebisacetamide-induced erythro-leukemia cell differentiation. Proc. Natl. Acad. Sci. USA. **84:** 5282-5286.

Moolenaar, W. H., Kruijer, W., Tilly, B. C., Verlaan, I., Bierman, A. J. and de Laat, S. W. (1986). Growth factor-like action of phosphatidic acid. Nature. **323:** 171-173.

Morris, S. A., Ahle, S. and Ungewickell, E. (1989). Cathrin-coated vesicles. Curr. Opinion Cell Biol. **1:** 684-690.

Mostov, K. E., Breitfeld, P. and Harris, J. M. (1987). An anchor-minus form of the polymeric immunoglobulin receptor is secreted predominantly apically in Madin-Darby canine kidney cells. J. Cell Biol. **105:** 2031-2036.

Nishizuka, Y. (1984). The role of protein kinase C in cell surface signal transduction and tumour promotion. Nature. **308:** 693-698.

Overath, P., Czichos, J., Stock, U. and Nonnengasser, C. (1983). Repression of glycoprotein synthesis and release of surface coat during transformation of *T. brucei* EMBO. J. **2:** 1721-1730.

Pays, E. and Steinert, M. (1988). Control of antigenic gene expression in African trypanosomes. Ann. Rev. Genet. **22:** 107-126.

Rifkin, M. R. and Landsberger, F. R. (1990). Trypanosome variant surface glycoprotein transfer to target membranes: A model for the pathogenesis of trypanosomes. Proc. Natl. Acad. Sci. USA. **87:** 801-805.

Rivas, L., Kahl, L., Manson, K. and McMahon-Pratt, D. (1991). Biochemical characterization of the protective glycoprotein GP46/M-2 of *Leishmania amazonensis*. Mol. Biochem. Parasitol. (in press).

Rothberg, K. G., Ying, Y.-S., Kamen, B. A. and Anderson, R. G. W. (1990)a. The glycophosphlipid-linked folate receptor internalizes folate without entering the clathrin-coated pit endocytic pathway. J. Cell Biol. **110:** 637-649.

Rothberg, K. G., Ying, Y.-S., Kamen, B. A. and Anderson, R. G. W. (1990)b. Cholesterol controls the clustering of the glycophospholipid-anchored membrane receptors for 5-methyltetrahydrofolate. J. Cell Biol. **111:** 2931-2938.

Roditi, I., Schwartz, H., Pearson, T. W., Beecroft, R. P., Lui, M. K., Richardson, J. P., Bühring, H.-J., Pleiss, J., Bülow, R., Williams, R. O. and Overath, P. (1989). Procyclin gene expression and loss of the variant surface glycoprotein during differentiation of *Trypanosoma brucei*. J. Cell Biol. **108:** 737-746.

Rodrigues-Boulan, E. (1983). Membrane biogenesis, enveloped RNA viruses and epithelial polarity. Mol. Cell Biol. **100:** 119-170.

Schell, D., Evers, R., Pries, D., Ziegelbauer, K., Kiefer, H., Lottsprich, F., Cornelissen, A. W. C. A. and Overath, P. (1991). A transferrin-binding protein of *T. brucei* is encoded by one of the two genes in the variant surface glycoprotein gene expression site. EMBO. J., **10:** 1061-1066.

Schenkman, S., Yoshida, N. and Cardoso de Almeida, M.-L. (1988). Glycophosphatatidylinositol anchored proteins in metacyclic *Trypanosoma cruzi*. Mol. Biochem. Parasitol. **29:** 141-151.

Shapiro, S. Z. (1986). *Trypanosoma brucei:* release of variant surface glycoprotein during the parasite life cycle. Exp. Parasitol. **61:** 1721-1729.

Simons, K. and Fuller, S. D. (1985). Cell surface polarity in epithelia. Ann. Rev. Cell Biol. **1:** 243-288.

Simons, K. and van Meer, G. (1988). Lipid sorting in epithelial cells. Biochemistry. **27:** 6197-6202.

Stevenson, B. R. and Paul, D. L. (1989). The molecular constituents of intercellular junctions. Current Opin. Cell Biol. **1:** 884-891.

Teixeira, A. R. L. and Santana, J. M. (1989). *Trypanosoma cruzi*: Endocytosis and degradation of specific antibodies by parasitic forms. Am. J. Trop. Med. Hyg. **40:** 165-170.

Vooheis, H. P., Gale, J. S., Owen, M. J. and Edwards, W. (1979). The isolation and partial characterization of the plasma membrane from *Trypanosoma brucei*. Bichem. J. **180:** 11-24.

Webster, P. (1989). Endocytosis by African trypanosomes. I Three-dimentional structure of the organelles in *Trypanosoma brucei* and *T. congolense*. Europ. J. Cell Biol. **49:** 295-302.

Webster, P. and Fish, W. R. (1989). Endocytosis by African trypanosomes. II Occurrence in different life cycle stages and intracellular sorting. Europ. J. Cell Biol. **49:** 303-310.

Webster, P. and Grab, D. J. (1988). Intracellular colocalization of variant surface glycoprotein and transferrin-gold in *Trypanosoma brucei*. J. Cell Biol. **106:** 279-288.

Webster, P., Russo, D. C. W. and Black, S. J. (1990). The interaction of *Trypanosoma brucei* with antibodies to variant surface glycoproteins. J. Cell Sci. **96:** 249-255.

PROBING THE SIGNAL FOR GLYCOPHOSPHATIDYLINOSITOL ANCHOR ATTACHMENT USING DECAY ACCELERATING FACTOR AS A MODEL SYSTEM

***Ingrid W Caras**

Genentech Inc, San Francisco, California, USA.

* Corresponding author.

INTRODUCTION

Although all glycophosphatidylinositol (GPI)-anchored proteins are presumably processed by a common pathway in mammalian cells they contain no recognizable consensus sequence or obvious homology, suggesting that the GPI-signal is independent of primary sequence, possibly relying instead on conformational features. The only obvious property common to all GPI-anchored proteins is the presence of a COOH-terminal hydrophobic domain, predicted by the cDNA but absent from the processed GPI-anchored protein. It is not surprising that the hydrophobic domain is a crucial part of the GPI signal. This article summarizes the approaches used to decipher the GPI signal and what was learned, using the GPI-linked protein, decay accelerating factor (DAF), as the model system.

LOCALIZATION OF THE GPI SIGNAL TO THE COOH-TERMINUS OF DAF

Early experiments localized the signal for GPI attachment to the COOH-terminus of DAF. Using gene manipulation techniques we transferred the last 37 amino acids predicted by the DAF cDNA to the COOH-terminus of a truncated, secreted form of the herpes glycoprotein, gD-1, lacking its normal transmembrane and cytoplasmic domains (Fig.1). Following expression in either COS or CHO cells the chimeric gD-1-DAF protein was targeted to the plasma membrane by means of a GPI anchor (Caras et al., 1987). Thus, transmembrane gD-1 could be converted to a GPI-anchored membrane protein by switching the transmembrane-cytoplasmic domains with the COOH-terminus of DAF. Similarly, the normally secreted protein, human growth hormone (hGH) could be converted to a GPI-anchored cell surface protein by the addition of 37 COOH-terminal residues of DAF (Caras et al., 1989). In these experiments GPI anchoring was assessed by two independent tests: 1) release of the protein from the cell surface upon digestion of the cells with purified PI-specific phospholipase C (PIPLC) from *Bacillus thuringiensis* and 2) metabolic labelling of the protein with [^{3}H]ethanolamine (a component of the GPI anchor) followed by analysis by immunoprecipitation. Similar experiments using sequences derived from the Qa-2 antigen (Waneck et al., 1988) or Thy-1 (Crise et al., 1989) confirmed that the COOH-terminal domains of these GPI anchored proteins contain the signals for GPI attachment.

Further analysis of the information contained within these COOH-terminal signalling domains has come from two basic experimental approaches: 1) the introduction of mutations (deletions or point mutations) that highlight essential regions or residues required for correct functioning of the GPI signal, and 2) the construction of chimeric proteins containing portions of the DAF COOH-terminus fused to a secreted protein, hGH.

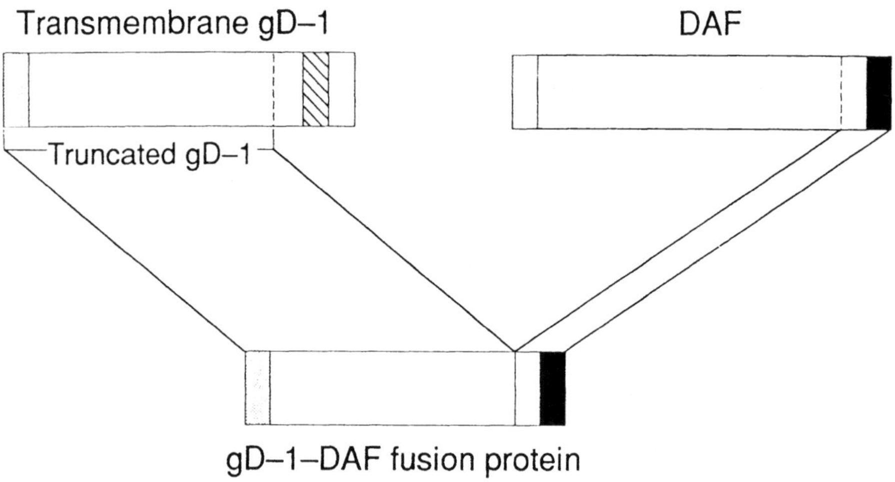

Figure 1.
Schematic diagram showing the regions of gD-1 and DAF that are present in the gD-1-DAF fusion protein. Truncated (secreted) gD-1, comprising the signal peptide (grey box) and residues 1-300, was fused to the last 37 residues of DAF, which include the 17-residue COOH-terminal hydrophobic domain (shown as a black box). The transmembrane domain of gD-1 is indicated by a hatched box (from Caras et al, 1987).

THE COOH-TERMINAL HYDROPHOBIC DOMAIN

As mentioned above, all GPI anchored proteins contain a short hydrophobic domain (15-20 residues) at the COOH terminus of the protein (Ferguson and Williams, 1988; Cross, 1990). Deletion of the DAF COOH-terminal hydrophobic domain (the last 17 amino acids predicted by the cDNA) abolished GPI attachment and produced a secreted form of DAF (Caras et al., 1989) indicating that the hydrophobic domain plays a crucial role in anchor attachment. Replacement of the COOH-terminal hydrophobic domain with a signal peptide that normally functions in membrane translocation, or with a random hydrophobic sequence, resulted in efficient and correct processing, producing GPI-anchored DAF on the cell surface (Caras and Weddel, 1989). This result suggests that the overall hydrophobicity of the COOH-terminal domain rather than its precise sequence is important for signalling GPI attachment. The signal peptide used in these experiments was derived from the NH_2-terminus of the hGH gene and contains a hydrophobic region of only 13 residues (compared with the 17-residue hydrophobic domain normally found at the DAF COOH-terminus). This is apparently sufficient to trigger anchor attachment. Using placental alkaline phosphatase (PLAP) Berger et al. (1988) found that a minimum hydrophobic sequence of 14 residues was required for anchor attachment, the precise sequence being unimportant. Taken together these results support the conclusion that the function of the COOH-terminal domain in GPI-anchoring is dependent only on its overall hydrophobicity and its length.

DISCOVERY OF A SECOND COMPONENT OF THE GPI SIGNAL

Analysis of the 37-residue, COOH terminal sequence of DAF (containing all the information necessary to signal anchor attachment) indicated that whereas deletion of the 17-residue COOH-terminal hydrophobic domain abolished GPI attachment and led to secretion of the protein, deletion of the adjacent sequence (20 residues) had no effect on anchor attachment, suggesting that the COOH-terminal hydrophobic domain alone might be responsible for signalling GPI attachment (Caras et al., 1989). However, when we tested this hypothesis by fusing the 17-residue hydrophobic domain directly to the COOH-terminus of either hGH or the truncated herpes gD-1 protein, the dimeric proteins failed to become GPI-anchored and instead were secreted, suggesting that the hydrophobic domain, although necessary, is in fact insufficient to trigger anchor attachment. A second element, contained within the adjacent 20 residues of DAF sequence, was required together with the hydrophobic domain to target chimeric proteins to the plasma membrane by means of a GPI-anchor (Caras et al., 1989). We speculated and later showed, that this second element provides the cleavage/attachment site for the anchor, which is preassembled in the endoplasmic reticulum (Masterson et al., 1989) and then added to the protein 'en bloc' in a processing event that involves coordinated proteolytic removal of 17-31 residues from the COOH-terminus of the nascent polypeptide (Boothroyd et al., 1981; Tse et al., 1985). The paradox as to why the DAF sequence adjacent to the hydrophobic domain is apparently required to signal GPI-addition in the context of a chimeric protein but can be deleted from DAF itself without affecting anchor addition, was resolved by the finding that DAF contains a number of cryptic cleavage sites NH_2-terminal to the natural processing site. These cryptic sites are able to serve as substitute cleavage/attachment sites when brought into juxtaposition with the hydrophobic domain by deletion of the natural site (Caras et al., 1989).

DETERMINATION OF THE CLEAVAGE/ATTACHMENT SITE OF DAF

To assess the importance of the cleavage/attachment site in GPI anchoring it was necessary to first locate it, which required determining the COOH-terminal residue of processed, GPI-linked DAF. We approached this by first labelling DAF metabolically with [^{3}H]ethanolamine, a component of the GPI anchor, thus providing a means of specifically tagging the COOH-terminus of the protein. The labelled DAF was then purified and cleaved with trypsin, giving rise to a single labelled (COOH-terminal) peptide which we isolated by HPLC and sequenced (Moran et al., 1991). These experiments pinpointed Ser319 as the COOH-terminal, GPI-linked residue of DAF (Fig.2) and suggested that a 28-residue peptide is removed in the processing event. In agreement with earlier predictions the cleavage/attachment site is located 12 residues NH_2-terminal to hydrophobic domain, within the 20-residue sequence identified as containing a second element necessary for another addition.

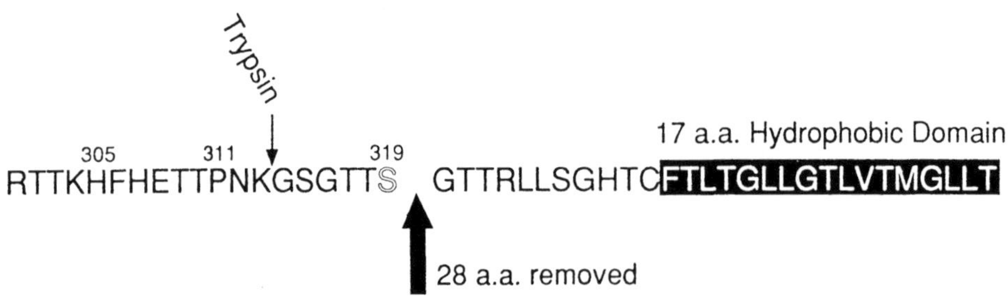

Figure 2.
COOH-terminal sequence of DAF showing the cleavage/attachment site for GPI addition (large arrow). The COOH-terminal hydrophobic domain is shown in black. A small arrow indicates the tryptic-cleavage site that produced a 6-residue, GPI-linked COOH-terminal peptide that was isolated by HPLC and sequenced (from Moran et al, 1991). aa, amino acid.

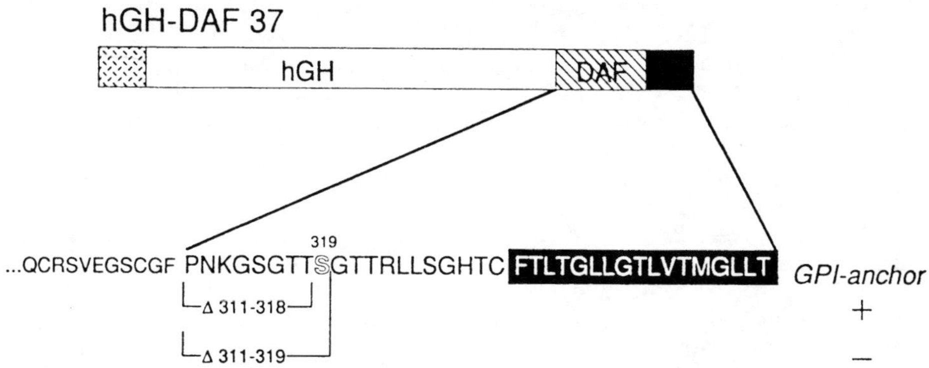

Figure 3.
Schematic diagram showing the GPI-anchored fusion protein hGH-DAF37 and deletion mutants derived from it. HGH-DAF37 contains the last 37 residues of DAF fused to the COOH-terminus of hGH. The 17-residue COOH-terminal hydrophobic domain of DAF is shown in black and the GPI-addition site (ser319) is indicated in the sequence below. Deletion of DAF residues 311-318, NH_2-terminal to the attachment site, had no effect on anchor attachment. Deletion of residues 311-319 which includes the attachment site Ser, abolished anchor addition (from Moran et al, 1991).

STRUCTURAL REQUIREMENTS OF THE CLEAVAGE/ATTACHMENT SITE

Since DAF contains multiple cryptic cleavage sites, the structural requirements of the cleavage/attachment site were studied using a GPI-anchored fusion protein, hGH-DAF27, containing the last 37 residues of DAF (including the 17-residue hydrophobic domain and the natural cleavage site) fused to the COOH-terminus of hGH (Fig.3). GPI-anchoring of this protein is absolutely dependent on the single cleavage site provided by the DAF sequence, making it an ideal substrate for mutational analysis. The DAF sequences immediately NH_2-terminal to the anchor addition site (8 residues) were found not to be important for GPI-anchoring and could be deleted from the fusion protein with no effect (Fig.3). In contrast, deletion of these 8 residues together with the attachment site Ser totally abolished anchor attachment. This established that the COOH-terminal 29 residues encoded by the DAF cDNA (consisting of the attachment site Ser and the 28-residue peptide that is removed during processing) comprise the minimum sequence that will trigger anchor attachment. Further, a critical role for the residue at the attachment site was suggested (Moran et al., 1991).

To assess the importance of the amino acid at the anchor addition site, we investigated the consequences of replacing this residue with alternative amino acids. Starting with the GPI-anchored fusion protein, hGH-DAF29, containing the minimal DAF sequence that will trigger anchor addition, 'in vitro' mutagenesis by the polymerase chain reaction (Saiki et al., 1988) was used to replace Ser319 of DAF (the natural acceptor for the anchor) with all possible alternative amino acids. Following expression of these mutants in COS cells, we tested for targeting of hGH to the cell surface and for the presence of a GPI anchor. We found that only Ser, Gly, Ala, Asp or Asn could serve as acceptors for GPI-addition (Moran et al., 1991). Other amino acids at the attachment site abolished or severely reduced both cleavage of the nascent polypeptide and anchor attachment, suggesting that these events might be tightly coupled. In addition, transport of the uncleaved fusion proteins to the cell surface was blocked. Similar experiments using placental alkaline phosphatase suggested that besides the 5 residues listed above, Cys can also serve as an acceptor for GPI addition (Micanovic et al., 1990). In the context of the hGH-DAF sequence Cys was an inefficient acceptor, producing barely detectable levels of cell-surface (presumably GPI-anchored) protein. In accordance with these results, these six amino acids and only these, have been found at the known cleavage/attachment sites of natural GPI-anchored proteins. The general rule that emerges from these observations is that the residue at the anchor attachment site must be small.

A similar restriction appears to apply to the residue found on the carboxyl side of the cleavage point. This was first suggested by inspection of the known processing sites of natural GPI-anchored proteins which show Ser, Gly, Ala, Asn, Asp and occasionally Arg at this position (Cross, 1990). This concept is supported by experiments showing that a non-functional sequence for GPI attachment (see below), consisting of a fragment of the low density lipoprotein receptor (LDLR) fused to the DAF COOH-terminal hydrophobic domain could be converted to a strong GPI signal by introducing a 'pair' of small residues at a position corresponding to the normal cleavage/attachment site in the DAF sequence (Moran and Caras, 1991). The LDLR sequence contained Val and Glu at this position, a sequence that is unable to function as a cleavage site (fusion protein HLD, Fig.4). A single mutation, converting Val at the anchor addition site to Ser, one of the 6 amino acids that can function as an acceptor for anchor addition, was insufficient to produce GPI anchoring. A second mutation was required, converting the Glu-residue on the COOH-terminal side of the cleavage point to a small residue such as Gly, the Ser-Gly combination now allowing efficient GPI attachment. In contrast, mutation of the Glu-residue to Phe (a bulky residue), producing the sequence Ser-Phe at the cleavage/attachment site, failed to produce GPI-linked protein (Fig.4). These observations suggest that in addition to a requirement for a small residue at the anchor addition site itself, there is a similar requirement for a small residue at the position immediately COOH-terminal to the cleavage point.

	hGH		DAF Hydrophobic Domain	GPI–Anchor
hGH–DAF37	...EGSCGF	PNKGSGTTSGTTRLLSGHTC	**FTLTGLLGTLVTMGLLT**	+
	hGH	LDL Receptor		
HLD	...EGSCGF	LTTVEIVTMSHQALG	**FTLTGLLGTLVTMGLLT**	–
HLD.S^{12}	...EGSCGF	LTTSEIVTMSHQALG	**FTLTGLLGTLVTMGLLT**	–
HLD.S^{12}G	...EGSCGF	LTTSGIVTMSHQALG	**FTLTGLLGTLVTMGLLT**	+
HLD.S^{12}F	...EGSCGF	LTTSFIVTMSHQALG	**FTLTGLLGTLVTMGLLT**	–

Figure 4.
COOH-terminal sequences of the hGH-DAF and HLD (hGH-LDLR-DAF) fusion proteins. Small print, hGH COOH-terminus; boxed sequence, DAF hydrophobic domain. The LDLR sequence of HLD is indicated. An arrow denotes the normal cleavage point in the DAF sequence of hGH-DAF37 and the anchor addition site, Ser319, is underlined. HLD.S^{12}, HLD.S^{12}G and HLD.S^{12}F were derived from HLD by oligonucleotide-directed in vitro mutagenesis and the residues that were altered are indicated by an underline (from Moran and Caras, 1991).

HGH-DAF37 contains the COOH-terminal 37 residues of DAF and is GPI anchored. In HLD, the DAF sequence NH_2-terminal to the hydrophobic domain has been replaced with a fragment of the LDLR, and GPI attachment is abolished. A Val to Ser mutation at the anchor addition site (producing HLD.S^{12}) does not restore anchor attachment, whereas a *pair* of mutations, converting Val and Glu at the cleavage site to Ser and Gly (as in HLD.S^{12}G) produces a GPI anchored protein. If the Gly is replaced with Phe (as in HLD.S^{12}F), anchor attachment is again abolished.

ADDITIONAL FEATURES OF THE GPI-SIGNAL

Thus far we have discussed two basic elements needed to constitute a GPI signal: a COOH-terminal hydrophobic domain and a cleavage/attachment site located NH_2-terminal to it, composed of a pair of small residues flanking the cleavage point. Are there additional structural or conformational features necessary for anchor attachment? As mentioned above, sequences immediately NH_2-terminal to the attachment site do not appear to be important for GPI anchoring. What about the sequences COOH-terminal to the cleavage site, between the cleavage site and the hydrophobic domain? While problems such as these are usually approached using substitution or deletion mutagensis to pinpoint critical residues or sequences, the general nature of the GPI signal suggested that the results of such a study might be ambiguous and difficult to interpret. We therefore approached the problem from the opposite direction by starting with a sequence that does not signal anchor attachment and asking what changes are required to convert it to a GPI signal. We began with an hGH-DAF fusion protein in which all of the DAF sequences adjacent to the COOH-terminal hydrophobic domain (including the cleavage/attachment site) were replaced with a non-specific, non-functional sequence derived from the extracellular domain of the LDLR. When expressed in COS cells this fusion protein (hGH-LDLR-DAF17, abbreviated HLD, Fig.4) failed to become GPI anchored and was not transported to the cell surface, remaining trapped inside the cells. We then asked what changes are required to convert HLD to a GPI-anchored protein? We found that the insertion of a cleavage/attachment site consisting of a pair of small residues (Ser-Gly) was both necessary and sufficient to produce a strong GPI signal that could target HLD to the plasma membrane via a GPI anchor (Moran and Caras, 1991). Since the LDLR sequence of HLD and DAF sequence that it replaced show no homology and are presumably also conformationally different, and since the insertion of a cleavage site into HLD is sufficient to induce GPI-anchoring, we concluded that the sequences surrounding the cleavage site, including the sequence between the cleavage site and the hydrophobic domain, do not contribute significantly to the GPI signal other than serving as a hydrophilic spacer. The complete absence of homology or recognizable consensus patterns within this region when different GPI-anchored proteins are compared, is consistent with this view.

The 'positioning' of the cleavage/attachment site relative to the COOH-terminal hydrophobic domain appears to be an important feature of the GPI signal. We tested this by constructing a series of HLD proteins containing an inserted cleavage/attachment site sequence (Ser-Gly) positioned such that the attachment site Ser was located either 12, 9, 6, 4, 2 or 16 residues NH_2-terminal to the hydrophobic domain (Fig.5). Following expression in COS cells the fusion protein HLD.S^{12}G, containing the cleavage/attachment site positioned 12 residues NH_2-

terminal to the hydrophobic domain, was efficiently processed to a GPI-anchored cell surface protein. The HLD.S^{9}G protein also became GPI-anchored but with reduced efficiency, and all of the remaining proteins, containing cleavage/attachment sites positioned closer to or further from the hydrophobic domain, failed to become GPI anchored at all (Moran and Caras, 1991). These data suggest that for optimal anchor attachment the cleavage/attachment site must be positioned 10-12 residues NH_2-terminal to the hydrophobic domain. As the processing site is moved beyond these constrains, the efficiency of anchor attachment drops off sharply, the GPI signal eventually becoming essentially non-functional.

	15 10 5 1	DAF Hydrophobic Domain	GPI–Anchor
HLD.S^{12}G	...EGSCGF LTTSGIVTMSHQALG	FTLTGLLGTLVTMGLLT	+
HLD.S^{9}G	...EGSCGF LTTVEISGMSHQALG	FTLTGLLGTLVTMGLLT	±
HLD.S^{6}G	...EGSCGF LTTVEIVTMSGQALG	FTLTGLLGTLVTMGLLT	—
HLD.S^{4}G	...EGSCGF LTTVEIVTMSHSGLG	FTLTGLLGTLVTMGLLT	—
HLD.S^{2}G	...EGSCGF LTTVEIVTMSHQASG	FTLTGLLGTLVTMGLLT	—
HLD.S^{16}G	...EGSCG SGTTVEIVTMSHQALG	FTLTGLLGTLVTMGLLT	—

Figure 5.
COOH-terminal sequences of HLD fusion protein containing a cleavage/attachment site (underlined) introduced at various positions NH_2-terminal to the hydrophobic domain. Boxed sequences, DAF hydrophobic domain; large print, unboxed, LDLR sequence; small print, hGH COOH-terminal sequence (from Moran and Caras, 1991).

AN INTERNALLY POSITIONED SIGNAL CAN DIRECT ATTACHMENT OF A GLYCOPHOSPHOLIPID MEMBRANE ANCHOR

All known GPI-anchored proteins contain the signal for GPI attachment at the COOH-terminus of the protein, and it has been suggested that the absence of a hydrophilic tail (beyond the hydrophobic domain) is an obligatory feature of the GPI signal (Berger et al., 1989; Cross, 1990). To examine this requirement we constructed a chimeric protein, DAF-hGH, in which the human growth hormone coding sequence (191 amino acids) was fused to the COOH-terminus of DAF, thereby placing the GPI signal in the middle of the chimeric protein between two large hydrophilic domains (Fig.6). Following expression in COS cells this fusion protein was processed at or near the normal DAF processing site, producing authentic GPI-anchored DAF on the cell surface (Caras, 1991). Thus, despite the presence of a long hydrophilic extension of 191 amino acids, DAF molecules can be targeted to the plasma membrane via a GPI anchor. This result suggests that the absence of a hydrophilic tail is not a critical feature of the GPI signal. A fraction of the DAF-hGH fusion molecules escaped processing however and were secreted as intact fusion protein, suggesting that an internal GPI signal may be recognized less efficiently than a COOH-terminal signal.

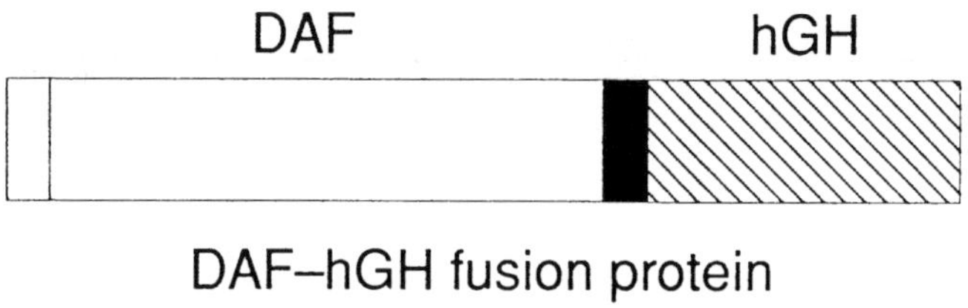

Figure 6.
Schematic diagram of the DAF-hGH fusion protein containing mature hGH (residues 1 - 191) (hatched box) fused to the COOH-terminus of DAF (unshaded box). The DAF signal peptide is denoted by the grey box and a black box indicates the COOH-terminal hydrophobic domain (from Caras, 1991).

The absence of hydrophilic extensions in natural GPI-anchored proteins may reflect an evolutionary adaptation rather than a functional constraint. Since the COOH-terminal sequences are removed and presumably degraded during anchor addition, there is no obvious advantage to synthesizing more than the minimum COOH-terminal sequence required to contain the GPI signal.

REFERENCES

Berger, J., Howard, A. D., Brink, L., Gerber, L., Hauber, J., Cullen, B.R. and Udenfriend, S. (1988) COOH-terminal requirements for the correct processing of a phosphatidylinositol-glycan anchored membrane protein. J. Biol. Chem. **263:** 10016-10021.

Berger, J., Micanovic, R., Greenspan, R. J., and Udenfriend, S. (1989) Conversion of placental alkaline phosphatase from a phosphatidylinositol-glycan-anchored protein to an integral transmembrane protein. Proc. Natl. Acad. Sci. USA **86:** 1457-1460.

Boothroyd, J. C., Paynter, C. A., Cross, G. A. M., Bernards, A. and Borst, P. (1981) Variant surface glycoproteins of *Trypanosoma brucei* are synthesized with cleavable hydrophobic sequences at the carboxy and amino termini. Nucl. Acids Res. **9:** 4735-4743.

Caras, I. W. and Weddell, G. N. (1989) Signal peptide for protein secretion directing glycophospholipid membrane anchor attachment. Science **243:** 1196-1198.

Caras, I., Weddell, G. N. and Williams, S. R. (1989) Analysis of the signal for attachment of a glycophospholipid membrane anchor attachment. J. Cell Biol. **108:** 1387-1396.

Caras, I. W., Weddell, G. N., Davitz, M. A., Nussenzweig, V. and Martin, D. W. Jr. (1987) Signal for attachment of a phospholipid membrane anchor in delay accelerating factor. Science **238:** 1280-1283.

Caras, I. W. (1991) An internally positioned signal can direct attachment of a glycophospholipid membrane anchor. J. Cell Biol. **113:** in press.

Crise, B., Ruusala, A., Zagouras, P., Shaw, A. and Rose, J. K. (1989) Oligomerization of glycolipid-anchored and soluble forms of the vesicular stomatitis virus glycoprotein. J. Virol. **63:** 5328-5333.

Cross, G. A. M. (1990) Glycolipid anchoring of plasma membrane proteins. Ann. Rev. Cell Biol. **6:** 1-39.

Ferguson, M. A. J. and Williams, (1988) Cell surface anchoring of proteins via glycosyl-phosphatidylinositol structures. Annu. Rev. Biochem. **57:** 285-320.

Masterson, W., Doering, T. L., Hart, G. W. and Englund, P. T. (1989) A novel pathway for glycan assembly: biosynthesis of the glycosyl-phosphatidylinositol anchor of the trypanosome variant surface glycoprotein. Cell **56:** 793-800.

Micanovic, R., Gerber, L. D., Berger, J., Kodukula, K. and Udenfriend, S. (1990) Selectivity of the cleavage/attachment site of phosphatidylinositol-glycan-anchored membrane proteins determined by site-specific mutagenesis at Asp-484 of placental alkaline phosphatase. Proc. Natl. Acad. Sci. USA **87:** 157-161.

Moran, P., Raab, H., Kohr, W. J. and Caras, I. W. (1991) Glycophospholipid membrane anchor attachment. Molecular analysis of the cleavage/attachment site. J. Biol. Chem. **266:** 1250-1257.

Moran, P. and Caras, I. W. (1991) A non-functional sequence converted to a signal for phosphatidylinositol membrane anchor attachment. Personal communication.

Saiki, R. K., Gelfand, D. H., Stoffel, S., Scharf, S. J., Higuchi, R., Horn, G. T., Mullis, K. B. and Erlich, H. A. (1988) Primer-directed enzymatic amplification of DNA with a thermostable DNA polymerase. Science **239:** 487-491.

Tse, A. G. D., Barclay, A. N., Watts, A. and Williams, A. F. (1985) A glycophospholipid tail at the carboxyl terminus of the Thy-1 glycoprotein of neurons and thymocytes. Science **230:** 1003-1008.

Waneck, G. L., Sherman, D. H., Kincade, P. W., Low, M. G. and Flavell, R. A. (1988) Molecular mapping of signals in the Qa-2 antigen required for attachment of the phosphatidylinositol membrane anchor. Proc. Natl. Acad. Sci. USA. **85:** 577-581.

INOSITOLGLYCANS AND CELLULAR SIGNALLING

***Guillermo Romero**

Department of Pharmacology, University of Virginia Health Sciences Center, Charlottesville, VA 22908, USA.

* Corresponding author. Present address: Department of Pharmacology, University of Pittsburgh School of Medicine, Pittsburgh, PA 15261, USA.

INTRODUCTION

Glycosylphosphatidylinositols (GPI) are lipids found covalently linked to many membrane proteins attached to the outer leaflet of the cell membrane. The structural features of GPIs have been well established during the last few years and have been the subject of many reviews (Ferguson and Williams, 1988; Low and Saltiel, 1988; Low, 1989; Thomas et al, 1990). However, the functional relevance of these lipids other than providing an alternative mechanism of anchoring of cell membrane proteins remains obscure. GPI anchored proteins are susceptible to cleavage by specific phospholipases C and D and exhibit high lateral mobility in membranes (Ferguson et al, 1988; Homans et al, 1988; Roberts et al, 1988; Ishihara, A et al, 1987). The rather complex structure of the GPI anchor shows a remarkable degree of evolutionary conservation suggestive of specialized functional roles (Ferguson and Williams, 1988; Low, 1989). Furthermore, the cleavage of these anchors by phospholipases C or D generates well known modulators of cell function, such as diacylglycerol and phosphatidic acid. Recent observations suggest that GPI-derived oligosaccharides may play important roles in cell signalling as well.

The GPI anchors of membrane proteins may participate in cellular signalling by many different hypothetical mechanisms. The four most obvious are the following: a) GPI anchored proteins may serve as precursors of inositolglycans, diacylglycerol, phosphatidic acid and other second messengers; b) GPI anchored enzymes and regulatory proteins may become "activated" by their release to the extracellular medium in a manner analogous to the activation of proenzymes by hydrolysis or simply by allowing delivery of the enzyme to a specific target different from its site of attachment in the cell membrane; c) GPI anchored membrane proteins may form aggregates inducing endocytosis followed by digestion and release mediators to the inside of the cell; and d) GPI anchored proteins may form clusters capable of direct interactions with other cell components such that the cellular response is a consequence of the activation of other as yet undetermined signalling systems. These mechanisms are independently supported by experimental evidence. Furthermore, these mechanisms do not exclude each other and, obviously, there may be significant differences in the relative importance of these signalling pathways in different systems. Mechanisms (a) and, to a lesser extent, (b) appear to be important in insulin and growth factor signalling; mechanisms (c) and (d) have been invoked to explain the activation of immune cells by antibodies directed against GPI-anchored antigens (for a review, see Robinson, 1991). To date, however, none of these models has been fully tested and these ideas remain largely speculative.

This article deals primarily with the role of GPIs and their inositolglycan derivatives in the mediation of insulin and growth factor function. The function of other compounds derived from GPI metabolism (such as diacylglycerol, alkyl-acyl-glycerol and phosphatidic acid) have not been considered but cannot be ignored when attempting to reconstruct the series of events involved in the complete signalling processes of these growth factors.

INSULIN MEDIATORS

Insulin binding to a target cell initiates a pleiotropic response: the transport of glucose and amino acids is rapidly augmented; lipid and glucose metabolism are altered; gene regulation, protein synthesis and, eventually, cell division are induced; and, concomitantly with these effects, endocytosis and degradation of the hormone proceed at a fast rate. The complexity of the cellular responses to insulin implies a complicated signalling mechanism in which regulation of different functions may require the existence and cross-interaction of diverse signalling pathways. Despite great efforts, our present understanding of the mechanisms by which insulin may regulate such a wide spectrum of cellular functions is still poor.

In 1974, Larner et al. (Larner et al, 1974) first proposed that at least some of the actions of insulin could be mediated by small molecules. This hypothesis was based on observations indicative of the existence of an insulin inducible, heat and acid stable compound that selectively inhibited the activity of CAMP dependent protein kinase. This compound was further purified by gel chromatography and shown to inhibit adenylate cyclase (Larner, 1979; Malchoff et al, 1987) and to activate mitochondrial pyruvate dehydrogenase (PDH) (Jarett and Seals, 1979).

Initial reports on the chemical composition of insulin mediators were conflicting for several years. In 1982, Larner et al reported finding amino acids and carbohydrates in a purified preparation of mediator (Larner et al, 1982). A few years later, Saltiel et al discovered that related compounds could be metabolically labelled with [^{3}H]-inositol and [^{3}H]-glucosamine and released from cell membranes by the action of inositol-specific phospholipase C (Saltiel et al, 1986; Saltiel et al, 1987). These results were rapidly confirmed by Mato et al (1987) and supported by the chemical composition data reported by Larner et al (1988). Further work in Larner's laboratory led to the isolation of two different putative mediators (Larner et al, 1988; Larner et al, 1989). These compounds differ in their chemical compositions as well as in their specific actions: one of them is a potent activator of PDH phosphatase and contains *chiro*-inositol and galactosamine; the second inhibits the cAMP-dependent protein kinase and contains *myo*-inositol and glucosamine. To this date, the complete chemical structure of insulin mediators remains undetermined.

In addition to chemical composition data, the evidence indicative that these putative mediators are related to the GPI anchors of membrane proteins arises from several sources. Chemical studies have shown that the biological activity of these compounds is destroyed by treatments known to break down the oligosaccharide moiety of GPI protein anchors, such as nitrous acid deamination, periodate oxidation and TFA hydrolysis (Saltiel, 1987; Mato et al, 1987; Larner et al, 1988). Mediator-like substances have been obtained from isolated cell lipids by treatment with phosphatidylinositol-specific phospholipase C (Saltiel et al, 1986; Saltiel et al, 1987; Mato et al, 1987). Immunochemical studies have shown that antibodies raised against the inositolglycan linkers of α-galactosidase-treated trypanosome variant surface glycoproteins (VSG) inhibit the in vitro action of purified mediator fractions (Romero et al, 1990). Putative insulin mediators can be selectively immunoprecipitated by or selectively bound to immunoaffinity columns prepared with anti-inositolglycan antibodies (Romero et al, 1990) (Figure 1). Inositolglycans purified from trypanosome VSGs mimic some of the in vitro biological effects of insulin mediators (Romero et al, 1990) (Figure 1).

The physiological role of these mediators is being actively investigated. Initial data indicated that these compounds could mimic in vitro the effects of insulin in the activation of PHD phosphatase and the inhibition of cAMP-dependent protein kinase (Jarett and Seals, 1979; Larner et al 1979). Other biological effects of insulin have been added to this original list (summary Table 1).

How can inositolglycans regulate such a broad spectrum of biochemical events? The current hypothesis is that inositolglycans regulate protein phosphorylation by direct interaction with protein kinases and phosphatases. Detailed information on the mechanisms of action of mediators has been obtained in the cases of the cAMP-

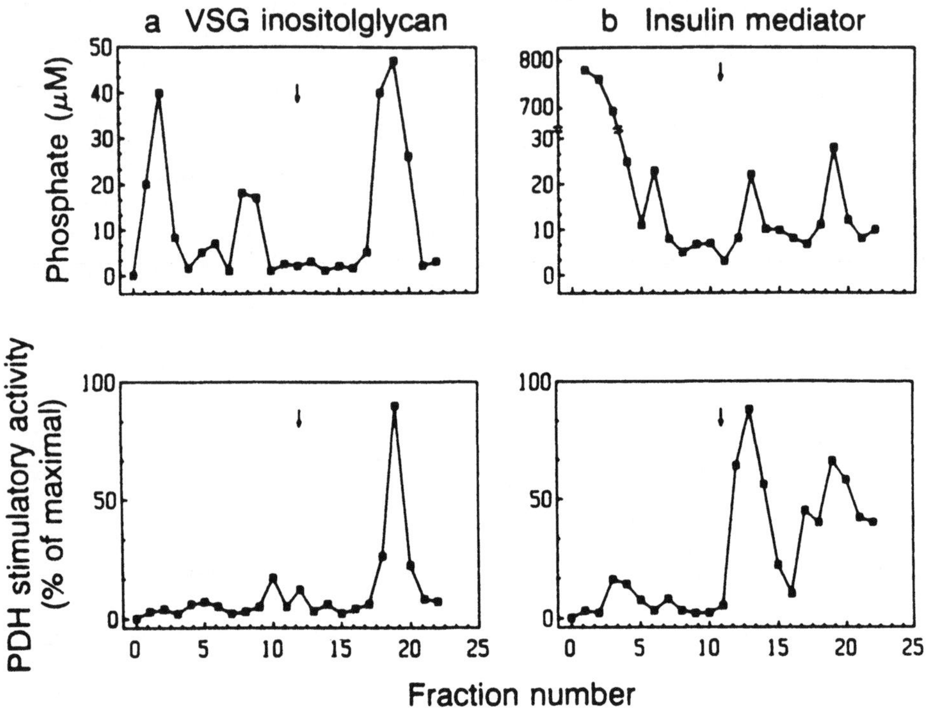

Figure 1. Elution profile of inositolglycan compounds chromatographed through an immunoaffinity column prepared with purified anti-inositolglycan antibodies. Affinity purified anti-VSG antibodies were purified as described by Romero et al (1990) and immobilized in CNBr-activated Sepharose (0.9 mg of antibody in 0.85 ml of packed gel). Samples were applied to the gel and equilibriated for 12-18 h at 4°C. The column was eluted first with 20 mM Tris (pH 7.5)/0.1 M NaCl and, where indicated by the arrows, elution of the bound material was started by addition of 0.1 M glycine (pH 2.8). Fractions (1 ml) were collected, neutralized, and assayed for organic phosphatase activation (lower). Panel (a) shows the elution profile of inositolglycans purified from trypanasome VSG. Panel (b) shows the results obtained with a crude preparation of rat liver insulin mediators. Reprinted from Romero et al (1990) with permission.

TABLE 1

Biological effects of inositolglycan mediators

Cell free assays:

Enzyme	Effect	Reference
Pyruvate dehydrogenase (PDH)	Activation	Jarett & Seals, 1979 Kiechle et al, 1980
PDH phosphatase	Activation	Larner et al, 1989
Adenylate cyclase	Inhibition	Malchoff et al, 1987
Casein kinase II	Biphasic	Alemany et al, 1990
cAMP-dependent protein kinase	Inhibition	Larner et al, 1979
cAMP phosphodiesterase	Activation	Saltiel, 1987
Phospholipid methyltransferase	Inhibition	Kelly et al, 1986
Acetyl-CoA carboxylase	Activation	Saltiel et al,1983
Glucose-6-phosphatase	Inhibition	Suzuki et al, 1984

Intact cell assays:

Effect	Cell system	Reference
Stimulation of PDH	Rat adipocytes	Jarett et al, 1985
Inhibition of glycogen phosphorylase	Rat adipocytes	Alvarez et al, 1987
Inhibition of pyruvate kinase	Rat adipocytes	Alvarez et al, 1987
Inhibition of hormone induced cAMP accumulation	Rat adipocytes Rat adipocytes	Machicao et al, 1990 Martiny et al, 1990
Inhibition of hormone induced lipolysis	Rat adipocytes	Machicao et al, 1990 Martiny et al, 1990
Stimulation of lipid synthesis	Rat hepatocytes Rat adipocytes	Caro et al, 1983 Saltiel & Sorbara-Cazán, 1987
Stimulation of glucose oxidation	Rat adipocytes	Saltiel & Sorbara-Cazán, 1987
Stimulation of glucose uptake	Rat adipocytes	Machicao et al, 1990
Stimulation of aminoacid transport	Rat hepatocytes	Varela et al, 1990
Stimulation of p33 mRNA production	H4 hepatoma cells	Sato et al, 1988
Stimulation of cell proliferation	Cochleovestibular ganglion	Varela-Nieto et al, 1991
Insulin-like changes in peptide phosphorylation	Rat adipocytes	Alemany et al, 1987

dependent protein kinase and PDH phosphatase. In the former, mediators act as inhibitors of the kinase reaction, apparently by competition with the polypeptide substrate (Malchoff et al, 1987). In the latter, mediator decreases the divalent metal requirements of PDH phosphatase (Larner et al, 1989) in a manner that mimics the effect of insulin.

THE MECHANISM OF GENERATION OF INSULIN MEDIATORS

The insulin receptor is a heterotetramer composed of two different subunits (Kasuga et al, 1982). The α-subunit faces the outside of the cell and interacts directly with insulin. The β-subunit is a tyrosine-specific protein kinase that spans the membrane (Kasuga et al, 1983; Ebina et al, 1985). Tyrosine kinase activity appears to be absolutely required for insulin receptor function. The question of how the tyrosine kinase activity of the receptor is linked to the generation of inositolglycan mediators has not been investigated in depth.

Much of the information available on the mechanisms by which insulin stimulates the generation of inositolglycan mediators has been obtained using BC_3H1 murine myocytes. Using metabolic labelling procedures, Farese and coworkers (1984; 1985) demonstrated initially that insulin induces a very rapid increase in the concentration of diacylglycerol in these cells. These results were later confirmed and extended by Saltiel et al (1987) and Luttrell et al (1988) who demonstrated that the initial phase of insulin induced generation of diacylglycerol was accompanied by the release of a putative mediator labelled with inositol and glucosamine and active in the modulation of phosphodiesterase and PDH phosphatase. It was therefore suggested that insulin stimulated a specific phospholipase C by an unknown mechanism and that this enzyme released simultaneously diacylglycerol and inositolglycan mediators from a specific lipid precursor (Saltiel et al, 1986).

Recent work has shown that pertussis toxin (PT) treatment inhibits insulin action in several systems (Goren et al, 1985; Heyworth et al, 1986; Luttrell et al, 1988; Romero et al, 1988; Rothenberg and Kahn, 1988; Luttrell et al, 1990). Among the PT sensitive effects of insulin are the generation of diacylglycerol and the formation of inositolglycan mediators (Luttrell et al, 1988; Romero et al, 1988). The effects of PT on the inhibition of cellular responses to insulin are apparently correlated with the ADP-ribosylation of a 40 kD protein (Luttrell et al, 1988), a fact that has been interpreted as indicative of the involvement of a heterotrimeric G protein in the insulin signal transduction system. Recent work in our laboratory has shown that nonhydrolyzable GTP analogues stimulate the release of a PDH phosphatase activator from isolated liver plasma membranes (Figure 2). Thus, although the precise role of this G protein remains to be established, the data suggest that it might be involved in the regulation of a specific phospholipase C activity required for the generation of inositolglycan mediators.

FIG. 2.

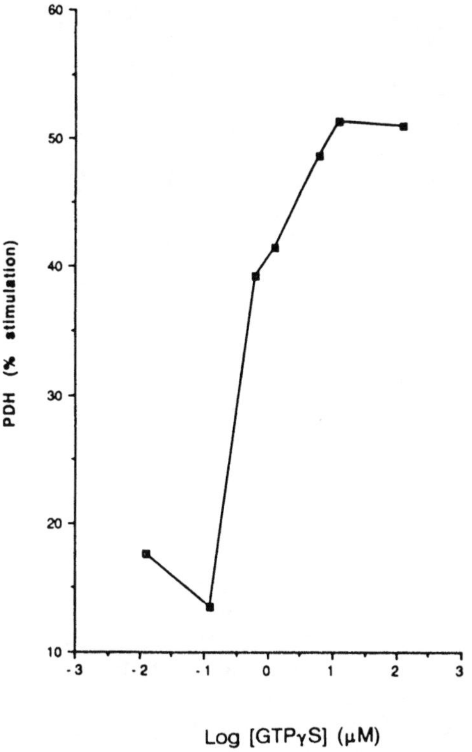

Figure 2. Dose dependence for the release of mediator activity from liver plasma membranes in response to GTPτS. Liver plasma membranes were incubated in the presence of 5mM $MgCl_2$ with the indicated concentrations of GTPτS. After 5 minutes at 37°C incubations were terminated and mediator fractions were prepared and assayed using a PDH phosphatase activation system. Kilgour, E. et al, in preparation.

The tyrosine kinase activity of the insulin receptor appears to be important in the mechanism of generation of insulin mediators. Earlier work by Suzuki et al (1987) showed that the insulin-stimulated release of mediator from liver plasma membranes was significantly increased by conditions known to stimulate the tyrosine kinase activity of the receptor, such as ATP in the presence of Mn^{2+}. More recent work by Villalba et al (1990) has shown that cells transfected with human insulin receptors mutated in the ATP binding site (and therefore without tyrosine kinase activity) have an impaired hydrolysis of GPI lipids in response to insulin. How is the tyrosine kinase activity of the insulin receptor linked to the generation of inositolglycan mediators? One possibility is the direct phosphorylation of phospholipase C. However, in contrast with the cases of the PDGF and EGF receptor systems (Meisenhelder et al, 1989; Nishibe et al, 1990), there is no evidence of the phosphorylation of phospholipase C upon insulin action that has been reported to date. A second alternative is the activation of phospholipase C via the phosphorylation of a specific G protein. However, our observations suggest that the insulin receptor does not phosphorylate G proteins in intact cells and that receptor-G protein interactions occur even in the absence of tyrosine phosphorylation (Luttrell et al, 1990). Thus, the role of the tyrosine kinase activity of the insulin receptor in the generation of mediators might not be associated to the activation of a phospholipase C and its precise function remains to be established.

MEDIATOR PRECURSORS: LIPIDS OR PROTEINS?

One of the most controversial issues related to the mechanism of generation of insulin mediators is that of the nature of the physiological precursors of these molecules. Initially, inositolglycan mediators were proposed to originate from insulin-sensitive lipids located in the inner leaflet of the cell membrane (Saltiel et al, 1986). Shortly thereafter, it was observed that insulin stimulated the release of GPI anchored proteins from the surface of BC_3H1 myocytes to the extracellular medium with a time course consistent with the generation of mediators (Romero et al, 1988; Figure 3). Furthermore, mediator activity was found in the incubation medium within 30 seconds of the stimulation of cells with insulin (Figure 3). Both PT and the protease inhibitor p-aminobenzamidine were shown to block the effects of insulin on mediator generation and PDH stimulation in BC_3H1 cells although by apparently different pathways: PT blocked the generation of diacylglycerol, suggestive of the inhibition of phospholipase C, but had no effect on the release of alkaline phosphatase to the medium; p-aminobenzamidine blocked the release of alkaline phosphatase to the medium with only a small inhibitory effect on the generation of diacylglycerol (Figure 3).

Figure 4 shows a model for the generation of insulin mediators from the GPI anchors of membrane proteins. Although this model is mostly hypothetical, it is strongly supported by experimental evidence. As shown, two insulin sensitive hydrolytic enzymes should be required to generate insulin mediators from the GPI anchors of membrane proteins. One of these is a p-aminobenzamidine sensitive hydrolase, most likely a protease; the second is a G protein regulated phospholipase, probably a phospholipase C. The insulin receptor is proposed to activate directly a PT sensitive heterotrimeric G protein which in turn stimulates the action of the phospholipase. By an independent mechanism, possibly protein phosphorylation, the receptor also stimulates a proteolytic enzyme that cleaves off the protein from the GPI lipid. The phospholipase C then releases the inositolglycan mediator to the culture medium. Since the target enzymes of the released mediators are

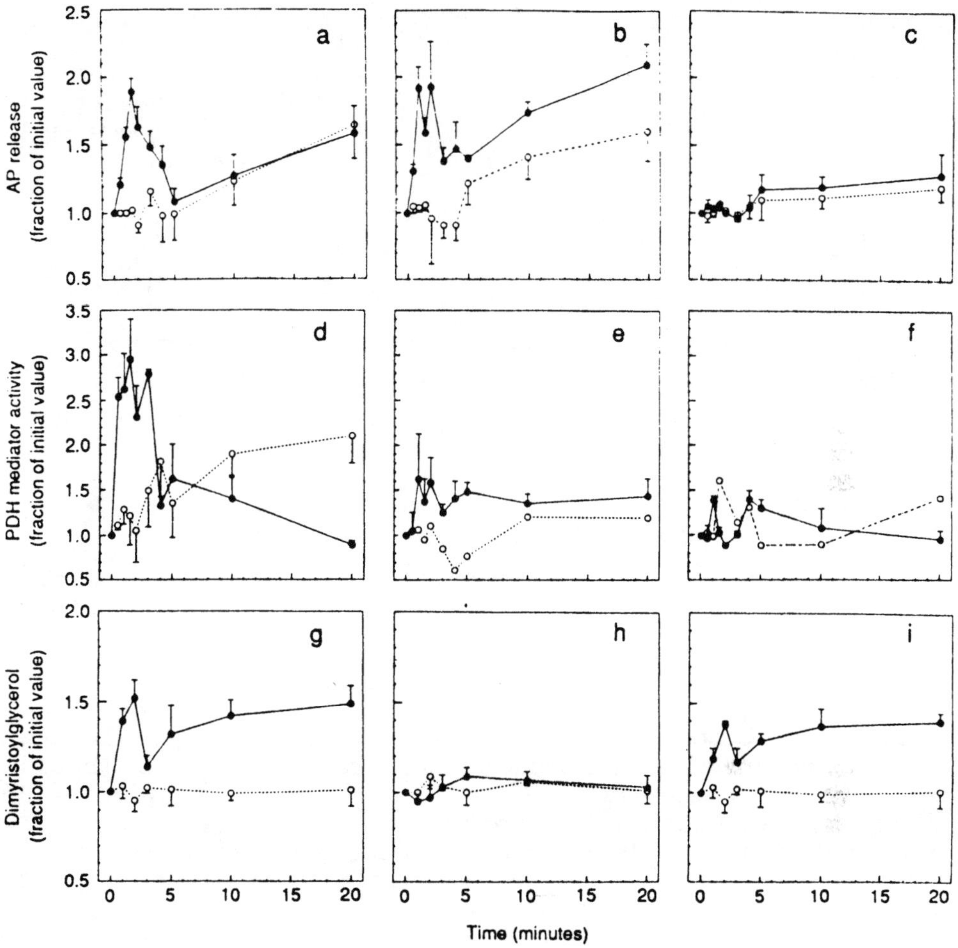

Figure 3. Early effects of insulin in BC_3H1 monocytes. Panels (a) through (c) show the time course of release of alkaline phosphatase to the incubation medium in the presence (●) or absence (○) of 100 nM insulin under the following experimental conditions: (a) insulin alone; (b) insulin after exposure of the cells (18-20 h) to 100 ng/ml pertussis toxin; (c) insulin in the presence of 0.8 mM p-aminobenzamidine. Panels (d), (e) and (f) show the contents of mediator activity of the same medium samples ((a), (b) and (c), respectively). Mediator activity was determined by using PDH phosphatase activation assay after preliminary purification of the samples by anion exchange chromatography. Panels (g), (h) and (i) show the accumulation of myristate labelled diacylglycerol in cells treated under identical conditions ((g): insulin alone; (h): insulin after pertussis treatment; (i): insulin and p-aninobenzamidine). Reprinted with permission from Romero et al (1988).

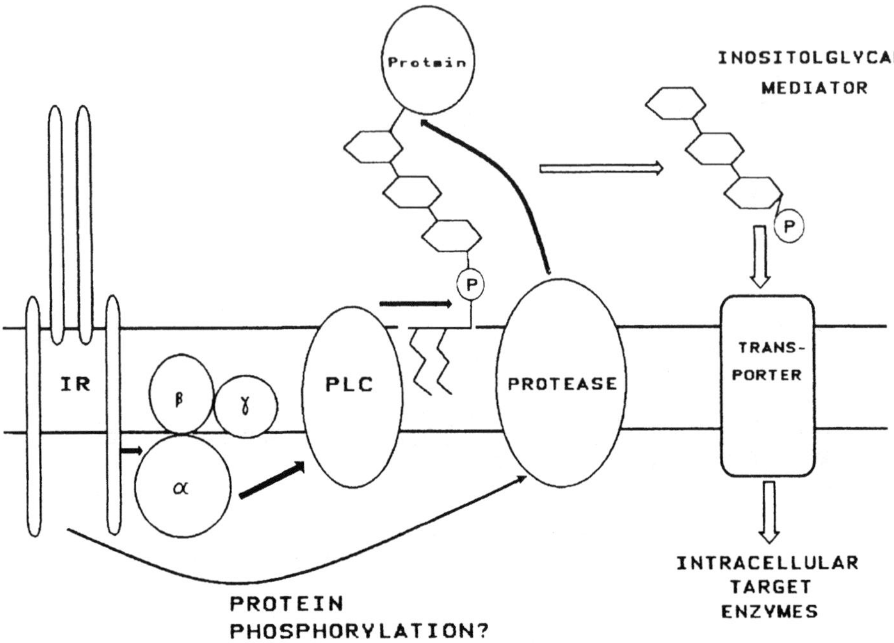

Figure 4. A model for the generation of inositolglycan mediators from GPI anchors of membrane proteins. Insulin activates, presumably by a protein phosphorylation process, a proteolytic enzyme that cleaves the GPI anchored protein, leaving GPI proteolipid attached to the membrane. A GPI binding protein is activated at the same time by direct interaction with the insulin receptor and stimulates the action of a phospholipase C which in turn releases inositolglycan to the extracellular medium. The extracellular mediator then binds a specific receptor/uptake system and is translocated to the intracellular medium where it may interact directly with specific protein kinases and/or phosphatases to exert its biological function.

intracellular, the existence of a receptor/transporter system responsible for the uptake of mediator is hypothesized.

Some of the evidence supporting this mechanism has been reviewed above. A significant amount of additional data provides further support for the proposed model. For instance, this model explains the insulinomimetic effects of proteolytic enzymes (Whipple and Larner, 1989), the anti-insulin effects of protease inhibitors (Muchmore et al, 1982; Begum et al, 1985), the reported stimulation of proteolytic activity (Cherqui et al, 1985) and the release of proteolytic enzymes to the extracellular medium (Kitagawa et al, 1989) by insulin. It is also consistent with the finding that trypsin treatment stimulates the generation of a mediator like substance from cell membranes (Rinaudo et al, 1988). The effects of insulin on the acute release of GPI anchored proteins have been observed in several laboratories (Romero et al, 1988; Chan et al, 1988; Lisanti et al, 1989). Furthermore, there is compelling evidence of the existence of a mediator receptor/uptake/transport system that will deliver extracellular inositolglycans to their intracellular target enzymes: purified mediators can mimic the effects of insulin in intact cells (see Table 1).

GPI anchored membrane proteins might serve as precursors of inositolglycan mediators by an alternative hypothetical mechanism in which agonist binding induces internalization of GPI anchored membrane proteins followed by attack by proteolytic enzymes and phospholipases in the endocytic vesicle resulting in the release of inositolglycans and other second messengers to the inside of the cell (Ferguson and Williams, 1988; Robinson, 1991). This model is not widely supported by experimental evidence in the insulin signalling system but may explain the mitotic effects of antibodies against GPI anchored proteins in T lymphocytes (reviewed by Robinson, 1991). The recent detection of a lysosomal GPI specific phospholipase C (Fouchier et al, 1990) provides some indirect support to this hypothetical mechanism.

Nevertheless, the nature of the physiological precursors of inositolglycan mediators remains an important issue. The contents of GPI lipids in BC_3H1 myocytes (Saltiel and Cuatrecasas, 1986; Luttrell et al, 1988) and other systems (Mato et al, 1987) decreases very rapidly after exposure to insulin. Several possible interpretations of these observations consistent with the model shown in Figure 4 may be envisioned. For instance, lipid "precursors" might arise from a constitutive or artifactual proteolytic release of GPI linked proteins, such that their hydrolysis represents only one half of the complete process. This may be the case of animal cell preparations in which collagenase treatment is used to break down the tissue and separate individual cells. Alternatively, these GPI lipids might be mostly the precursors of GPI anchored proteins: the hydrolysis of the protein precursors from the cell membrane might induce a rapid utilization of available lipids to regenerate GPI anchored proteins lost from the surface. Finally, these lipids may be, as suggested (Saltiel et al, 1986), true physiological precursors of inositolglycan mediators. For the moment, the issue remains unresolved.

INSULIN MEDIATORS ARE GENERATED OUTSIDE THE CELL

A controversial feature of the model depicted in Figure 4 is that insulin mediators are generated *outside* the cell. This was originally proposed based on the observation that mediators appeared in the extracellular medium very shortly after insulin treatment (Romero et al, 1988). Further evidence in favour of this hypothesis arises from the finding of mediator-like substances in material purified from the bloodstream (Sinha and Caro, 1985; Machicao et al, 1990) and the well established fact that treatment of intact cells with purified mediators induces insulin-like responses: activation of PDH (Jarett et al, 1985), inhibition of hormone-induced accumulation of cAMP (Zhang et al, 1983; Martiny et al, 1990), glycogen synthesis (Saltiel and Sorbara-Cazán, 1987), inhibition of lipolysis (Martiny et al, 1990), expression of specific mRNAs (Sato et al, 1988) and others. Even glucose uptake, an effect that was not believed to be mediated by lipid-derived inositolglycans (Kelly et al, 1987), appears to be modulated by mediator-like substances isolated from hemodialysates (Machicao et al, 1990). Thus, given the hydrophillic nature of inositolglycans, a specific receptor/uptake/transport mechanism for these substances must exist.

In 1988, Alvarez et al (1988) published elegant chemical labelling experiments showing that most of the insulin sensitive GPI lipids were exposed to the extracellular medium. We used a different approach: if mediators were primarily generated in the extracellular medium, than it should be possible to specifically block at least some of the actions of insulin by removing mediator from the medium. This can be accomplished by the use of specific antibodies. We thus prepared anti-inositolglycan antibodies using the soluble form of trypanosome VS pretreated with immobilized α-galactosidase in order to maximize cross reactivity with inositolglycans of mammalian origin. These antibodies were found to inhibit the stimulation of PDH by insulin in intact BC_3H1 cells by a mechanism that was consistent with the prevention of the entry of mediator into the cells (Figure 5). These observations thus demonstrated that the effects of insulin on PDH activation are mediated by inositolglycans and that these compounds are generated outside the cell.

One of the predictions of the proposed mechanism is the following: if mediators are generated at the external surface of the cell membrane and are free to diffuse to the extracellular medium, then the response of cells to insulin must be a function of the cell density of the system. Experiments to test this prediction are difficult to design since insulin induces relatively small responses in most systems. However, the synthesis of progesterone by swine ovary granulosa cells is very responsive to insulin (Veldhuis and Kolp, 1985) thus providing a system with the sensitivity required to test this hypothesis. Figure 6 shows the cell density dependence of the prog-esterone accumulation in response to insulin in these cells.

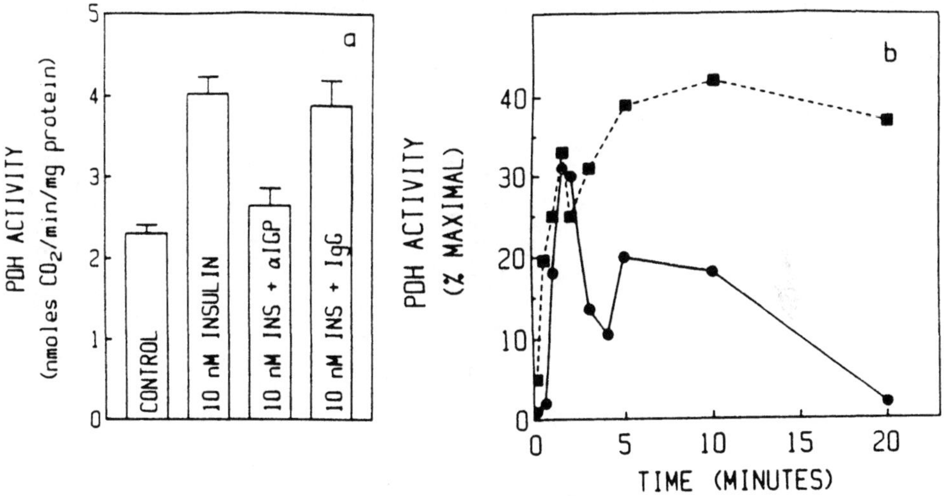

Figure 5. The effects of anti-inositolglycan antibodies on the responses of BC_3H1 myocytes to insulin. (a): Anti-inositolglycan antibodies block the insulin-induced stimulation of PDH in intact BC_3H1 cells. In these experiments, cells were exposed to insulin in the presence of 20 μg/ml anti-inositolglycan antibodies (αIPG) or non-immune IgG (IgG). The response to insulin was measured by the stimulation of PDH activity. In the experiments shown in (b), cells were exposed to insulin in the presence (■ or absence (●) of antibodies (20 μg/ml). Aliquots (0.5ml) of the medium were withdrawn when indicated and the antibodies were precipitated with protein A. Mediator activity was then eluted from the immunoprecipitates by acidification with 50 mM HCl and the eluted material was neutralized and assayed for PDH phosphatase activation. In control experiments (●) antibody was added to the same final concentration after removal of the medium from the tissue culture plate and processed in an identical manner. Reprinted with permission from Romero et al (1990).

As predicted, the response increases with cell density in a manner that is consistent with the effects of insulin being mediated by a paracrine factor released to the culture medium (Figure 6b, solid line). In agreement with the proposed model, we have recently detected inositolglycan mediators in the culture medium and shown that anti-inositolglycan antibodies block the insulin induced generation of progesterone by these cells. This suggests that the paracrine factor is probably an inositolglycan.

Very little is known about the hypothetical mediator receptor/uptake system. A few years ago, Ishihara, M et al (1987) demonstrated that proteoheparan sulfate, a GPI

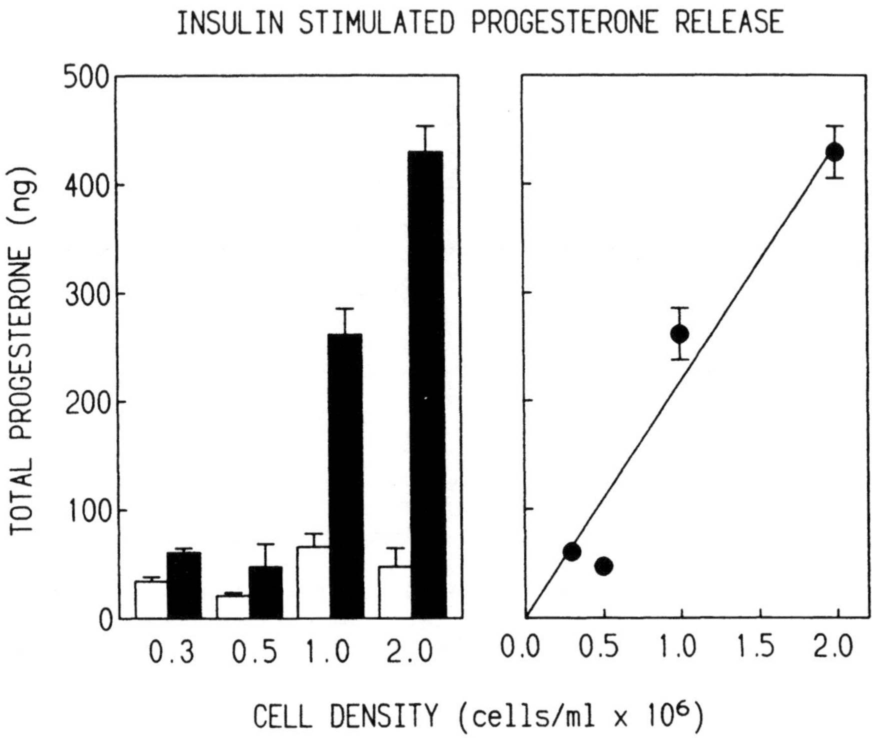

Figure 6. Insulin induces the synthesis of progesterone by swine ovary granulosa cells by a mechanism that depends on the cell density of the system. The solid line shown in (b) represents the best fit of the data to a simple paracrine model of action in which the paracrine factor is allowed to diffuse and equilibriate in the culture medium. Data taken and adapted from Veldhuis and Kolp (1985).

anchored polypeptide released from the surface of hepatocytes by the action of insulin, was slowly taken up into the cells. This uptake could be blocked specifically with inositol phosphates. More recent work has shown that the effects of purified inositolglycans on intact cell systems can be blocked by inositol phosphates and other sugars (Saltiel and Sorbara-Cazán, 1987; Machicao et al, 1990). These observations suggest that a specific inositolglycan receptor/transporter exists.

Do inositolglycan mediators mimic the actions of insulin in vivo? There are some indications that this might be the case. Fully active mediator-like substances have been found in the bloodstream or in hemodialysates (Sinha and Caro, 1985; Machicao et al, 1990). Simultaneous intraperitoneal administration of [^{14}C]-glucose and mediator results in a dose dependent increase in the incorporation of [^{14}C]-glucose into glycogen in a manner analogous to that observed with insulin (Huang et al, 1990). Furthermore, intravenous injection of purified mediators reduces the blood glucose concentration of streptozotocin treated diabetic rats (Huang et al, 1990). These observations, although still at a preliminary stage, may open new perspectives in the treatment of certain forms of diabetes.

DIABETES AS A MOLECULAR DISEASE

Diabetes is a severe disease characterized by impaired production of and/or response to insulin. Two main clinical forms of the disease are known. Type 1 or insulin dependent diabetes mellitus (IDDM) is characterized by destruction of the ß cells of the pancreatic islet which results in the absence of secretion of insulin. Patients are usually controlled with insulin. Type 2 or non-insulin dependent diabetes mellitus (NIDDM) is a syndrome in which patients may secrete large quantities of insulin but have impaired responses at the tissue level (Bogardus et al, 1984; Mandarino et al, 1986; Kida et al, 1990). The etiology of NIDDM is unknown.

Given the complex nature of the mechanism of generation of insulin mediators, multiple defects that impair cellular response to insulin might be envisioned. There have been several attempts to correlate diabetes to defects in the insulin signalling mechanism. It has been shown that mutations in the insulin receptor may lead to impaired responses to insulin (for a review see Taylor et al, 1990). Other observations point in the direction of the existence of insulin antagonistic substances, such as anti-insulin or anti-insulin receptor antibodies. These forms of NIDDM, however, are rare, accounting for only a small proportion of the cases and a unified theory for the pathogenesis of NIDDM is not within reach as yet.

Recently, Larner's laboratory has engaged in the analysis of possible defects in the metabolism of insulin mediators in NIDDM patients. Previous work by Larner and coworkers had shown the presence of *chiro*-inositol in one of the putative insulin mediators (Larner et al, 1988). It was then hypothesized that at least some of the NIDDM patients might have impaired mechanisms for the generation of mediators as a result of a defect in the metabolism of this rare inositol. In order to examine this hypothesis, urine from NIDDM patients and normal individuals were collected and analyzed for the presence of *chiro*-inositol using gas chromatography/mass spectrometry techniques (Kennington et al, 1990). A surprising finding followed: the daily urinary *chiro*-inositol excretion was reduced by 30-50 fold in the case of the patients with NIDDM (Figure 7).

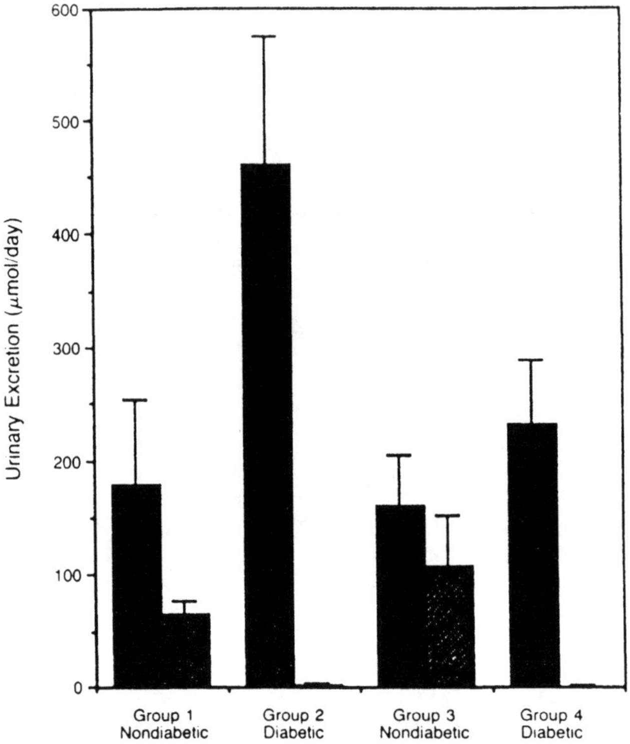

Figure 7. Mean 24-hour urinary excretion of *chiro*-inositol (hatched bars) and *myo*-inositol (solid bars) in four groups of subjects. Group 1 consisted of non-diabetic whites from Virginia; group 2, diabetic blacks from Virginia; group 3, nondiabetic whites and Pima Indians from Arizona; group 4, diabetic Pima Indians. T bars denote standard errors. The bars corresponding to the *chiro*-inositol excretion of diabetic subjects are barely visible. Reproduced from Kennington et al (1990) with permission.

Careful analysis of the data failed to find any correlations between the low excretion of *chiro*-inositol and any variables other than the diabetic state. Further studies confirmed the absence of *chiro*-inositol in mediator samples purified from muscle biopsies obtained from NIDDM patients; samples from normal individuals obtained under identical conditions, in contrast, contained detectable quantities of *chiro*-inositol (Kennington et al, 1990). A similar pattern was found in a primate *(Macaca mulatta)* model of NIDDM (Kennington et al, 1990).

The picture that emerges from these findings is the following. A large proportion of NIDDM patients owe their condition to one or several metabolic defects that prevent or impair the formation of *chiro*-inositol. In the absence of this compound, the patients can only generate *myo*-inositol-containing mediators and, as a result of this, their responses to insulin are severely impaired. Several corollaries may be derived from this hypothetical model. One of these is the theoretical possibility to bypass insulin by administration of mediator analogues or derivatives. A second consequence of interest is the potential to diagnose or predict the onset of NIDDM at early stages by means of the detection of *chiro*-inositol or possibly by genetic analysis of the defective enzyme involved in the process. A considerable amount of additional research is necessary to explore the viability of these ideas.

INOSITOLGLYCAN MEDIATORS MAY BE INVOLVED IN THE SIGNAL TRANSDUCTION MECHANISMS OF OTHER GROWTH FACTORS

Parallel to the explosion in the research on GPI anchored proteins of the last few years, great advances have been made in the research for other functions of inositolglycan mediators. To date, inositolglycan production and/or hydrolysis of GPI lipids and GPI anchored proteins have been directly or indirectly involved in the mechanism of action of several growth factors (see Table 2). Most of the work has been patterned on the initial studies of the insulin signalling systems and the results obtained bear striking resemblance to those reported with insulin. What is then the complete spectrum of events controlled by inositolglycans? This is a very complex issue since the generation of inositolglycan mediators is accompanied by the production of diaclyglycerol, alkyl-acyl-glycerol or possibly phosphatidic acid, all of which are potent modulators of cellular metabolism. Furthermore, GPI lipids may have physiological functions of their own, as suggested by the recently described inhibitory effects of the GPI antigens of the protozoan parasite *Leishmania donovani* on protein kinase C activity in vitro (McNeely et al, 1989).

The current approach to analyze the role of inositolglycans in signalling pathways is based on the fact that isolated inositolglycans mimic some of the

TABLE 2

Cell regulation systems that alter GPI metabolism

Agent	Cell system	Effects on GPI lipids	Effects on GPI anchored protein	Reference
Growth Factors:				
Insulin	BC_3H1 myocytes	yes	yes	Saltiel et al, 1986; Luttrell et al, 1988; Romero et al, 1988; Lisanti et al, 1989.
	H35 cells	yes	?	Kelly et al, 1987
	lymphocytes	yes	?	Gaulton et al, 1988
IGF-1	BC_3H1 myocytes	yes	?	Farese et al, 1988
	Balb/c 3T3 cells	yes	?	Kojima et al, 1990
EGF	BC_3H1 myocytes	yes	?	Farese et al, 1988
NGF	PC12 cells	yes	yes	Chan et al, 1989; Margolis et al, 1988
Cytokines:				
IL-2	CTLL2 (T cells)	yes	?	Mérida et al, 1990
Miscellaneous:				
ACTH	adrenal cells	yes	?	Cozza et al, 1988
TSH	thyroid cells	yes	?	Martiny et al, 1990

actions of the growth factor of interest. This approach has met with some success but presents some serious drawbacks. In most cases, the inositolglycans used are derived from cell membrane lipids purified and then treated with phospholipase C. Such studies therefore use a mixture of several inositolglycan species which may induce processes that are not necessarily involved in the specific signalling pathway of interest. Furthermore, some of these compounds may have no physiological role in signalling or may turn out to be inhibitory of the function of others. Second, the effects of the inositolglycans alone is unlikely to reflect the complete scope of action of the growth factor. This appears to be the case of the function of inositolglycans in 1L-2 signalling: the isolated oligosaccharide cannot stimulate the proliferation of CTLL-2 cells by itself, although it reduces considerably the threshold required for 1L-2 induced proliferation (Mérida et al, 1990). Likewise, inositol oligosaccharides isolated from different sources may have dissimilar properties and/or different actions in specific cell systems. For instance, inositolglycans prepared from whole cell lipids failed to stimulate glucose uptake in rat hepatocytes (Kelly et al, 1987), whereas related compounds obtained from a different source were found to be fully active in the stimulation of glucose uptake in rat adipocytes (Machicao et al, 1990). The issue is further complicated by the fact that inositolglycans of different structure may have very different functions, as shown for the *chiro*- and *myo*-inositol containing insulin mediators. Finally, different growth factors may generate specific inositolglycans and therefore produce different cellular responses. It is thus apparent that significant care must be exercised in the assignment of specific signalling functions to these compounds. Detailed chemical structures are needed in order to avoid unwarranted speculation. Given the structural diversity of GPI and GPI anchors, the analysis of the detailed regulatory functions of these compounds emerges as a formidable task.

ACKNOWLEDGEMENTS

Supported by Grants from the National Institutes of Health (DK40753), the Diabetes and Endocrinology Research Center of the University of Virginia and the Jeffress Trust Fund (J129). The author thanks Dr Joseph Larner for many useful discussions and for his critical evaluation of this manuscript.

REFERENCES

Alemany, S., Mato, J. M. and Stalfors, P. (1987). Phospho-dephospho-control by insulin is mimicked by a phospho-oligosaccharide in adipocytes. Nature. **330:** 77-79.

Alemany, S., Puerta, J., Guadaño, A. and Mato, J. M. (1987). Modulation of casein kinase II activity by the polar head group of an insulin-sensitive glycosyl-phosphatidylinositol. J. Biol. Chem. **265:** 4849-4852.

Alvarez, J. F., Cabello, M. A., Feliu, J. E. and Mato, J. M. (1987). A phospho-oligosaccharide mimics insulin action on glycogen phosphorylase and pyruvate kinase activities in isolated rat hepatocytes. Biochem. Biophys. Res. Commun. **147:** 765-771.

Alvarez, J., Varela, I., Ruiz-Albusac, J. M. and Mato, J.M. (1988). Localization of the insulin-sensitive phosphatidylinositol-glycan at the outer surface of the cell membrane. Biochem. Biophys. Res. Commun. **152:** 1445-1462.

Begum, N., Tepperman, H. M., and Tepperman, J. (1985). Studies on the effects of protease substrate analogues on some of the actions of insulin. Biochem. Biophys. Res. Commun. **126:** 489-495.

Bogardus, C., Lillioja, S., Stone, K. and Mott, D. (1984). Correlation between muscle glycogen synthase activity and in vivo action of insulin in man. J. Clin. Invest. **73:** 1185-1190.

Chan, B. L., Chao, M. V. and Saltiel, A. R. (1989). Nerve growth factor stimulates the hydrolysis of glycosyl-phosphatidylinositol in PC-12 cells: a mechanism of protein kinase C regulation. Proc. Natl. Acad. Sci. USA **86:** 1756-1760.

Chan, B. L., Lisanti, M. R., Rodríguez-Boulan, E. and Saltiel, A. R. (1988). Lipoprotein lipase is anchored to the cell surface by glycosyl-phosphatidylinositol. Science. **241:** 1670-1672.

Cherqui, G., Caron, M., Capeau, J. and Picard, J. (1985). Further evidence for the involvement of a membrane proteolytic step in insulin action. Biochem. J. **227:** 137-147.

Cozza, E. N., Vila, M. del C., Gómez-Sánchez, C. E., and Farese, R. V. (1988). ACTH stimulates turnover of the phosphatidylinositol-glycan. Biochem. Biophys. Res. Commun. **157:** 585-589.

Ebina, Y., Ellis, L., Jarnagin, K., Edery, M., Graf, L., Clauser, E Ou, J-h., Masiarz, F., Kan, Y. W., Goldfine, I. D., Roth, R. A. and Rutter, W. J. (1985). The human insulin receptor cDNA: The structural basis for hormone-activated transmembrane signalling. Cell. **40:** 747-758.

Farese, R V., Barnes, D. E., Davis, J. S., Standaert, M. L. and Pollet, R. J. (1984). Effects of insulin and protein synthesis inhibitors on phospholipid metabolism, diacylglycerol levels, and pyruvate dehydrogenase activity in BC_3H1 cultured myocytes. J. Biol. Chem. **259:** 7094-7100.

Farese, R. V., Davis, S. J., Barnes, D. E., Standaert, M. L., Babishkin, J. S., Hock, R., Rosic, N. K. and Pollet, R. J. (1985). The de novo phospholipid effect of insulin is associated with increases in diacylglycerol, but not inositol phosphates or cytosolic Ca^{2+}. Biochem. J. **231:** 269-278.

Farese, R. V., Nair, G. P., Standaert, M. L. and Cooper, R. (1988). Epidermal growth factor and insulin-like growth factor-1 stimulate the hydrolysis of the insulin-sensitive phosphatidylinositol-glycan in BC_3H1 myocytes. Biochem. Biophys. Res. Commun. **156:** 1346-1352.

Ferguson, M. A., Homans, S. W., Dwek, R. A. and Rademacher, T. W. (1988). Glycosyl-phosphatidylinositol moiety that anchors *Trypanosoma brucei* variant surface glycoprotein to the membrane. Science. **239:** 753-759.

Ferguson, M. A. J. and Williams, A. F. (1988). Cell-surface anchoring of proteins via glycosyl-phosphatidylinositol structures. Annu. Rev. Biochem. **57:** 285-320.

Fouchier, F., Baltz, T. and Rougon, G. (1990). Identification of glycosylphosphatidylinositol-specific phospholipases C in mouse brain membranes. Biochem. J. **269:**321-327.

Gaulton, G. N., Kelly, K. L., Powlowski, J., Mato, J. M. and Jarett, L. (1988). Regulation and function of an insulin-sensitive glycosyl-phosphatidylinositol during T lymphocyte activation. Cell. **53:** 963-970.

Goren, J. J., Northup, J. K. and Hollenberg, M. D. (1985). Action of insulin is modulated by pertussis toxin in rat adipocytes. Can. J. Physiol. Pharmacol. **63:** 1017-1022.

Heyworth, C. M., Grey, A.-M., Wilson, S. R., Hanski, E. and Houslay, M. D. (1986). The action of islet activating protein (pertussis toxin) on insulin's ability to inhibit adenylate cyclase and activate cyclic AMP phosphodiesterases in hepatocytes. Biochem. J. **235:** 145-149.

Homans, S. W., Ferguson, M. A., Dwek, R. A., Rademacher, T. W., Anand, R. and Williams, A. F. (1988). Complete structure of the glycosyl-phosphatidylinositol membrane anchor of rat brain Thy-1 glycoprotein. Nature. **333:** 269-272.

Huang, L. C., Houston, D. B., Zhang, C. and Larner, J. (1990). In vivo effects of two phosphoinositol glycan insulin mediators. FASEB J. **4:** A1829.

Ishihara, A., Hou, Y. and Jacobson, K. (1987). The Thy-1 antigen exhibits rapid lateral diffusion in the plasma membrane of rodent lymphoid cells and fibroblasts. Proc. Natl. Acad. Sci. USA. **84:** 1290-1293.

Ishihara, M., Fedarko, N. S. and Conrad, H. E. (1987). Involvement of phosphatidylinositol and insulin in the coordinate regulation of proteoheparan sulfate metabolism and hepatocyte growth. J. Biol. Chem. **262:** 4708-4716.

Jarett, L. and Seals, J. R. (1979). Pyruvate dehydrogenase activation in adipocyte mitochondria by an insulin-generated mediator from muscle. Science. **262:** 1407-1408.

Jarett, L., Wong, E. H. A., Macaulay, S. L. and Smith, J. A. (1985). Insulin mediators from rat skeletal muscle have different effects on insulin-sensitive pathways of intact adipocytes. Science. **227:** 533-535.

Kasuga, M., Hedo, J.-A., Yamada, K. M. and Kahn, C. R. (1982). The structure of the insulin receptor and its subunits: evidence for multiple nonreduced forms and a 210,000 kD possible proreceptor. J. Biol. Chem. **257:** 10392-10399.

Kasuga, M., Fujita-Yamaguchi, Y., Blithe, D. L. and Kahn, C. R. (1983). Tyrosine specific protein kinase activity is associated with the purified insulin receptor. Proc. Natl. Acad. Sci. USA. **80:** 2137-2141.

Kelly, K. L., Mato, J. M., Mérida, I. and Jarett, L. (1987). Glucose transport and antilipolysis are differentially regulated by the polar head group of an insulin sensitive glycophospholipid. Proc. Natl. Acad. Sci. USA. **84:** 6404-6407.

Kiechle, F. L., Jarett, L., Popp, D. A. and Kotagel, N. (1980). Isolation from rat adipocytes of a chemical mediator for insulin activation of pyruvate dehydrogenase. Diabetes. **29:** 852-855.

Kitagawa, K., Rosen, B.S., Spegelman, B. M., Lienhard, G. E. and Tanner, L. I. (1989). Insulin stimulates the acute release of adipsin from 3T3-L1 adipocytes. Biochim. Biophys. Acta. **1014:** 83.89.

Kojima, I., Kiaoka, M. and Ogata, E. (1990). Insulin-like growth factor-1 stimulates diacylglycerol production via multiple pathways in Balb/c 3T3 cells. J. Biol. Chem. **265:** 16846-16850.

Kennington, A. S., Hill, C. R., Craig, J., Bogardus, C., Raz, I., Ortmeyer, H.K., Hansen, B. C., Romero, G. and Larner, J. (1990). Low urinary *chiro*-inositol excretion in non-insulin dependent diabetes mellitus. N. Engl. J. Med. **323:** 373-378.

Kida, Y., Esposito-Del Puente, A., Bogardus, C. and Mott, D. M. (1990). Insulin resistance is associated with reduced fasting and insulin-stimulated glycogen synthase phosphatase activity in human skeletal muscle. J. Clin. Invest. **85:** 476-481.

Larner, J., Huang, L. C., Brooker, G., Murad, F. and Miller, T. B. (1974). Inhibitor of protein kinase formed in insulin treated muscle. Fed. Proc. **33:** 261.

Larner, J., Galasko, G., Cheng, K., DePaoli-Roach, A. A., Huang, L. C., Daggy, P. and Kellog, J. (1979). Generation by insulin of a chemical mediator that controls protein phosphorylation and dephosphorylation. Science. **206:** 1408-1410.

Larner, J., Cheng, K., Schwartz, C., Kikuchi, K., Tamura, S., Creacy, S., Dubler, R., Galasko, G., Pullin, C. and Katz, M. (1982). A proteolytic mechanism for the action of insulin via oligopeptide mediator formation. Fed. Proc. **41:** 2724-2729.

Larner, J., Huang, L. C., Schwartz, C. F. W., Oswald, A. S., Shen, T.-Y., Kinter, M., Tang, G. and Zeller, K. (1988). Rat liver insulin mediator which stimulates pyruvate dehydrogenase phosphatase contains galactosamine and D-*chiro*-inositol. Biochem. Biophys. Res. Commun. **151:** 1416-1426.

Larner, J., Huang, L. C., Sizuki, S., Tang, G., Zhang, C., Schwartz, C. F. W., Romero, G., Luttrell, L. and Kennington, A. S. (1989). Insulin mediators and the control of pyruvate dehydrogenase complex. Annals NY Acad. Sci. **573:** 297-305.

Lisanti, M. P., Darnell, J. C., Chan, B. L., Rodríguez-Boulan, E. and Saltiel, A. R. (1989). The distribution of glycosyl-phosphatidylinositol anchored proteins is differentially regulated by serum and insulin. Biochem. Biophys. Res. Commun. **164:** 824-832.

Low, M. G. (1989). Glycosyl-phosphatidylinositol: a versatile anchor for cell surface proteins. FASEB J. **3:** 1600-1608.

Low, M. and Saltiel, A. R. (1988). Structural and functional roles of glycosyl-phosphatidylinositol in membranes. Science. **239:** 268-275.

Luttrell, L. M., Hewlett, E. L., Romero, G. and Rogol, A. D. (1988). Pertussis toxin attenuates some effects of insulin in BC_3H1 murine myocytes. J. Biol. Chem. **263:** 6131-6141.

Luttrell, L., Kilgour, E., Larner, J. and Romero, G. (1990). A pertussis toxin-sensitive G protein mediates some aspects of insulin action in BC_3H1 murine myocytes. J. Biol. Chem. **265:** 16873-16879.

Machicao, F., Mushack, J., Seffer, E., Ermel, B. and Haring, H.U. (1990). Mannose, glucosamine and inositolmonophosphate inhibit the effects of insulin in lipogenesis. Further evidence for a role for inositolphosphate oligosaccharides in insulin action. Biochem. J. **266:** 909-916.

Malchoff, C. D., Huang, L., Gillespie, N., Villar-Palasi, C., Schwartz, C. F. W., Cheng, K., Hewlett, E. L. and Larner, J. (1987). A putative mediator of insulin action which inhibits adenylate cyclase and adenosine 3′,5,-monophosphate dependent protein kinase: Partial purification from rat liver, site and kinetic mechanism of action. Endocrinology. **120:** 1327-1237.

Mandarino, L. J., Madar, Z., Kolterman, O. G., Bell, J. M. and Olefsky, J. M. (1986). Adipocyte glycogen synthase and pyruvate dehydrogenase in obese and type II diabetes subjects. Am. J. Physiol. **251:** E489-496.

Margolis, R. K., Goossen, B. and Margolis, R. K. (1988). Phosphatidylinositol-anchored glycoproteins of PC12 pheochromocytoma cells and brain. Biochemistry. **27:** 3454-3458.

Mato, J. M., Kelly, K. L., Abler, A. and Jarett, L. (1987). Identification of a novel insulin-sensitive glycophospholipid from H35 hepatoma cells. J. Biol. Chem. **262:** 2131-2137.

McNeely, T. B., Rosen, G., Londner, M. V. and Turco, S. J. (1989). Inhibitory effects of protein kinase C by lipophosphoglycan fragments and glycosylphosphatidylinositol antigens of the protozoan parasite *Leishmania*. Biochem. J. **259:** 601-604.

Meisenhelder, J., Suh, P.-G., Rhee, S. G. and Hunter, T. (1989). Phospholipase D-τ is a substrate for the PDGF and EGF receptor protein kinases in vivo and in vitro. Cell. **57:** 1109-1122.

Mérida, I., Pratt, J. C. and Gaulton, G. N. (1990). Regulation of interleukin 2-dependent growth responses by glycosylphosphatidylinositol molecules. Pro. Natl. Acad. Sci. USA. **87:** 9421-9425.

Muchmore, D. B., Raess, B. U., Bergstrom, R. W. and de Haen, C. (1982). On the mechanism of inhibition of insulin action by small molecular weight trypsin inhibitors. Diabetes. **31:** 976-984.

Nishibe, S., Wahl, M. I., Hernández-Sotomayor, S. M. T., Tonks, N. K., Rhee, S. G. and Carpenter, G. (1990). Increase in the catalytic activity of phospholipase C-τ1 by tyrosine phosphorylation. Science. **250:** 1256-1256.

Rinaudo, M. T., Curto, M., Bruno, R., Marino, C., Rossetti, V. and Mostert, M. (1987). Evidence of an insulin generated pyruvate dehydrogenase stimulating factor in rat brain plasma membranes. Int. J. Biochem. **19:** 909-913.

Roberts, W. L., Santikarn, S., Reinhold, V. N. and Rosenberry, T. L. (1988). Characterization of the glycoinositol phospholipid anchor of human erythrocyte acetylcholinesterase by fast atom bombardment mass spectrometry. J. Biol. Chem. **263:** 18776-18784.

Robinson, P. (1991). Phosphatidylinositol membrane anchors and T-cell activation. Immunol. Today. **12:** 35-41.

Romero, G., Luttrell, L., Rogol, A., Zeller, K., Hewlett, E. and Larner, J. (1988). The phosphatidylinositol anchors of membrane proteins: potential precursors of insulin mediators. Science. **240:** 509-511.

Romero, G., Gámez, G., Huang, L. C., Lilley, K. and Luttrell, L. (1990). Anti-inositolglycan antibodies selectively block some of the actions of insulin in intact BC_3H1 cells. Proc. Natl. Acad. Sci. USA. **87:** 1476-1480.

Rothenberg, P. L. and Kahn, C. R. (1988). Insulin inhibits pertussis toxin-catalyzed ADP ribosylation of G-proteins. J. Biol. Chem. **263:** 15546-15552.

Saltiel, A. R. (1987). Insulin generates an enzyme modulator from hepatic plasma membranes: regulation of cAMP phosphodiesterase, pyruvate dehydrogenase and adenylate cyclase. Endocrinology. **120:** 967-972.

Saltiel, A. R. and Cuatrecasas, P. (1986). Insulin stimulates the generation from hepatic plasma membranes of modulators derived from an inositol glycolipid. Proc. Natl. Acad. Sci. USA. **83:** 5793-5797.

Saltiel, A. R., Fox, J. A., Sherline, P. and Cuatrecasas, P. (1986). Insulin stimulates the release from liver plasma membranes of a chemical modulator of pyruvate dehydrogenase. Science. **233:** 967-972.

Saltiel, A. R., Sherline, P. and Fox, J. A. (1987). Insulin stimulated diaglycerol production results from the hydrolysis of a novel phosphatidylinositol glycan. J. Biol. Chem. **262:** 1116-1121.

Saltiel, A. R. and Cobara Cazán, L. R. (1987). Inositolglycan mimics the action of insulin on glucose utilization in rat adipocytes. Biochem. Biophys. Res. Commun. **149:** 1084-1092.

Sato, T., Villar Palasi, C., Huang, L., Tang, G., Larner, A. C. and Larner, J. (1988). Insulin and a putative insulin metabolic mediator fraction from liver and muscle stimulate p33 mRNA accumulation by apparently different mechanisms. Endocrinology. **123:** 1559-1564.

Sinha, M. K. and Caro, J. F. (1985). Presence of an activator of pyruvate dehydrogenase in human circulation: elevation following a glucose load and possible relation to insulin mediator. Biochem. Biophys. Res. Commun. **129:** 41-45.

Suzuki, S., Toyota, T., Sukuki, H. and Goto, Y. (1984). A putative second messenger of insulin action regulates hepatic microsomal glucose-6-phosphatase. Biochem. Biophys. Res. Commun. **118:** 40-46.

Suzuki, S., Toyota, T., Tamura, S., Kikuchi, K., Tsuiki, S., Huang, L., Villar-Palasi, C., Larner, J. and Goto, Y. (1987). ATP-Mn^{2+} stimulates the generation of a putative mediator of insulin action. J. Biol. Chem. **262:** 3199-3204.

Taylor, S. I., Kadowski, T., Kadowski, H., Accili, D., Cama, A. and McKeon, C. (1990). Mutations in insulin-receptor gene in insulin-resistant patients. Diabetes Care. **13:** 257-279.

Valera, I., Avila, M., Mato, J. M. and Hue, L. (1990). Insulin-induced phospho-oligosaccharide stimulates aminoacid transport in isolated rat hepatocytes. Biochem. J. **257:** 541-544.

Valera-Nieto, I., Represa, J., Avila, M. A., Miner, C., Mato, J. M. and Giraldez, F. (1991). Inositol phospho-oligosaccharide stimulates cell proliferation in the early developing inner ear. Develop. Biol. **143:** 432-435.

Veldhuis, J. D. and Kolp, L A. (1985). Mechanisms subserving insulin's differentiating actions on progestin biosynthesis by ovarian cells: studies with swine granulosa cells. Endocrinology. **116:** 651-659.

Villalba, M., Alvarez, J. F., Russell, D. S., Mato, J. M. and Rosen, O. M. (1990). Hydrolysis of glycosyl-phosphatidylinositol in response to insulin is reduced in cells bearing kinase deficient insulin receptors. Growth Factors. **2:** 91-97.

Whipple-Leef, J. and Larner, J. (1987). Insulin-mimetic effect of trypsin on the insulin receptor tyrosine kinase in intact adipocytes. J. Biol. Chem. **262:** 14837-14842.

Zhang, S.-R., Shi, Q.-H. and Ho, R.-j. (1983). Cyclic AMP-lowering mediator of insulin. Generation, quantitation and properties. J. Biol. Chem. **258:** 6471-6476.

ELECTROSPRAY MASS SPECTROMETRY OF THE GLYCOSYLINOSITOL PHOSPHOLIPID OF THE SCRAPIE PRION PROTEIN

[1]Neil Stahl, [1]Michael A Baldwin and *[1,2]Stanley B Prusiner.

Department of [1]Neurology and [2]Biochemistry and Biophysics, University of California, San Francisco, California 94143, USA.

* Corresponding author.

INTRODUCTION

Among the most enigmatic of glycosylinositol phospholipid (GP1)-linked proteins is the scrapie prion protein, denoted PrP^{Sc}. This host-encoded protein forms a major component of the prion - a transmissible pathogen for which a nucleic acid component has not been found (Prusiner, 1989). In addition to scrapie, prions containing PrP are implicated in the transmissible human neurodegenerative diseases - kuru, Creutzfeldt-Jakob disease (CJD), and Gerstmann-Sträussler-Scheinker (GSS) disease. Several varieties of CJD and GSS are unique: the disease occurs in families with an autosomal dominant pattern of inheritance, yet brain tissue from the afflicted individuals contains transmissible prions (Masters et al., 1981; Hsiao et al., 1989). All of the inherited prion diseases appear to be associated with mutations in the human PrP gene. Inadvertent transmission of sheep scrapie to cattle through contaminated diet supplements has given rise to a new disease called bovine spongiform encephalopathy (BSE), which threatens to decimate the British beef industry and has caused widespread concern among the beef-eating public (Wilesmith et al., 1988).

Although the mechanism of formation and the structure of the prion is not yet known, a large body of information has accumulated that implicates PrP^{Sc} as an important participant. PrP^{Sc} is a disease-specific isoform of a normal cellular protein called PrP^{c} (Bolton et al., 1982; Oesch et al., 1985). PrP^{Sc} slowly accumulates in the brain in a protease-resistant form that aggregates; criteria which distinguish it from PrP^{c} (Meyer et al., 1986).

Experimental evidence (Borchelt et al., 1990) confirms earlier assertions (Basler et al., 1986) that PrP^{c} or a precursor is transformed into PrP^{Sc} by a posttranslational event. Thus PrP^{c} and PrP^{Sc} appear to differ by either a chemical modification, a stable conformational change, or tight binding to other cellular components. We are analyzing the chemical structure of PrP^{Sc} to look for posttranslational modifications that might be responsible for its unusual physical properties; PrP^{c} structure is less well characterized. These studies have confirmed that the primary structure of PrP 27-30, which is the protease-resistant core of PrP^{Sc}, corresponds to that predicted by the gene sequence from normal animals (N. Stahl, M. Baldwin, S. Prusiner, manuscript in preparation). Our findings contrast with those in the genetic forms of prion diseases such as GSS, where mutations in the amino acid sequence can lead to neurodegenerative diseases (Hsiao et al., 1989; Hsiao and Prusiner, 1990). This has been experimentally reproduced in animals; transgenic mice harbouring a mouse PrP gene with an engineered mutation corresponding to one found in human GSS spontaneously develop a neurodegenerative disease (Hsiao et al., 1990).

The major posttranslational modifications identified for PrP^{Sc} are the GP1 anchor (Stahl et al., 1987) and two N-linked carbohydrates whose structures are partially known (Endo et al., 1989). Pulse-chase radiolabelling experiments in cultured cells have shown that a mutant protein with ablated N-link consensus sequences can still become protease resistant, demonstrating that this carbohydrate is probably not required to form PrP^{Sc} (Taraboulos et al., 1990a). The possible roles of a truncated form of PrP^{Sc} (Stahl et al., 1990a), or poorly characterized modifications of N-terminal arginines of PrP^{Sc} (Hope et al., 1988; Turk et al., 1988) and PrP^{c} (Turk et al., 1988) are not yet known.

We have thus focused on determining the structure of the GP1 anchor to reveal any unusual features that might contribute to the distinctive physical properties of PrP^{Sc} and permit this protein to become a prion component. Previous studies have shown that PrP^{c} and PrP^{Sc} are differentially released from cellular membranes by PIPLC (Stahl et al., 1990b; Caughey et al., 1990). Even though both were accessible to added reagents, PrP^{c} was released by PIPLC while PrP^{Sc} was resistant. In contrast, the glycolipid anchor of PrP^{Sc} is sensitive to PIPLC once the protein is denatured (Stahl et al., 1990b). Thus the conformation or aggregation of PrP^{Sc} appears to block access of PIPLC to the C-terminus. The GPI is attached to Syrian hamster PrP^{Sc} at Ser_{231} (Stahl et al., 1990a); the attachment site to PrP^{c} is unknown, but there is no reason to believe that it is different.

We describe in this report electospray mass spectrometry of a C-terminal peptide purified from the scrapie isoform of the scrapie prion protein. This technique has proven useful for obtaining the mass of the intact GPI anchor including additional phosphoethanolamines.

EXPERIMENTAL

The reverse phase HPLC separation and purification of K12 (the C-terminal peptide containing the GPI) from endoproteinase Lys C digests of PrP 27-30, a N-terminally truncated form of PrP^{Sc}, has been described previously (Baldwin et al., 1990b).

Electrospray Mass Spectrometry: This was carried out using a VG Bio-Q mass spectrometer (VG Bio Tech Ltd, Altrincham, UK). The sample buffer was 1:1 $H_2O:CH_3CN$ containing 1% acetic acid and 2pmol/µL gramicidin S, pumped by an ABI model 130 pump at 4 µL/min. This was sprayed into the atmospheric pressure ionization chamber from a nebulizer consisting of a needle held at 4 kV with a coaxial nitrogen supply. The chamber was heated to 50°C and additional nitrogen was blown through it causing the solvent to evaporate from the droplets, leaving isolated ions. These were focused by electrostatic lenses through a small aperture into a quadrupole mass analyzer, where differential pumping maintained a pressure of $<10^{-4}$ Torr. Samples were injected into the liquid flow via a Rheodyne microinjector fitted with a 10 µL loop. The mass spectrometer was set for repetitive 10 s scans and its performance was optimized by monitoring the doubly protonated molecular ion of gramicidin S at m/z 571.7. The mass scale was calibrated by injecting 100 pmol myoglobin (mol. mass 16250.6 Da) and monitoring the multiply charged peaks in range 600-1400 m/z. An aliquot of an HPLC fraction containing approximately 300 pmol of K12 was reduced to a volume of 5 µL on a speedvac. The sample was injected following the addition of 5.92% acetic acid in acetonitrile. Multiple spectra were acquired over a period of 3 min. and averaged under computer control. The data were smoothed using Savitsky-Golay algorithms to give the spectrum shown in Figure 1. In electrospray mass spectrometry, molecular ions having *n* excess protons and thus *n* positive charges are observed at m/z values corresponding to $(M_r+n)/n$. By fitting different possible integer values of *n* to this equation, series of ions can be identified originating from a common molecular mass M_r. In the case of K12, three separate molecular species were identified.

RESULTS

Our initial structural analysis of the PrP^{Sc} GPI involved reverse phase HPLC purification of the C-terminal peptide-GPI after digestion of PrP 27-30 with Endoproteinase Lys-C and PIPLC (Stahl et al., 1990a). Incubation of this fraction in 50% aqueous HF selectively cleaves the phosphodiester bonds and allows repurification of both the peptide and the glycan portions of the GPI (Ferguson et al., 1988; Baldwin et al., 1990b). Permethylation of the glycan introduces a positive charge by forming a quaternary ammonium ion on the conserved glucosamine, which increases the sensitivity of analysis by soft ionization mass spectrometry (Baldwin et al., 1990b). We then used high resolution tandem mass spectrometry to uncover the order and branching of the GPI glycan. The results of this procedure revealed 3 forms of the GPI. During these studies, we found two additional glycoforms with masses that differed from the original forms by the addition of a sialic acid. Modification of the permethylation procedure allowed these new glycan species to be reproducibly observed (Baldwin et al., 1990a).

We then sought other methods to confirm the structure of the GPI from PrP 27-30. Electrospray mass spectrometry provides the means to measure the molecular mass of proteins and large peptides with an accuracy of up to 1 in 10,000, yet typically requires only 50-300 pmoles of sample. This method gives multiply charged ions with a mass/charge ratio in a range that can be measured with a relatively simple quadruple mass spectrometer. Figure 1 shows an electospray spectrum of purified K12. The spectrum reveals 3 molecular species of the GPI present at both the +4 and +3 charge states. Calculation of the molecular mass from this data gives valued of 2832.1, 2995.0, and 3285.4, with an estimated accuracy of better than ± 1 Da. Although PIPLC is thought to leave an inositol phosphodiester as product, these masses indicate that the phosphate attached to the inositol is a monoester. It is likely that conversion to the monoester occurred during sample purification and mass spectrometry under acidic conditions. This data confirms our earlier analysis, and is consistent with the presence of sialic acid on the species with the highest mass. The data also confirms the presence of a second phosphoethanolamine present on all of the GPI species. Figure 2 shows the proposed structure for the middle of these three GPI species with a calculated mass of 2994.6 Da. The smaller species corresponds to the loss of a hexose, while the larger corresponds to the addition of a sialic acid. These calculations are based on average atomic masses of the constituent elements rather than monisotopic atomic masses since the quadruple mass analyzer is a low resolution device incapable of separating the adjacent peaks within an isotopic cluster. Placement of the phosphoethanolamines is based on the locations proposed for rat brain Thy-1 (Homans et al., 1989).

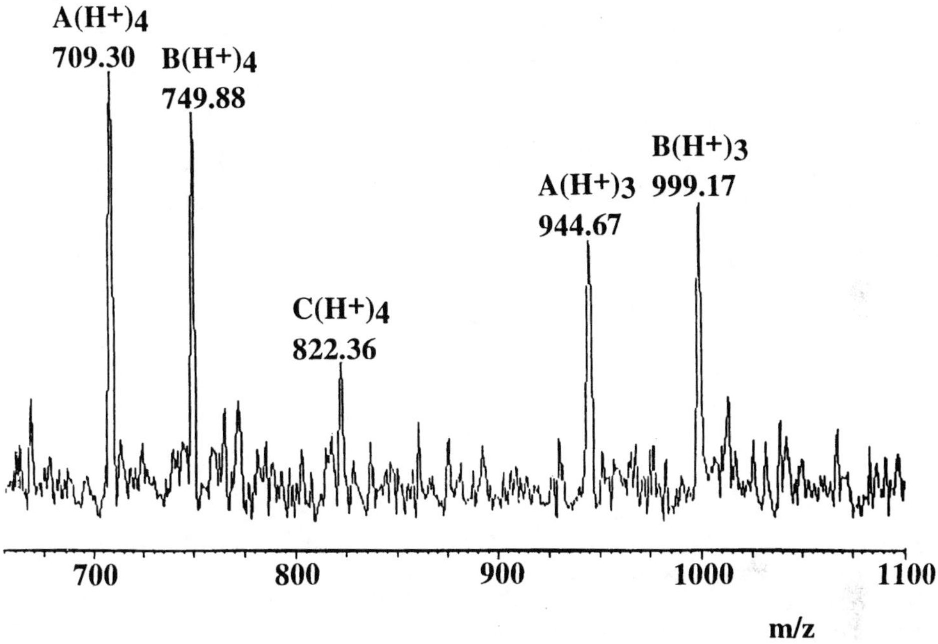

Figure 1.
Electrospray mass spectrum of K12 allowing the identification of three distinct molecular species having the following M_r values. A=2833.2, 2831.0, mean 2832.1; B=2995.5, 2994.5, mean 2995.0; c=3285.4. For details and for explanation of the calculations see Experimental.

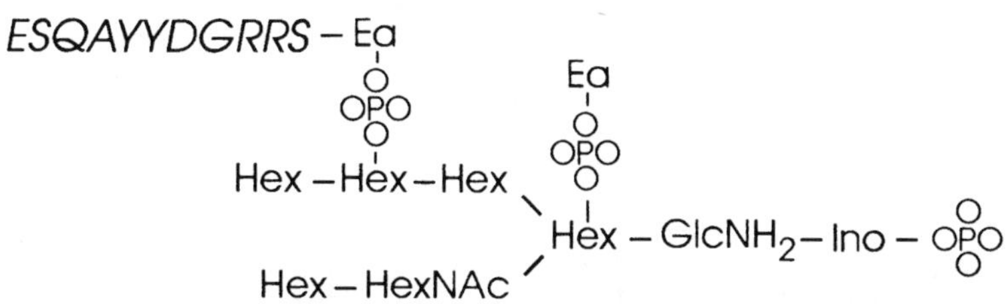

Figure 2.
Proposed structure of the C-terminal Lys-C peptide and GPI anchor from the scrapie isoform of the prion protein. The italic letters correspond to the 1 letter amino acid codes, Ea = ethanolamine, Hex = hexose, HexNAc = N-acetlyhexosamine, $GlcNH_2$ = glucosamine, Ino = inositol.

DISCUSSION

Although a great deal has been learned about the chemical structure of the scrapie prion protein, the nature of the difference between PrP^c and PrP^{Sc} has not yet been uncovered. Other results have shown that the primary sequence of PrP 27-30, the protease-resistant core of PrP^{Sc} that purifies with prion infectivity, exactly matches that predicted by the gene sequence (Stahl et al., 1990c). Pulse-chase radiolabelling experiments indicate that PrP^{Sc} is formed from PrP^c or a related precursor by a post-translational event (Borchelt et al., 1990). We now have evidence indicating that all posttranslational modifications of the majority of PrP^{Sc} molecules have been identified (N. Stahl, M. Baldwin, and S. Prusiner, manuscript in preparation). The two Asn-linked oligosaccharides do not appear to play a critical role since a mutant protein lacking these sites can still become protease resistant and can probably form PrP^{Sc} (Taraboulos et al., 1990a). Our efforts have now turned toward deciphering the structures of the GPI anchors from both PrP^c and PrP^{Sc} to evaluate whether different GPI anchors could play a role in the development of scrapie (Baldwin et al., 1990b).

Analysis of the GPI of PrP^{Sc} by a variety of techniques indicates the presence of at least five glycoforms. Other results indicate that the largest two forms contain neuraminidase-sensitive sialic acid (N. Stahl, M. Baldwin, and S. Prusiner, manuscript in preparation). Fragmentation patterns observed in tandem mass spectrometry of the permethylated GPI glycans indicate that the sialic acid is attached in a trisaccharide branch Sia-Hex-HexNAc. Compositional analysis of the PrP 27-30 GPI reveals the presence of galactose and N-acetylgalactosamine (N. Stahl, M. Baldwin, and S. Prusiner, manuscript in preparation), and we suggest that these sugars are found in this trisaccharide branch. This portion of the GPI would have a structure analogous to relatively common O-linked trisaccharides with the sequence Sia-Gal-GalNAc. Whether the same glycosyltransferases are active on both types of substrates is unknown, as are the linkages between the sugars in this GPI branch. While the presence of sialic acid in other GPI anchors has apparently not been reported, this may change as structural analysis is completed for the GPI from other proteins.

It is not yet known whether the PrP^c GPI anchor also contains sialic acid. If not, then this would constitute the first known difference between the normal and scrapie-specific isoforms. Although it is conceivable that such a difference could play a role in the replication and pathogenesis of the scrapie prion, it is also possible that any differences in the structure of the GPI could reflect secondary changes resulting from the different cellular localizations of the two isoforms (Taraboulos et al., 1990b). PrP^{c} is largely localized to the cell surface, while at least a portion of PrP^{Sc} accumulates intracellularly in secondary lysosomes (McKinley et al., 1990). Answers to these questions must await structural analysis of the GPI of PrP^{c}, as well as experiments with site-directed mutants that inquire whether the GPI is required to form PrP^{Sc}.

These results attest to the usefulness of electrospray mass spectrometry for the analysis of GPI anchors. Electrospray mass spectrometry provides a method for obtaining the overall mass of the GPI. The spectrum illustrated in Figure 1 was obtained on a GPI-linked peptide derived from the C-terminus of PrP^{Sc} that was purified by reverse-phase HPLC (Stahl et al., 1990a). We believe that the peptide attachment is required to obtain reasonably good ionization of the sample for analysis in the positive ion mode, but this proposal remains untested at this time. We have not tried negative ion electrospray mass spectrometry but this might be favoured by the phosphates and the acidic sugar residue. Attempts to obtain a spectrum of this sample by liquid secondary ion mass spectrometry were unsuccessful regardless of whether the peptide-GPI had been permethylated (Stahl and Baldwin, unpublished results). Electrospray mass spectrometry also revealed the presence of a second phosphoethanolamine, which had been predicted on the basis of amino acid analysis (Stahl et al., 1987). We had previously suggested a stoichiometry of 2.8 ethanolamine per mole of PrP 27-30, but this value was adjusted from the observed value of 1.8 due to the loss noted upon hydrolysis of an ethanolamine standard (Stahl et al., 1987). Although ethanolamine is attached to glutamate in another protein (Rosenberry et al., 1989), we have found no evidence for the presence of ethanolamine in PrP at sites other than the GPI (Stahl et al., 1990c), suggesting that an overall stoichiometry of 2 is more likely correct.

ACKNOWLEDGMENTS

Mass spectrometry was provided by the UCSF Mass Spectrometry Facility (A. L. Burlingame, Director) supported by RR01614. We wish to acknowledge the assistance of David Maltby with the electrospray mass spectrometry. SBP was supported by research grants from the National Institutes of Health (AG02132, NS14069, AG08967 and NS22786) and American Health Assistance Foundation as well as by gifts from Sherman Fairchild Foundation and National Medical Enterprises.

REFERENCES

Baldwin, M. A., Stahl, N., Burligame, A. L. and Prusiner, S. B. (1990a). Structure determination of glyco-inositol phospholipid anchors by permethylation and tandem mass spectrometry. METHODS; A Companion to methods in Enzymology **1:** 306-314.

Baldwin, M. A., Stahl, N., Reinders, L. G., Gibson, B. W., Prusiner, S. B. and Burlingame, A. L. (1990b). Permethylation and tandem mass spectrometry of oligosaccharides having free hexosamine: analysis of the glycoinositol phospholipid anchor glycan from the scrapie prion protein. Anal. Biochem. **191:** 174-182.

Basler, K., Oesch, B., Scott, M., Westaway, D., Walchli, M., Groth, D. F., McKinley, M. P., Prusiner, S. B. (1986). Scrapie and cellular PrP isoforms are encoded by the same chromosomal gene Cell **46:** 417-428.

Bolton, D. C., McKinley, M. P. and Prusiner, S. B. (1982). Identification of a protein that purifies with the scrapie protein. Science **218: 1309-1311.**

Borchelt, D. R., Scott, M., Taraboulos, A., Stahl, N. and Prusiner, S. B. (1990). Scrapie and cellular prion protein differ in their kinetics of synthesis and topology in cultured cells. J. Cell Biol. **110:** 743-752.

Caughey, B., Neary, K., Butler, R., Ernst, D., Perry, L., Chesebro, B. and Race, R. (1990). Normal and scrapie -associated forms of prion protein differ in their sensitivities to phospholipase and proteases in intact neuroblastoma cells. J. Virol. **64:** 1093-1101.

Endo, T., Groth, D., Prusiner, S. B. and Kobata, A. (1989). Diversity of oligosaccharide structures linked to asparagines of the scrapie prion protein. Biochemistry **28:** 8380-8388.

Ferguson, M. A. J., Homans, S. W., Dwek, R. A., Rademacher, T. W., Anand, R., and Williams, A. F. (1988). Glycosyl-phosphatidylinositol moiety that anchors *Trypanosoma brucei* variant surface glycoprotein to the membrane. Science **239:** 753-759.

Homans, S. W., Ferguson, M. A., Dwek, R. A., Rademacher, T. W., Anand, R., and Williams, A. F. (1988). Complete structure of the glycosyl phosphatidyl inositol membrane anchor of rat brain Thy-1 glycoprotein. Nature **333:** 269-272.

Hope, J., Morton, L. J. D., Farquhar, C. F., Multhaup, G., Beyreuther, K. and Kimberlin, R. H. (1986). The major polypeptide of scrapie-associated fibrils (SAF) has the same size, charge distribution and N-terminal protein sequence as predicted for the normal brain protein (PrP) EMBO J. **5:** 2591-2597.

Hope, J., Multhaup, G., Reekie, L. J. D., Kimberlin, R. H. and Beyreuther, K. (1988). Molecular pathology of scrapie-associated fibril protein (PrP) in mouse brain affected by the ME7 strain of scrapie. Eur. J. Biochem. **172:** 271-277.

Hsiao, K. and Prusiner, S. B. (1990). Inherited human prion diseases. Neurology **40:** 1820-1827.

Hsiao, K., Baker, H. F., Crow, T. J., Poulter, M., Owen, F., Terwilliger, J. D., Westaway, D., Ott, J. and Prusiner, S. B. (1989). Linkage of a prion protein missense variant to Gerstmann-Sträussler syndrome. Nature **338:** 342-345.

Hsaio, K. K., Scott, M., Foster, D., Groth, D. F., DeArmond, S. J. and Prusiner, S. B. (1990). Spontaneous neurodegeneration in transgenic mice with mutant prion protein. Science **250:** 1587-1590.

Masters, C. L., Gajdusek, D. C. and Gibbs, C. J. Jr. (1981). Creutzfeldt-Jakob disease virus isolations from the Gerstmann-Sträussler syndrome. Brain **104:** 559-588.

McKinley, M. P., Taraboulos, A., Kenaga, L., Serban, D., DeArmand, S. J., Stieber, A. and Prusiner, S. B. (1990). Ultrastructural localization of scrapie prion proteins in secondary lysosomes of infected cultured cells. J. Cell Biol. **111:** 316a.

Meyer, R. K., McKinley. M. P., Bowman, K. A., Braunfield, M. B., Barry, R. A. and Prusiner, S. B. (1986). Separation and properties of cellular and scrapie prion proteins. Proc. Natl. Acad. Sci. USA **83:** 2310-2314.

Oesch, B., Westaway, D., Walchli, M., McKinley, M. P., Kent, S. B. H., Aebersold, R., Barry, R. A., Tempst, P., Teplow, D. B., Hood, L. E., Prusiner, S. B. and Weissman, C. (1985). A cellular gene encodes scrapie PrP 27-30 protein. Cell **40:** 735-746.

Prusiner, S. B. (1989). Scrapie prions. Annu. Rev. Microbiol. **43:** 345-374.

Rosenberry, T. L., Krall, J. A., Dever, T. E., Haas, R., Loubard, D. and Merrick, W. C. (1989). Biosynthetic incorporation of [^{3}H]ethanolamine into protein synthesis elongation factor 1 alpha reveals a new post-translational protein modification. J. Biol. Chem. **264:** 7069-7099.

Stahl, N., Borchelt, D. R., Hsiao, K. and Prusiner, S. B. (1987). Scrapie prion protein contains a phosphatidyl-inositol glycolipid. Cell **51:** 229-240.

Stahl, N., Baldwin, M. A., Burlingame, A. L. and Prusiner, S. B.(1990a). Identification of glycoinositol phospholipid linked and truncated forms of the scrapie prion protein. Biochemistry **29:** 8879-8884.

Stahl, N., Borchelt, D. R. and Prusiner, S. B. (1990b). Differential release of cellular and scrapie prion proteins from cellular membranes by phosphatidylinositol-specific phospholipase C. Biochemistry **29:** 5405-5412.

Stahl, N., Baldwin, M., Gibson, B., Burlingame, A. and Prusiner, S. B. (1990c). Mass spectrometric mapping of the scrapie prion protein. VIIIth International Congress of Virology, Berlin: 69 (abstr).

Taraboulos, A., Rogers, M., Borchelt, D. R., McKinley, M. P., Scott, M., Serban, D. and Prusiner, S. B. (1990a). Acquisition of protease resistance by prion proteins in scrapie-infected cells does not require asparagine-linked glycosylation. Proc. Natl. Acad. Sci. (USA) **87:** 8262-8266.

Taraboulos, A., Serban, D. and Prusiner, S. B. (1990b). Scrapie prion proteins accumulate in the cytoplasm of persistently infected cultured cells. J. Cell Biol. **110:** 2117-2132.

Turk, E., Teplow, S. B., Hood, L. E. and Prusiner, S. B. (1988). Purification and properties of the cellular and scrapie hamster prion proteins. Eur. J. Biochem. **176:** 21-30.

Wilesmith, J. H., Wells, G. A. H., Cranwell, M. P. and Ryan, J. B. M. (1988). Bovine spongiform encephalopathy: epidemiological studies. Vet. Rec. **123:** 638-644.

BIOSYNTHESIS OF GLYCOPHOSPHOINOSITOL ANCHORS IN *SACCHAROMYCES CEREVISIAE*

***Andreas Conzelmann, Christoph Fankhauser, Alessandro Puoti and Chantal Desponds**

Institute of Biochemistry, University of Lausanne, CH-1066 Epalinges, Switzerland.

* Corresponding author: Institut de Biochimie, Perolles, 1700 Fribourg, Switzerland.

ABSTRACT

Numerous membrane glycoproteins of *Saccharomyces cerevisiae* are posttranslationally modified by the addition of a glycophosphatidylinositol (GPI). These proteins can be detected most easily by metabolic labelling of yeast cells with ^{3}H-myoinositol or ^{3}H-palmitate. This report summarizes what little is known about the identity, biosynthesis and cellular localization of GPI-modified glycoproteins in *Saccharomyces cerevisiae* as well as what could be learned from the system with respect to the biosynthesis of GPI's in general.

INTRODUCTION

Numerous membrane glycoproteins of protozoan and mammalian origin are anchored in the lipid bilayer by a glycophosphatidylinositol (GPI) anchor and the potential functional roles of these anchors have been discussed in recent reviews (Ferguson and Williams, 1988; Low and Saltiel, 1988; Thomas et al, 1990; Doering et al, 1990; Cross 1990). The complete structural characterization of the GPI anchor of a variant surface glycoprotein from *Trypanosoma brucei*, of the major surface protease of *Leishmania major* (Schneider et al, 1990) as well as of the mammalian Thy-1 glycoprotein have recently been reported (Ferguson et al, 1988; Homans et al, 1988). All evidence argues that the anchor is synthesized as a precursor glycolipid (Krakow et al, 1986, 1989; Menon et al, 1988, Mayor et al, 1990a,b) which is transferred "en bloc" onto glycoproteins shortly after their synthesis (Bangs et al, 1985, Ferguson et al, 1986, Conzelmann et al, 1987). The biosynthesis of the precursor glycolipid by the stepwise addition of monosaccharides to phosphatidyl-inositol in crude microsomes of *Trypanosoma brucei* has been reported (Masterson et al, 1989, 1990; Menon et al, 1990a).

Yeast cells contain a considerable number of glycoproteins which contain similar GPI anchors (Conzelmann et al, 1988, 1990). Yeast seems therefore to be an attractive organism to identify genes involved in the biosynthesis of GPI anchors. However the exact functional significance of this modification is not well defined and all the more so because no enzymatic or other activity of any GPI modified protein of yeast has been reported yet.

MATERIALS AND METHODS

The following haploid strains of *Saccharomyces cerevisiae* were used: HMSF176, *a, sec18-1* (Novick et al, 1980, 1981; Esmon et al, 1981); DBY1790, *a, alg2-1, ade2, ura3;* DBY1791 *a, alg3-1, ade2, ura3* (Huffaker and Robbins, 1983). Labelling of cells, extraction of proteins and analysis by gel electrophoresis were done as recently described (Conzelmann et al, 1990).

RESULTS

1. Identification and structural analysis of yeast GPI-modified proteins. Many proteins of *S. cerevisiae* are readily labelled with ^{3}H-myoinositol and, as judged by co-migration on sodium dodecyl polyacrylamide gel electrophoresis (SDS-PAGE), the same proteins are strongly labelled with ^{3}H-palmitate (Fig. 1). The ^{3}H-myoinositol-labelled peptides obtained by protease treatment from such proteins are amphiphilic but become completely hydrophilic upon treatment with phosphatidylinositol-specific phospholipase C (PI-PLC) (Conzelmann et al, 1990). Moreover, nitrous acid treatment of such proteins generates labelled lipid moieties which migrate closely to the phosphatidylinositol standard when analyzed by TLC (Conzelmann et al, 1990). The cross-reacting determinant (CRD) which is generated by the pI-PLC-catalyzed removal of diacylglycerol has been found on a prominent GPI-modified glycoprotein of yeast (Conzelmann et al, 1988). These findings suggest that yeast contains GPI anchors of similar description as mammals and protozoan organisms. Recent analysis of the anchors from a pool of all GPI-modified yeast proteins by NMR, methylation analysis and fast atom bombardment reveals the presence of the same carbohydrate core structure as in protozoan and mammalian organisms while the lipid moiety consists predominantly of a ceramide (in preparation).

2. General features of GPI-modified proteins of *S. cerevisiae*. Most GPI-modified proteins of yeast are N-glycosylated since they bind to Concanavalin A and since their apparent molecular weight is drastically reduced by endoglycosidase H treatment. The increase in apparent molecular weight during maturation in the Golgi is quite dramatic in most of the GPI-modified yeast proteins (Conzelmann et al, 1990; Fig. 1), a phenomenon which is due mainly to elongation of N-glycans and which is almost absent in *mnn9* cells (Tsai et al, 1984), a mutant in which the elongation of N-glycans in the Golgi is deficient (not shown).

PI-PLC treatment of GPI-modified proteins of mammalian as well as protozoan organisms usually abolishes the amphiphilic properties of these proteins and renders them hydrophilic. This can be monitored by a change in the partitioning of proteins in the detergent Triton X-114 (Bordier, 1981) or by the release of proteins from cell surfaces. These phenomena indicate that the GPI-modification is solely responsible for the membrane attachment of these proteins and justifies the term GPI-"anchor". Although PI-PLC completely removes the ^{3}H-palmitate label from most of the GPI-modified proteins of yeast even if they are in their native state, the PI-PLC treated proteins retain some detergent binding capacity as measured by partitioning of these proteins in Triton X-114 (Conzelmann et al, 1990; Orlean, 1990). This indicates that most GPI-modified proteins of yeast contain other hydrophobic domains outside the GPI moiety or that they interact strongly with other membrane binding proteins. The one striking exception to this rule is a major surface 125kD glycoprotein which readily becomes hydrophilic upon treatment with phospholipases (Conzelmann et al, 1988).

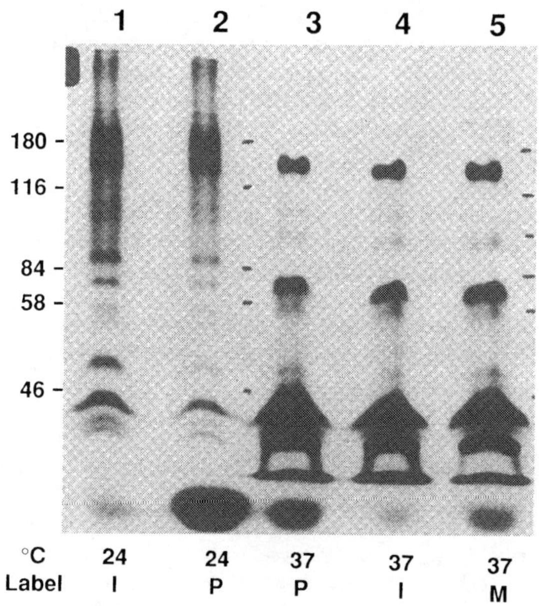

Figure 1. Biosynthetic labelling of GPI moieties of yeast glycoproteins

Aliquots of 5 OD_{600}units of *sec18* cells were preincubated for 15 min and labelled with either ^{3}H-myoinositol (I, 50µCi/tube, lanes 1,4) or ^{3}H-palmitate (P, 300µCi/tube, lanes 2,3) or ^{3}H-myristate (M, 150µCi, lane 5) for 2h at 24°C (lanes 1,2) or 37°C (lanes 3,4,5). Cells were broken in SDS (Conzelmann et al, 1990, procedure C) and the extracts incubated with Concanavalin A-Sepharose. The bound material was eluted by boiling the SDS and analysed by SDS-PAGE and fluorography. Positions of molecular weight standard proteins are indicated on the left in kD. Exposure: 3 days.

3. Biosynthesis of GPI-modified proteins in yeast. ^{3}H-myoinositol, when added to growing yeast cells, is almost completely taken up and incorporated into phosphatidylinositol within a few minutes. ^{3}H-myoinositol appears in proteins as early as 3 minutes after addition. The first proteins to be labelled migrate to the same positions in SDA-PAGE as the proteins observed after labelling secretion mutants in which the vesicular transport to the Golgi is blocked (Novick et al, 1980, 1981; Esmon et al, 1981). Addition of cycloheximide rapidly stops the incorporation of ^{3}H-myoinositol into proteins (Conzelmann et al, 1990).

This clearly indicates that the anchors are added rapidly after translation and while proteins are still in the ER. It also indicates that the pool of preformed glycolipid anchors ready to be attached to proteins must be quite small. Both, ^{3}H-palmitate and ^{3}H-myristate are incorporated into the very same proteins (Fig. 1) as get labelled with ^{3}H-myoinositol whereas ^{3}H-oleate is not incorporated into proteins under the same conditions (not shown). Attempts to label these proteins with $^{32}PO_4$ or ^{3}H-glycerol have been unsuccessful (Wen and Schlesinger, 1984).

4. Biosynthetic studies with mutant cells. The mature GPI ready to be transferred onto proteins either made in intact cells or by microsomes cannot easily be observed in yeast cells. A possible biosynthetic intermediate of GPI's has been observed in a mutant with a temperature dependent lesion in dolicholphosphomannose synthase (Orlean, 1990). At nonpermissive temperatures this mutant accumulates a highly nonpolar insositol- and glucosamine-containing lipid. This result (together with direct biochemical studies in microsomes of trypanosomes (Menon et al, 1990b)) firmly establishes that in vivo dolicholphosphomannose is an obligatory donor substrate for at least the first mannose residue which gets added to the glucosamine of GPI anchors. An interesting feature of this potential intermediate is that it probably carries an acyl chain on the myoinositol. In contrast, the mature anchors do not contain this additional fatty acid, since the ^{3}H-myoinositol label of anchor peptides is completely hydrophilic after PI-PLC (Conzelmann et al, 1990). This situation is reminiscent of *Trypanosoma brucei* in which acylation of the myoinositol can be found on part of the free GPI's but not on protein-bound GPI's (Krakow et al, 1986, 1989; Menon et al, 1988; Mayor et al, 1990a,b). The same lipid also accumulates in *sec*53, a temperature sensitive mutant affecting synthesis of GDP-mannose (Kepes and Schekman, 1988). This mutant stops adding GPI's to proteins within minutes after a shift to the nonpermissive temperature (Conzelmann et al, 1990). This is not unexpected and underscores the fact that the pool of mature GPI's is small. Other mutants with potential defects in the mannose transfer to the high mannose core of N-glycans have no effect on the incorporation of ^{3}H-myoinositol. Thus, *alg2* and *alg3* (Huffaker and Robbins, 1983), two mutants which transfer $GlcNAc_2Man_1$ and $GlcNAc_2Man_5$ respectively instead of $GlcNAc_2Man_9Glc_3$ onto glycoproteins do add GPI's to proteins normally (Fig. 2).

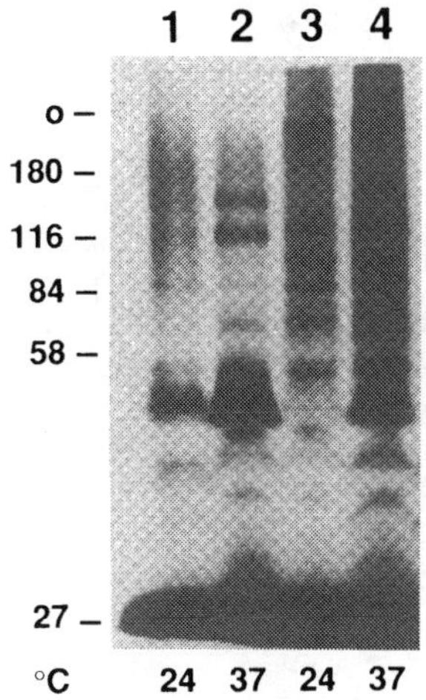

Figure 2. Incorporation of ^{3}H-myoinositol by *alg2* and *alg3* cells
Aliquots of $2OD_{600}$ units of *alg2* (lanes 1,2) or *alg3* (lanes 3,4) cells were precultured at 24°C in minimal medium supplemented with 1% casein hydrolysate and preincubated (20 min) and labelled with ^{3}H-myoinositol at either 24°C or 37°C for 2h. Detergent extracts of broken cells were analysed by SDS-PAGE and fluorography. o = top of running gel.

5. Characterization of GPI-modified proteins of yeast.
The only GPI-modified protein which has been purified so far is a 125kD glycoprotein of unknown function. The protein led to the discovery of GPI anchors of yeast in that it is the only yeast protein which becomes hydrophilic after PI-PLC treatment and because of its

relative abundance. The 125kD protein seems to be identical with a 115kD protein studied by others (Popolo et al, 1986, 1988; Vai et al, 1990; Grandori et al, 1990) the synthesis of which was found to be cell cycle controlled. The 100kD precursor form is made at higher rate in G_1 and its synthesis is stimulated by conditions which elevate intracellular cAMP levels. The protein is not required for progression through the cell cycle since the 125kD protein has been cloned and found to be encoded by a single gene which can be deleted without getting a cell cycle arrest (Nuoffer et al, 1991). Although the sequence of the 125kD protein does not bear any homologies to proteins of known function, the overall structure of the sequence is typical of GPI anchored proteins of other species. The protein contains an N-terminal signal sequence and a hydrophobic C-terminal peptide. Introduction of a charged residue into this C-terminal peptide abrogates the addition of GPI's. Anchors are added to an asparagine residue close to the C-terminus which is preceded by a serine-rich region (Nuoffer et al, 1991). The protein is very likely to contain a high amount of O-glycans since the N-deglycosylated form is about 30kD bigger (as judged by SDA-PAGE) than the chemically deglycosylated form (Fankhauser and Conzelmann, 1991). It is interesting to note that the precursor form leaves the ER and reaches the Golgi with a half time of about 15 min although in general most yeast proteins reach the cell surface within a few minutes. A large proportion of the 125kD protein is found on the surface and is removed by protease treatment of spheroplasts (Nuoffer et al, 1991). Yet, the use of trinitrobenzenesulfonic acid for tagging of surface molecules in order to monitor the surface appearance of the 125kD protein during pulse chase experiments showed that about half of newly made protein remains inaccessible for hours, thus suggesting the existence of an intracellular pool of this protein as well (Fankhauser and Conzelmann, 1991). Further studies are necessary to resolve this discrepancy.

A cAMP-binding protein of the inner mitochondrial membrane of *S. cerevisiae* has been reported to be anchored via an inositol-containing phospholipid and to be released by an endogenous phospholipase (Muller and Brandlow, 1989a,b). The data clearly indicate that a cAMP-binding protein can be labelled with ^{3}H-myoinositol and with ^{3}H-stearate and can be partially released from membranes by phospholipase D but not C. It is unlikely that this protein carries a classical GPI anchor, not only because of its intracellular localization but also because it cannot be labelled with ^{3}H-palmitate as other GPI-modified proteins of *S. cerevisiae*.

The *KRE1* protein which is required for the later biosynthetic steps of the (1-6)ß-glucan of the yeast cell wall has also been proposed to contain a GPI-modification based on the presence of a hydrophobic C-terminal sequence but further experimental evidence for a GPI anchor on this protein is not available (Boone et al, 1990).

PERSPECTIVES

In view of the fact that the GPI structure of yeast is the same as the one of the mammalian and protozoan systems, yeast might become an attractive system for the study of GPI-anchoring because of its well developed genetics and molecular biology. So far, however, our attempts to isolate mutants in the GPI biosynthetic pathway of yeast have met with little success although such mutants do exist in the mammalian system. Since there are so many GPI-modified proteins in yeast, we assume that a defect in this pathway would be lethal. Thus, if such mutants are to be obtained, it would seem necessary to find a way of selecting or screening for an abnormal phenotype which is specific for GPI-mutants and which becomes apparent within hours after a shift to the nonpermissive temperature.

ACKNOWLEDGEMENT

This work was supported by grants from the Swiss National Foundation for Scientific Research No. 31-9069.87 and 31-9418.88.

REFERENCES

Bangs, J. P., Hereld, D., Krakow, J. L., Hart, J. W. and Englund, P. T. (1985). Rapid processing of the carboxyl terminus of a trypanosome variant surface glycoprotein. Proc. Natl. Acad. Sci. USA **82:** 3207-3211.

Bordier, C. (1981). Phase separation of integral membrane proteins in Triton X-114 solution. J. Biol. Chem. **256:** 1604-1607.

Conzelmann, A., Spiazzi, A. and Bron, C. 1987). Glycolipid anchors are attached to Thy-1 glycoprotein rapidly after translation. Biochem. J. **246:** 605-610.

Conzelmann, A., Riezman, H., Desponds, C. and Bron, C. (1988). A major 125 kD membrane glycoprotein of *Saccharomyces cerevisiae* is attached to the lipid bilayer through an inositol-containing phospholipid. EMBO J. **7:** 2233-2240.

Conzelmann, A., Fankhauser, C. and Desponds, C. (1990). Myoinositol gets incorporated into numerous membrane glycoproteins of *Saccharomyces cerevisiae*: incorporation is dependent on phosphomannomutase (SEC53) EMBO J. **9:** 653-661.

Cross, G. A. M. (1990). Glycolipid anchoring of plasma membrane proteins. Annu. Rev. Cell Biol. **6: 1-39.**

Doering, T. L., Masterson, W. J., Hart, G. W. and Englund, P. T. (1990). Biosynthesis of glycosyl-phosphatidylinositol membrane anchors. J. Biol. Chem. **265:** 611-614.

Esmon, B., Novick, B. and Schekman, R. (1981). Compartmentalized assembly of oligosaccharides on exported glycoproteins in yeast. Cell **25:** 451-460.

Fankhauser, C. and Conzelmann, A. (1991). Purification, biosynthesis and cellular localization of a major 125-kDa glcophosphatidylinositol-anchored membrane glycoprotein of *Saccharomyces cerevisiae*. Eur. J. Biochem. **195:** 439-448.

Ferguson, M. A. J., Duszenko, M., Lamont, G. S., Overath, P. and Cross, G. A. M. (1986). Biosynthesis of *Trypanosoma brucei* variant surface glycoproteins N-glycosylation and addition of phosphatidylinositol membrane anchor. J. Biol. Chem. **261:** 356-362.

Ferguson, M. A. and Williams, A. F. (1988). Cell surface anchoring of proteins via glycosyl-phosphatidylinositol structures. Ann. Rev. Biochem. **57:** 285-320.

Ferguson, M. A. J., Homans, S. W., Dwek, R. A. and Rademacher, T. W. (1988). Glycosyl-phosphatidylinositol moiety that anchors *Trypanosoma brucei* variant surface glycoprotein to the membrane. Science **239:** 753-759.

Grandori, R., Popolo, L., Vai, M. and Alberghina, L. (1990). cAMP promotes the synthesis in early G1 of gp115, a yeast glycoprotein containing glycosyl-phosphatidylinositol. J. Biol. Chem. **265:** 14314-14320.

Homans, S. W., Ferguson, M. A. J., Dwek, R. A., Rademacher, T. W., Anand, R. and Williams, A. F. (1988). Complete structure of the glycosyl phosphatidylinositol membrane anchor of rat brain Thy-1 glycoprotein. Nature **333:** 269-272.

Huffaker, T. C. and Robbins, P. W. (1983). Yeast mutants deficient in protein glycosylation. Proc. Natl. Acad. Sci. USA **80:** 7466-7470.

Kepes, F. and Shekman, R. (1988). The yeast SEC53 gene encodes phosphomannomutase. J. Biol. Chem. **263:** 9155-9161.

Krakow, J. L., Doering, Tl L., Masterson, W. J., Hart, G. W. and Englund, P. T. (1989). A glycolipid isolated from *Trypanosoma brucei* related to the variant surface glycoprotein membrane anchor. Mol. Biochem. Parasit. **36:** 263-270.

Krakow, J. L., Hereld, D., Bangs, J. D., Hart, G. W. and Englund, P. T. (1986). Identification of a glycolipid precursor of the *Trypanosoma brucei* variant surface glycoprotein. J. Biol. Chem. **261:** 12147-12153.

Low, M. G. and Saltiel, A. R. (1988). Structural and functional role of glycosylphosphatidylinositol in membranes. Science **239:** 268-275.

Masterson, W. J., Doering, T. L., Hart, G. W. and Englund, P. T. (1989). A novel pathway for glycan assembly: biosynthesis of the glycosyl-phosphatidylinositol anchor of the trypanasome variant surface glycoprotein. Cell **56:** 793-800.

Masterson, W. J., Raper, J., Doering, T. L., Hart, G. W. and Englund, P. T. (1990). Fatty acid remodelling: a novel reaction sequence in the biosynthesis of trypanosome glycosyl phosphatidylinositol membrane anchors. Cell **62:** 73-80.

Mayor, S., Menon, A. K., Cross, G. A., Ferguson, M. A., Dwek, R. A. and Rademacher, T.W. (1990a). Glycolipid precursors for the membrane anchor of *Trypanosoma brucei* variant surface glycoproteins.I. Glycan structure of the phosphatidylinositol-specific phospholipase C sensitive and resistant glycolipids. J. Biol. Chem. **265:** 6164-6173.

Mayor, S., Menon, A. K. and Cross, G. A. (1990b). Glycolipid precursors for the membrane anchor of *Trypanosoma brucei* variant surface proteins.II. Lipid structures of phosphatidylinositol-specific phospholipase sensitive and resistant glycolipids. J. Biol. Chem. **265:** 6174-6181.

Menon, A. K., Mayor, S., Ferguson, M. A., Duszenko, M. and Cross, G. A. (1988). Candidate glycolipid precursor for the glycosylphosphatidylinositol membrane anchor of *Trypanosoma brucei* variant surface glycoproteins. J. Biol. Chem. **263:** 1970-1977.

Menon, A. K., Schwartz, R. T., Mayor, S. and Cross, G. A. (1990a). Cell-free synthesis of glycosyl-phosphatidylinositol precursors for the glycolipid membrane anchor of *Trypanosoma brucei* variant surface glycoproteins. Structural characterization of putative biosynthetic intermediates.J. Biol. Chem. **265:** 9033-9042.

Menon, A. K., Mayor, S. and Schwartz, R. T. (1990b). Biosynthesis of glycosyl-phosphatidylinositol lipids in *Trypanosoma brucei*: involvement of mannosyl-phosphoryldolichol as the mannose donor. EMBO J. **9:** 4249-4258.

Muller, G. and Bandlow, W. (1989a). An amphitropic cAMP-binding protein in yeast mitochondria.2.Phospholipid nature of the membrane anchor. Biochemistry **28:** 9968-9973.

Muller, G. and Bandlow, W. (1989b). An amphitropic cAMP-binding protein in yeast mitochondria. 3. Membrane release requires both Ca2(+)-dependent phophorylation of the cAMP-binding protein and a phospholipid-activated mitochondrial phospholipase. Biochemistry. **28:** 9974-9981.

Novick, P., Ferro, S. and Schekman, R. (1981). Order of events on the yeast secretory pathway. Cell **25:** 461-469.

Novick, P., Field, C. and Schekman, R. (1980). Identification of 23 complementation groups required for post-translational events in the secretory pathway. Cell **21:** 205.

Nuoffer, C., Jeno, P., Conzelmann, A. and Riezman,H. (1991). Determinants for glycophospholipid anchoring of the *Saccharomyces cerevisiae* GAS1 protein to the plasma membrane. Mol. Cell Biol. **11:** 27-37.

Orlean, P. (1990). Dolichol phosphate mannose synthase is required in vivo for glycosyl phosphatidylinositol membrane anchoring, O mannosylation and N glycosylation of protein in *Saccharomyces cerevisiae*. Mol. Cell Biol. **10:** 5796-5805.

Popolo, L., Grandori, R., Vai, M., Lacana, E. and Alberghina, L. (1988). Immunochemical characterization of gp115, a yeast glycoprotein modulated by the cell cycle. Eur. J. Cell Biol. **47:** 173-180.

Popolo, L., Vai, M. and Alberghina, L. (1986). Identification of a glycoprotein involved in cell cycle progression in yeast. J. Biol. Chem. **261:** 3479-3482.

Schneider, P., Ferguson, M. A. J., McConville, M. J., Mehlert, A., Homans, S. W. and Bordier, C. (1990). Structure of the glycosyl-phosphatidylinositol membrane anchor of the *Leishmania major* promastigote surface protease. J. Biol. Chem. **265:** 16955-16964.

Thomas, J. R., Dwek, R. A. and Rademacher, T. W. (1990). Structure, biosynthesis, and function of glycophosphatidylinositols. Biochemistry **29:** 5413 -5422.

Tsai, P., Frevert, J. and Ballou, C. E. (1984). Carbohydrate structure of *Saccharomyces cerevisiae* mnn9 mannoprotein. J. Biol. Chem. **259:** 3805-3811.

Vai, M., Popolo, L., Grandori, R., Lacana, E. and Alberghina, L. (1990). The cell modulated glycoprotein GP115 is one of the major yeast proteins containing glycosylphosphatidylinositol. Biochim. Biophys. Acta **1038:** 277-285.

Wen, D. and Schlesinger, M. J. (1984). Fatty acid-acylated proteins in secretory mutants of *Saccharomyces cerevisiae*. Mol. Cell Biol. **4:** 688-694.

CHARACTERIZATION OF THE PLASMA GLYCOSYLPHOSPHATIDYLINOSITOL-SPECIFIC PHOSPHOLIPASE D (GPI-PLD)

[1]Christine N Metz, [2]Sergio Schenkman and *[1]Michael A Davitz.

[1]Division of Immunology, Department of Pathology, New York University School of Medicine, New York, USA. and [2]Escola Paulista de Medecina, Sao Paulo, Brazil.

* Corresponding author.

INTRODUCTION

Proteins anchored to the cell membrane by glycosylphosphatidylinosital (GPI) are functionally diverse. They include: complement regulatory proteins, cell-adhesion molecules, ectoenzymes, lymphocyte differentiation antigens, tumor markers, both the normal and scrapie forms of the prion protein, as well as parasitic protozoan cell surface antigens (Cross, 1990). One proposed function for the GPI-anchor is that it facilitates the release of the protein from the membrane by serving as a target substrate for anchor-specific phospholipases (Low, 1990). Recently, we and others have discovered (Davitz et al., 1987; Cardoso de Almeida et al., 1988; and Low and Prasad, 1988), purified (Davitz et al., 1989; Huang et al., 1990), and cloned (Scallon et al., 1991) a GPI-specific phospholipase D (GPI-PLD) isolated from mammalian plasma. Based on a cleavage analysis of the GPI-biosynthetic precursors for the membrane form variant surface glycoprotein (mfVSG) of *Trypanosoma brucei*, the minimal carbohydrate structure recognized by the GPI-PLD consists of Mannose-Glucosamine-phosphatidylinositol (Man_1GlcN-PI) (Masterson et al., 1989) (Figure 1). This carbohydrate unit is part of a conserved glycan core backbone that is present on both trypanosomal as well as mammalian anchors (Cross, 1990). Thus it is likely that all GPI-anchors could serve as potential substrates for the GPI-PLD.

DISCOVERY

The plasma GPI-PLD was discovered simultaneously by three groups. While studying the spontaneous release of the GPI-anchored protein, decay accelerating factor (DAF) from HeLa cells, Davitz et al. (1987) found that both bovine and human serum contained high levels of a phospholipase D activity. Closer examination of the activity revealed that the phospholipase D was specific for the GPI-anchor, that is it did not cleave phosphatidylcholine, phosphatidylethanolamine, or phosphatidylinositol, but did cleave the intact GPI-anchor present on GPI-anchor present on

GPI-anchored proteins as well as one of the biosynthetic precursors of mfVSG, lipid A (Davitz et al., 1987).

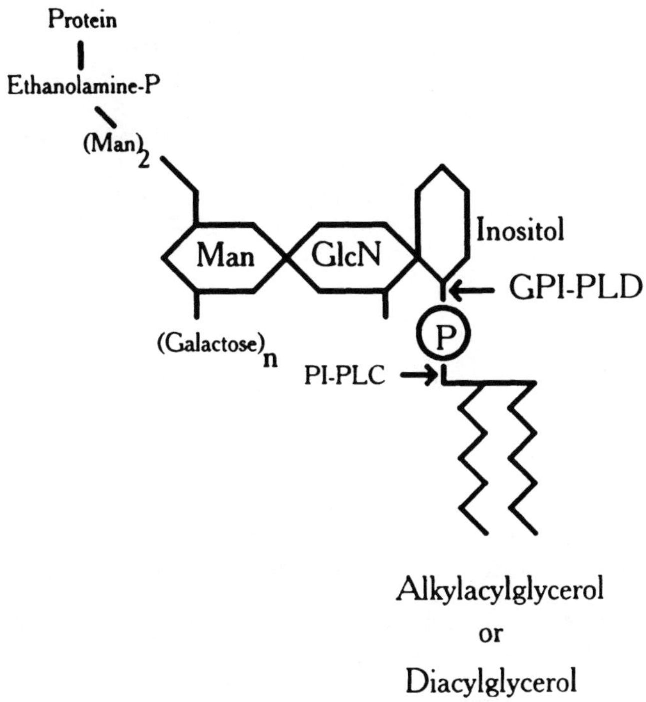

Figure 1. Structure of the GPI-anchor and site of GPI-PLD cleavage.

Low (Malik and Low, 1986) identified an alkaline phosphatase degrading activity in extracts of human placenta; this activity altered the mobility of alkaline phosphatase on gel filtration chromatography. On the basis of its molecular size as well as its sensitivity to inhibition with 1, 10-phenanthroline, this degrading activity appeared to be different from the cytoplasmic phosphatidylinositol (PI)-specific phospholipase C (Low and Prasad, 1988). Moreover, cleavage of [^{3}H]-labelled mfVSG by the degrading activity generated phosphatidic acid, rather than diacylglycerol (Malik and Low, 1986). They subsequently identified and characterized the enzyme as a plasma GPI-PLD, which was specific for the GPI-anchor, and inhibited by 1,10 phenanthroline as well as EGTA (Low and Prasad, 1988).

Cardoso de Almeida et al. (1988) identified an acid-lipase in human serum which converted mfVSG to a water-soluble form. In contrast to Pl-specific phospholipase C, cleavage of mfVSG with this enzyme did not expose the cross reacting determinant (CRD) (Cardoso de Almeida et al., 1988). The CRD epitope is a cryptic carbohydrate epitope present on the C-terminal of many GPI-anchored proteins (Cardoso de Almeida and Turner, 1983; Cross, 1990). The epitope is composed in part of a terminal inositol-phosphate group that is exposed after cleavage of the anchor with a Pl-specific phospholipase C (Shak et al., 1988). Treatment of the GPI-anchor with a phospholipase D eliminates the terminal phosphate on the inositol destroying the CRD epitope (Cardoso de Almeida et al., 1988; Shak et al., 1988). A similar observation was made by Davitz et al. (1987), that is, treatment of DAF with the GPI-PLD did not expose the CRD epitope.

PURIFICATION AND BIOCHEMICAL CHARACTERIZATION OF THE PLASMA GPI-PLD

Cardoso de Almeida et al. (1988) partially purified the GPI-PLD by chromatography over DEAE cellulose, Sephacryl S-300, and wheat germ lectin-Sepharose. The enzyme had a pH optimum around 5.4, eluted on gel filtration with an apparent molecular weight of 140 kD, and was inhibited by 1,10 phenanthroline. The discovery that the enzyme bound to wheat germ lectin proved to be crucial for the ultimate purification of the GPI-PLD.

Davitz et al. (1989) purified the human plasma GPI-PLD to homogeneity by a four step procedure involving a Mono Q and phenyl-5PW column, followed by a wheat germ lectin affinity column and finally another Mono Q column. A 4,500-wheat germ lectin affinity column and finally another Mono Q column. A 4,500-fold purification was achieved with a 5% yield. By sodium dodecyl sulfate-polyacrylamide gel electrophoresis, the homogenous enzyme had a Mr of 110,000 and consisted of a single polypeptide chain. It exhibited identical substrate specificity as compared with the crude preparation and was inhibited by 1,10 phenanthroline. The enzyme was active over a broad pH range (4.0-8.5), which may be explained by the use of a different assay.

Huang et al. (1990) purified the GPI-PLD from bovine serum. Initially the enzyme was purified from a 9% polyethylene glycol precipitate of plasma; this was followed by Q Sepharose anion-exchange chromatography, S-300 gel filtration, wheat-germ lectin-Sepharose anion-exchange chromatography, S-300 gel filtration, wheat-germ lectin-Sepharose, hydroxylapatite, zinc chelate matrix,

Mono-Q HPLC, and Superose (gel filtration). The purified material was used to generate monoclonal antibodies, which permitted the development of a simple immunoaffinity chromatographic scheme for the GPI-PLD. By SDS-PAGE, the purified enzyme had a Mr of 100,000 and a pI of 5.6. It exhibited similar substrate specificity and enzymatic activity as compared with the human enzyme. Davitz et al. (1989) and Huang et al. (1990) estimated that the plasma concentration of the GPI-PLD is 5-10 μg/ml.

Hoener et al. (1990) characterized and partially purified a GPI-PLD from bovine brain. This phospholipase exhibited similar substrate specificity as compared with the plasma GPI-PLD and was inhibited by both EGTA and 1,10-phenanthroline. Unexpectedly the GPI-PLD activity was barely detectable in crude homogenates of brain, while partial purification of the enzyme markedly increased the specific activity of the brain GPI-PLD. The authors attributed this observation to the presence of inhibitors in the crude extract. A similar increase of total activity was observed during the purification of the GPI-PLD from plasma (Davitz et al., 1989).

One interesting modification of the basic GPI-anchor structure is acylation of the inositol ring (Roberts et al., 1988). This modification renders the anchor resistant to cleavage with Pl-specific phospholipase C, but does not affect cleavage of the GPI-anchor by the GPI-PLD (Davitz et al., 1987; Roberts et al., 1988).

CLONING OF THE GPI-PLD

Twelve tryptic peptides were obtained from the purified enzyme and subjected to microsequencing. Two of the peptides shared extensive sequence homology (50%) indicating the existence of internal repeats. Using this information, Scallon et al. (1990) cloned a GPI-PLD from a bovine liver cDNA library. Consistent with the size of the protein, the GPI-PLD mRNA was approximately 3.0kb. The deduced amino acid sequence of the protein confirmed the existence of four internal regions of homology with 21-54% homology. These repeats had significant similarity to the metal ion binding domains of the integrin alpha subunits, that is they share an aspartate-rich core sequence flanked by short conserved segments. The identity of the cells in the liver producing the GPI-PLD, however, remains to be established as assays of liver microsomes, hepatocytes, kupffer, and Ito cells are all negative for GPI-PLD enzymatic activity (Metz and Davitz, unpublished observations).

FUNCTION OF THE GPI-PLD

Neither the crude, purified, nor recombinant GPI-PLD releases GPI-anchored proteins from intact cell membranes. This observation is not surprising in view of the fact that Low (1990) has estimated that sufficient catalytic activity is present in the plasma to remove most of the GPI-anchored proteins on the surfaces of a cell in less than one minute. Therefore the GPI-PLD must be under tight regulation to prevent the complete loss of GPI-anchored proteins on the cell membrane. Low (1990) has shown that degradation of the GPI-anchor on an intact cell membrane does occur if a detergent (e.g., 0.1% NP-40) is added to the reaction mixture, but the physiological relevance of this observation is unclear.

Scallon et al. (1990) made the intriguing observation that co-transfection of the cDNA encoding the GPI-PLD into COS cells together with the cDNA for a GPI-anchored protein (e.g., placental alkaline phosphatase, PLAP) greatly increased the release of PLAP from the cell. Interestingly, the cleavage of PLAP by the GPI-PLD appeared to occur inside the cell, rather than on the cell membrane. The broad pH range over which the GPI-PLD is active (Cardoso de Almeida et al., 1988; Davitz et al., 1989) could allow the enzyme to operate in an intracellular compartment such as an endosome where the pH is between 5.0 and 6.0 (Anderson and Orci, 1988). Thus the expression of the GPI-PLD in a particular cell may alter the regulation of GPI-anchored proteins in cells expressing the enzyme.

Soluble forms of many GPI-anchored proteins are known to exist normally in plasma (Cross, 1990), however whether these soluble molecules originate from a GPI-PLD cleavage of the intact protein or are the result of other cleavage events is unknown. A soluble form of Thy-1 has been identified in the cerebrospinal fluid (Almqvist and Carlsson, 1988). The analysis of its C-terminal structure is consistent with cleavage of the intact Thy-1 molecule by a GPI-PLD (Almqvist and Carlsson, 1988). The GPI-anchored protein DAF is released spontaneously from HeLa cells (Choi et al, in preparation). Analysis of the soluble protein by metabolic labelling indicates that a portion of the molecules contain inositol, but lack the CRD epitope, suggesting cleavage of DAF by a GPI-PLD.

Potentially, the cleavage of the GPI-anchor may generate second messages. Phosphatidic acid can elevate intracellular Ca^{2+} levels and induce mitogenesis in certain cell types (Moolenaar et al., 1986; Yu et al., 1988). In addition, the glycan may serve as a precursor for the putative GPI-derived insulin-mediator (Romero et al.,

1988). Nonetheless, the physiological connection between the GPI-PLD and the cleavage of GPI-anchored proteins remains to be established.

FUTURE DIRECTIONS

Since the initial discovery of the GPI-PLD rapid progress has been made on the biochemical characterization of the enzyme, but the crucial physiological questions remain to be answered. The specific cells expressing the GPI-PLD must be determined and the possible cellular factors regulating cleavage of GPI-anchored proteins by the GPI-PLD defined. One intriguing possibility is that co-factor proteins are required to facilitate cleavage of the GPI-anchored protein by the GPI-PLD. For example, in the case of glycolipid hydrolyses, such as ß-galactosidase and arylsulfatase A, the attack of these enzymes on their lipid substrates, Ganglioside G_{M1} and sulfatide, respectively, is mediated by activator proteins (Conzelmann and Sandhoff, 1987). These activator proteins bind to the glycolipid substrate, extract it from the membrane forming a water-soluble complex which is the actual substrate for the enzyme.

ACKNOWLEDGEMENT

Michael A Davitz is a Lucille P Markey Scholar in the Biomedical Sciences and this work was supported by a grant from the Lucille P Markey Charitable Trust (88-25).

REFERENCES

Almqvist, P. and Carlsson, S. R. (1988). Characterisation of a hydrophilic form of Thy-1 purified from human cerebrospinal fluid. J. Biol. Chem. **263:** 12709-12715.

Anderson, R. G. W. and Orci, L. (1988). A view of acidic intracellular compartments. J. Cell Biol. **106:** 539-543.

Cardoso de Almeida, M. L. and Turner, M. J. (1983). The membrane form of the variant surface glycoproteins of *Trypanosoma brucei*. Nature **302:** 349.

Cardoso de Almeida, M. L., Turner, M. J., Stambuk, B. B. and Schenkman, S. (1988). Identification of an acid-lipase in human serum which is capable of solubilising glycophosphatidylinositol-anchored proteins. Biochem. Biophys. Res. Comm. **150:** 476-482.

Choi, N. H., Davitz, M. A. Caras, I. and Nussenzweig, V. unpublished observations.

Conzelmann, E. and Sandhoff, K. (1987). Activator proteins for lysosomal glycolipid hydrolysis. Methods in Enzymology. **138:** 792-821.

Cross, G. A. M. (1990). Glycolipid anchoring of plasma membrane proteins. Ann. Rev. Cell Biol. **6:** 1-37.

Davitz, M. A., Hereld, D., Shak, S., Krakow, J., Englund, P. T. and Nussenzweig, V. (1987) A glycan-phosphatidylinositol-specific phospholipase D in human serum. Science **238:** 81-84.

Hoener, M. C., Steiger, S. and Brodbeck, U. (1990). Isolation and characterisation of a phosphatidylinositol-glycan-anchor-specific phospholipase D from bovine brain. Eur. J. Biochem. **190:** 593-601.

Huang, K. S., Fung, W. J. C., Hulmes, J. D., Reik, L., Pan, Y. C. E. and Low, M. G. (1990). Purification and characterisation of glycosylphosphatidylinositol specific phospholipase D. J. Biol. Chem. **265:** 17738-17745.

Low, M. G. and Prasad, A. R. S. (1988). A phospholipase D specific for the phosphatidylinositol anchor of cell-surface proteins is abundant in plasma. Proc. Natl. Acad. Sci. USA **85:** 980-984.

Low, M. G. (1990). Degradation of glycosyl-phosphatidylinositol anchors by specific phospholipases. In: Molecular and Cell Biology of Membrane Proteins: Glycolipid Anchors of Cell Surface Proteins. A. J. Turner Ed. Ellis Horwood Ltd. Chichester. England. pp35-63.

Malik, A. S. and Low, M. G. (1986). Conversion of human placental alkaline phosphatase from a high Mr. form to a low Mr form during butanol extraction. Biochem. J. **240:** 519-527.

Masterson, W. J., Doering, T. L., Hart, T. W. and Englund, P. T. (1989). A novel pathway for glycan assembly: biosynthesis of the glycosyl-phosphatidylinositol-anchor of the *Trypanosome* variant surface glycoprotein. Cell **56:** 793-800.

Moolenaar, W. H., Kruijer, W., Tilly, B. C., Verlaan, I., Bierman, A. J. and de Laat, S. W. (1986). Growth factor like action of phosphatidic acid. Nature **323:** 171-173.

Roberts, W. L., Myher, J. J., Kuksis, A., Low, M. G. and Rosenberry, T. L. (1988). Lipid analysis of the glycoinositol phospholipid membrane anchor of human erythrocyte acetylcholinesterase. J. Biol. Chem. **263:** 18766-18775.

Romero, G., Luttrell, L., Rogol, A., Zeller, K., Hewlett, E. and Larner, J. (1988). Phosphatidyl-glycan anchors of membrane proteins: potential precursors of insulin mediators. Science **240:** 509-511.

Scallon, S., Fung, W-E., Tsang, C., Li, S., Kado-fong, H., Huang, K. S. and Kochan, J. (1991). Phosphatidylinositol-glycan-specific phospholipase D provides a novel mechanism for protein secretion. Science **252:** 446-448.

Shak, S., Davitz, M. L., Wolinsky, M. L., Nussenzweig, V., Turner, M. J. and Gurnett, A. (1988). Partial characterisation of the cross reacting determinant, a carbohydrate epitope shared by decay accelerating factor and the variant surface glycoprotein of the African *Trypanosoma brucei*. J. Immunol. **140:** 2047-2050.

Yu,C. L., Tsai, M. H. and Stacey, D. W. (1988). Cellular ras activity and phospholipid metabolism. Cell **52:** 63-71.

EVOLUTIONARY ASPECTS OF GPI METABOLISM IN KINETOPLASTID PARASITES

***Michael A J Ferguson, Wayne J Masterson, Steve W Homans and Malcolm J McConville**

Department of Biochemistry, University of Dundee, Dundee, DD1 4HN, Scotland.

* Corresponding author.

SUMMARY

There is a growing, but still very patchy, data base of GPI structure, biosynthesis and function. In this article we speculate freely on the function of GPI anchors, and the origins of GPI-related molecules, primarily with reference to the protozoan parasites *Trypanosoma brucei* and the *Leishmania*. The views expressed draw on fairly wild extrapolations and some will, no doubt, not stand the tests of time. Several of the hypotheses presented should therefore be taken with a pinch of salt, some lemon, and large quantities of tequila!

INTRODUCTION

Glycosyl-phosphatidylinositol (GPI) anchors are probably ubiquitous throughout the eukaryotes. Their primary function is to afford the stable association of certain glycoproteins with the outer leaflet of the plasma membrane, or occasionally with the topologically equivalent inner leaflet of secretory granule membranes. The GPI anchor can be thought of as an alternative to the use of a transmembrane polypeptide anchor in type-1 membrane proteins (i.e. proteins which span the bilayer only once with their N-terminal domains as ectodomains). Throughout this article we will be comparing the distribution and function of GPI-anchored proteins and proteins which span the bilayer **only once** (type-1 and type-2 membrane proteins).

The addition of GPI to proteins is a post-translational event which occurs in the endoplasmic reticulum. Generally speaking, post-translational modifications are sometimes essential to protein bioactivity, sometimes mechanisms of fine-tuning or regulating protein function, and sometimes apparently irrelevant (at least to testable criteria). This much is certainly true of phosphorylation, acylation and glycosylation. However, just one **vital** function of a post-translational modification would constitute the selective pressure to retain the necessary biosynthetic and processing machinery. Once such machinery is in place it is perhaps inevitable that it will act on a number of proteins by chance, rather than design, provided that the effects are not detrimental to their function. However, as organisms evolve it is possible that some of these originally neutral modifications may be adopted for new functions.

GPI FUNCTION IN HIGHER EUKARYOTES

In all cases GPI anchors allow stable membrane association, but this does not explain why this mechanism of anchorage co-exists with the use of transmembrane domain polypeptide anchors. Some clues to specialised GPI functions have become available over the last few years (reviewed most recently by Low, 1989; Thomas et al, 1990; Cross, 1990; Ferguson, 1991 and several of the articles in this volume). For example, in mammals, GPI anchors appear to:

1. Target proteins to apical and axonal membranes in polarised cells such as epithelial cells and neurons (Lisanti et al, 1990; Dotti et al, 1991).

2. Exclude proteins from clathrin-dependent endocytosis (Lemansky et al, 1990).

3. Mediate specialised transmembrane signalling events in lymphoid and myeloid cells (Robinson, 1990).

4. Mediate small nutrient uptake via a non-clathrin dependent pseudo-endocytosis (in the case of the folate receptor, Rothberg et al, 1990).

Some of these functions may only have relevance in multicellular organisms and might therefore constitute new functions, not required earlier in evolution (i.e. in unicellular eukaryotes).

GPI FUNCTION IN LOWER EUKARYOTES

In the protozoa, cell surface proteins appear to be predominantly GPI anchored, at least for the most highly expressed cell surface proteins (e.g. in *Toxoplasma*, *Plasmodium*, *Paramecium*, *Trypanosoma* and *Leishmania*). In higher eukaryotes only a minority of proteins are anchored in this way. What is the origin of this evolutionary skew? Multicellular higher eukaryotes require complex communication between cells and these functions are best mediated through transmembrane proteins which can directly transmit information across the bilayer. However, the cell surfaces which must face particularly harsh environments (e.g. the apical membranes of gut and hepatic epithelia) are more involved with processing (and defending themselves against) the external milieu. These membranes are particularly rich in glycolipids and GPI anchored proteins. The same is broadly true of the plasma membranes of free living protozoa where plasma membrane function is primarily protection and nutrient uptake, rather than cell-cell communication. Thus for free living cells type-1 and type-2 membrane protein transmembrane polypeptide anchors are likely to be only rarely advantageous over GPI anchors. However, what could be the selective advantage for GPI anchors? One possible general advantage is that GPI anchored proteins have very high surface expression and very low turnover rates (Bülow et al, 1989; Lisanti et al, 1990; Lemansky et al, 1990). Thus for proteins whose function resides solely in the ectoplasmic domain (e.g. surface hydrolases and coat proteins) the overall biosynthetic burden may be reduced by using GPI anchors.

When considering the protozoa in general (free living, symbiotic and parasitic) it is important to remember that the various phyla and subphyla diverged at the earliest stages of eukaryotic evolution. The pre-eminence of GPI anchorage in members of the *Ciliophera* (e.g. *Paramecium*), *Sarcodina* (e.g. *Plasmodium*) and *Mastigophora* (e.g. the kinetoplastids *Trypanosoma* and *Leishmania*), which diverged hundreds of millions of years ago, does at least support the notion of a general advantage of GPI anchorage in single cell organisms. The abundance of GPI anchored proteins in yeast (Conzelmann et al, 1990) is also consistent with this hypothesis, although wider ranging studies are clearly required to test this idea.

GPI MOLECULES OF THE KINETOPLASTID PARASITES

The formation of the basic GPI anchor precursor (Figure 1) is believed to be highly conserved throughout the eukaryotes. All of the GPI protein anchors characterised so far from *Trypanosoma brucei, T.cruzi, Leishmania major,* rat, humans and yeast contain the same **ethanolamine-PO4-6Manα1-2Manα1-6Manα1-4GlcNH$_2$α1-6 *myo*-Inositol-1-PO4-lipid** core structure (Ferguson, 1991). The biosynthetic pathway of GPI precursor formation has been delineated by intensive studies in *T.brucei* (see article by Menon et al in this volume), and similar intermediates have been observed recently in mammalian thymocytes (De Gasperi et al, 1990). The known parasite specific adaptations to this conserved pathway (see Figure 1) include:

1. Specialised lipid remodelling of the VSG GPI anchor in *T.brucei* (Masterson et al, 1990).

2. The unique α-galactosylation of the VSG anchor (Ferguson et al, 1988).

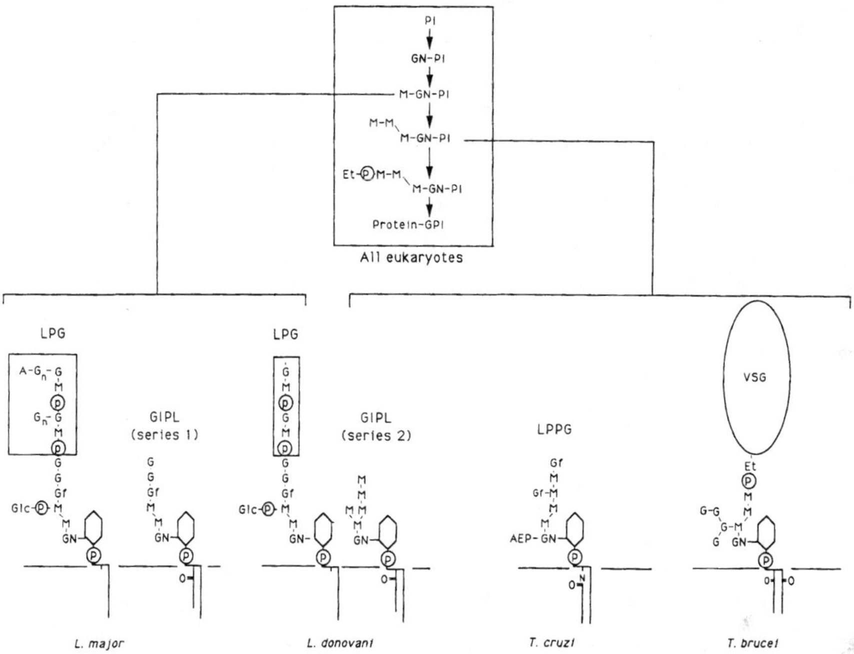

Fig 1. Examples of novel GPI structures produced by the kinetoplastid parasites.
The box shows the GPI anchor biosynthetic pathway which is believed to be ubiquitous throughout the eukaryotes. The structures outside the box represent parasite-specific GPI-related molecules or GPI anchor modifications. The kinetoplastid parasites utilise GPI glycolipids to anchor their major surface proteins to the plasma membrane. In some cases, the structure of these anchors (i.e in *T. brucei* VSG) are elaborated with additional glycan structures. Other kinetoplastid parasites (i.e *T. cruzi* and *Leishmania* spp) also produce a number of novel GPIs which are not linked to protein (i.e the LPGs, GIPLs and LPPGs). These structures may either be structurally similar to the protein anchors (i.e LPPG and GIPL (series 2)) or diverge completely from these anchors beyond the conserved core., Manα1-4GlcNα1-6*myo*inositol (i.e the LPGs and GIPL (series 1)). The structures of the parasite GPIs are also notable in the diversity of their lipid moieties: dimyrstylglycerol in the VSG, 1-O-alkylglycerol in the LPGs, alkylacylglycerols in the GIPLs and ceramide in the LPPG. Symbols: M,Mannose; G,galactopyranose; Gf,galactofuranose; GN,glucosamine; P,phosphate; A,arabinopyranose;AEP,2-aminoethylphosphonate.

3. The synthesis of the lipopeptidophosphoglycan (LPPG) by *T.cruzi* epimastigotes (Previato et al, 1990; Lederkremer et al, 1990).

4. The synthesis of glycoinositol phospholipids (GIPLs) and lipophosphoglycans (LPGs) in the *Leishmania* (McConville et al, 1990a,b, 1991; Turco, 1990).

Examples 3 and 4 are glycophospholipids which are not attached to protein and there are no known counterparts in other eukaryotes. As an operational definition we consider all molecules containing the structural motif **Manα1-4GlcNH$_2$α1-6*myo*-Inositol-1-PO4-lipid** to belong to the GPI family. This definition excludes glycosylated phosphoinositides of mycobacterial, yeast and plant origin which lack the characteristic non-N-acetylated glucosamine residue. The series-2 GIPLs of *Leishmania* and the LPPG of *T.cruzi* bear considerable structural homology to GPI protein anchors in their glycan moieties whereas the series-1 GIPLs/LPG sub-family are the most divergent in this respect (Figure 1). In contrast, all the *Leishmania* GIPLs and LPGs have similar lipid moieties to the promastigote protein-linked GPI anchor (Schneider et al, 1990), whereas the *T.cruzi* epimastigote LPPG lipid moiety (a ceramide) is different from the sn-1-alkyl-glycerolipid found in the *T.cruzi* metacyclic 1G7 antigen GPI anchor (Guther et al, unpublished data). However we must be cautious when comparing different developmental stages of the same organism since the bloodstream and procyclic forms of *T.brucei* are known to express different types of protein linked GPI glycerolipids (Field et al, 1991).

POSSIBLE ORIGINS OF PARASITE SPECIFIC GPI METABOLISM

Why should the kinetoplastid parasites either specifically modify their protein GPI anchors (e.g. the VSG anchor of T.brucei) or express novel but related glycophospholipids (e.g. *Leishmania* and *T.cruzi*). Perhaps high levels of GPI anchor expression in their ancestors has pre-adapted them for expanding their GPI metabolism for "unforeseen" functions that have complemented their evolution, firstly to monogenetic parasitism, and subsequently to digenetic parasitism. In the first instance there may be an inherent advantage to presenting a carbohydrate rich surface to the hostile environment of an insect digestive tract. Thus ancestral organisms which expanded and adapted their already active GPI anchor biosynthetic pathway might have achieved this phenotype relatively simply. The estimated copy numbers of LPPG in *T.cruzi* epimastigotes, and GIPLs plus LPG in *Leishmania* promastigotes, are extremely high (1-2 x 10^7 copies per cell) and it is likely that the glycan moieties of these molecules completely cover the otherwise fragile phospholipid bilayer. Consistent with this, *Crithidia fasiculata* (a monogenetic kinetoplastid parasite which colonises the mid- and hind-gut of mosquitoes) expresses, in addition to a GPI anchored protein (Zaretskaia et al, 1989), high levels of a GPI-related glycophospholipid (GPL) molecule containing an ~16 residue arabino-galactan chain containing 2-4 phosphate groups (Milne, McConville and Ferguson, preliminary observations). The surfaces of *C.fasiculata*, *T.cruzi* epimastigotes and *Leishmania* promastigotes will also be highly negatively charged by virtue of the phosphate, 2-aminoethylphosphonate (AEP), and phosphodiester groups of the GPL, LPPG and LPG molecules respectively. Such a feature may also be generally protective. At first sight this pattern seems to fall down when one considers the cell surface of the insect dwelling procyclic forms of T.brucei. We were unable to detect any significant levels of GIPL or LPG-like molecules in this species (McConville, Murray and Ferguson, preliminary observations). However we found that the GPI anchored procyclic acidic repetitive protein (PARP, also known as procyclin, Roditi et al, 1987; Clayton and Mowatt, 1989) has very similar physicochemical properties to *Leishmania* LPG. PARP behaves almost identically to LPG in terms of solvent extraction and hydrophobic interaction chromatography properties. It is present at similar levels to LPG (about $6x10^6$ copies per cell; Clayton and Mowatt, 1989) and it has a GPI anchor followed by an acidic $(Asp\text{-}Pro)_2\text{-}(Glu\text{-}Pro)_{22\text{-}29}$ repeat domain, which is predicted to take up an extended conformation (Roditi et al, 1989). The acidic phosphosaccharide repeat domain of LPG is predicted to adopt a "slinky spring" conformation where it can vary its length dramatically (Homans et al, 1991). However, the range of predicted inter-repeat distances for LPG overlaps with that predicted for PARP.

Thus one might speculate that the African trypanosomes and the *Leishmania* have undergone convergent evolution to express molecules of similar size, charge distribution and surface density on their insect dwelling forms (Figure 2). Unfortunately nothing is known about the function of PARP but it has been suggested that LPG plays an important role in the adhesion and development of *Leishmania* parasites in the insect vector (Davies et al, 1990), see below. It is conceivable that PARP plays a similar role in parasite-tsetse gut adhesion; a D-GlcNH$_2$ specific lectin has been implicated in the establishment of tsetse fly mid gut infections by *T.congolense* (Maudlin and Welburn, 1987). This parasite is a close relative of *T.brucei* and it expresses a PARP-like protein (J.D. Barry, personal communication). It has been suggested that the tsetse gut lectin could be recognising the D-GlcNH$_2$ residue of the PARP GPI anchor (Barry and Maudlin, personal communication). However more studies are required to tie

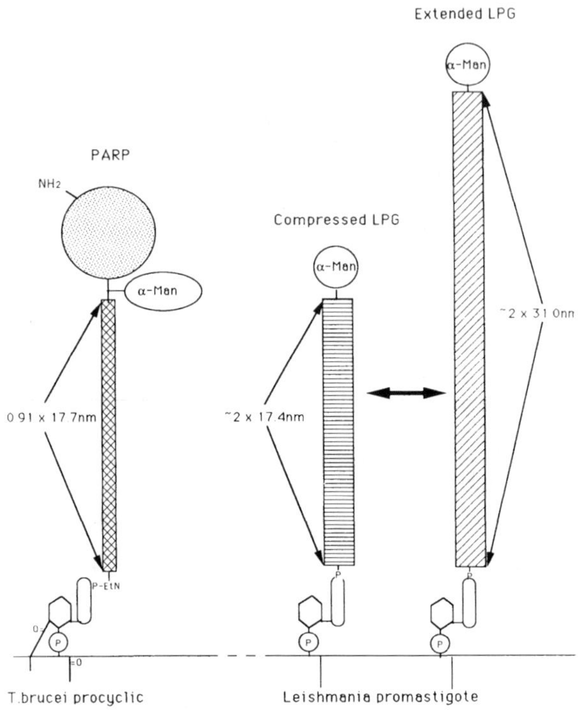

Fig 2. *T. brucei* PARP and *Leishmani* LPG: Convergent evolution of a polyanionic surface molecule?.
This diagram indicates the approximate sizes of *T. Brucei* PARP and *Leishmania* LPG. The PARP model is based on an (Asp-Pro)$_2$-(Glu-Pro)$_{29}$ repeat structure, adapted from Roditi et al(1989) showing the lipid structure according to Field et al (1991). The charge density is approximately one negative charge per 0.57nm along the polypeptide cylinder. For comparison, the theoretical minimum (fully compressed) and maximum (fully extended) models of a [-PO$_4$-6Galß1-4Manα1-]$_{31}$ repeat LPG[1] are shown, according to the 'slinky spring' model of Homans et al (1991). The charge density is approximately one negative charge per 0.56-1.0nm of phosphosaccharide helix. In addition to the charge density analogy, both molecules display a-Man containing structures at the terminus of the repeat structures, i.e. the "cap" structures of LPG and the N-linked glycan of PARP. Both molecules are expressed in the respective insect-dwelling stages of the parasites.
[1] The LPG molecule is shown without any of the repeat side-chains found in *L.major* and *L.mexicana* LPG. Although 31 phosphosaccharide repeats were chosen for direct comparison with PARP, this is in fact a typical average value for LPG repeat number.

up any relationships between these interesting observations.

EXAMPLES OF GPI STRUCTURE-FUNCTION RELATIONSHIPS IN PARASITE INFECTIVITY

a. The GPI anchor of *T.brucei* VSG

The VSG coat of the African trypanosomes is the parasites' first line of defence against both specific and non-specific immunity (reviewed in Ferguson and Homans, 1989). The parasites' ability to undergo antigenic variation, by sequentially expressing immunologically distinct VSG coat glycoproteins, enables the parasite to evade the hosts' humoral response (specific immunity). The different VSG coats pack into a dense monolayer which acts as a macromolecular diffusion barrier, allowing the access of small nutrient molecules to the plasma membrane but excluding host macromolecules. It is the diffusion barrier characteristic of the VSG coats which protects the parasite from destruction by the alternative complement pathway (non-specific immunity).

The VSG GPI anchor structure (Ferguson et al, 1988) distinguishes itself from all other known examples in two ways: (i) it contains exclusively sn-1,2-dimyristyl-glycerol in its Pl moiety and (ii) it contains a complex α-Gal branch in its glycan moiety (Figure 1).

(i) The dimyristyl-Pl moiety

The functional significance of the exclusively dimyristyl-Pl moiety is unknown, and has been the cause of much angst among Tryp anchor researchers. Presumably it has some inherent advantage to the trypanosome since the dimyristyl-Pl feature is common to several (and probably all) species of African trypanosomes (Lamont et al, 1986). The fatty acid remodelling steps, which result in the formation of the dimyristyl-Pl unit (Figure 3; Masterson et al, 1990), involves at least 4 enzymes, the expense of ATP, and the scavenging of relatively low abundance myristic acid. These features indicate that the selective advantage of the remodelling must be quite strong for the trypanosome. Consistent with this view, it has been shown that myristic acid analogues can be incorporated into VSG but that these compounds are trypanocidal (Doering et al, 1991), suggesting some crucial function for the dimyristyl-Pl unit. Myristic acid is a relatively short (14 carbon) saturated fatty acid and this would result in the anchor lipid being restricted to the outer leaflet of the membrane bilayer (c.f. other anchors which have alkyl or acyl chains of up to 26 carbons). Similarly the relatively low total lipid chain length of the VSG anchor (28 carbons versus an average of about 40 in other anchors) may have some significance. It is tempting to

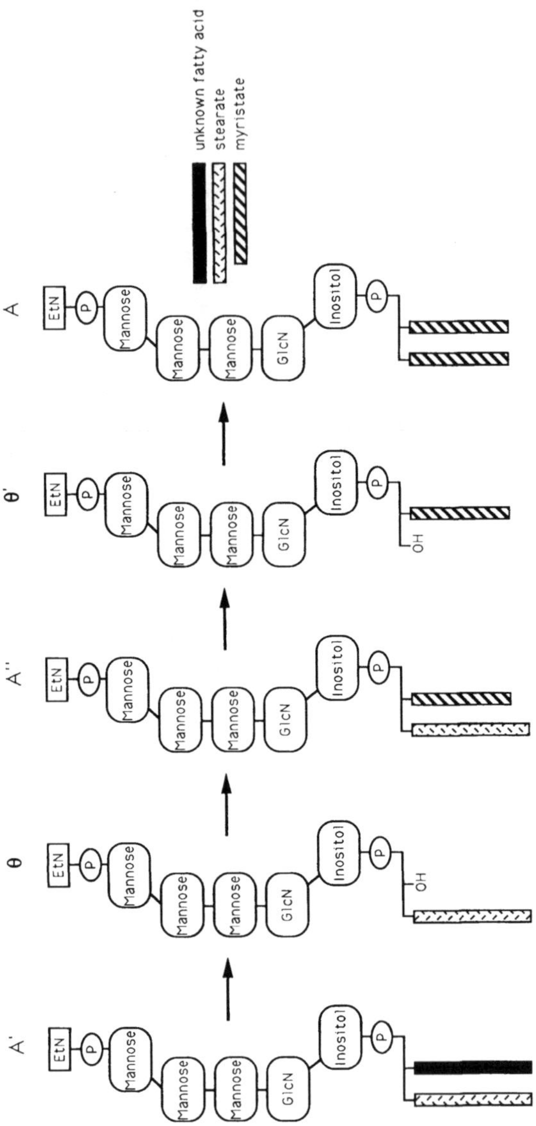

Figure 3. Fatty acid remodelling of the GPI precursor in *T. brucei*. The African trypanosomes express exclusively dimyristyl-GPI anchors on their VSG coat glycoproteins. This appears to be an African trypanosome-specific phenomenon and it arises through a complex series of remodelling reactions involving phospholipases and acyl-transferases. All of the intermediate species shown have been identified and characterized (except for θ), Masterson et al (1990).

speculate that the unusual lipid content has something to do with allowing the very close packing of the VSG coat molecules on the plasma membrane, or the transport vesicles which take pre-assembled VSG coat to the surface. If this were the case, one could argue that the trypanosome fatty acid remodelling pathway may have evolved to enable trypanosomes to assemble a tightly packed protein coat, and therefore to become competent for survival in the animals upon which their insect hosts were feeding.

(ii) The α-Gal sidechain

The α-Gal sidechain of the VSG GPI anchor is an unusual structure, containing glycosidic linkages which are either rare in nature (i.e. Galα1-2Gal and Galα1-6Gal) or unique (i.e. Galα1-3Man). The α-Gal branch appears to be restricted to the African trypanosome; it is not even found in the GPI anchors of the related parasites *L.major* (Schneider et al, 1990) and *T.cruzi* (Guther et al, unpublished data). Why have the African (VSG coated) trypanosomes evolved unilaterally the processing machinery (at least 3 distinct α-galactosyltransferases) to assemble this α-Gal sidechain? It is not clear from the primary structure what function this appendage might have. It can not be involved in anchorage per se since the linkage of the VSG polypeptide to the lipid moiety is via the trimannosyl core. Three dimensional modelling studies of the VSG variant MlTat1.4 anchor (Homans et al, 1989) indicate that the GPI glycan may adopt a plate-like conformation, lying parallel to the plane of the membrane, and covering a significant area ($6nm^2$) of plasma membrane. About half of this cross sectional is due to the α-Gal branch. This may be quite significant since the widest cross sectional area of the VSG polypeptide N-terminal domain is very similar (Metcalf et al, 1987). Thus the α-Gal branch may assist in ensuring the macromolecular diffusion barrier characteristics of the VSG coat by occupying space very close to the plasma membrane. Two other lines of evidence support this hypothesis:

(i) The extent of VSG GPI α-galactosylation correlates with VSG sub-class, which in turn is defined by C-terminal domain homology groups (Ferguson and Homans, 1989). This suggests that the extent of GPI galactosylation is dependent on the three dimensional structure of the C-terminal polypeptide domain to which it is attached.

(ii) Kinetic studies show that GPI α-galactosylation occurs when the VSG is passing through the Golgi apparatus (Bangs et al, 1988). This may indicate that the α-Gal transferases add residues at the time that the VSG molecules are being packaged into coat arrays for transport to the cell surface (in the trans-Golgi network).

Taken together, these findings could be consistent with the notion that the α-Gal transferases act as spatial probes, filling space close to the membrane according to the intra- and inter-molecular steric constraints imposed by the three dimensional structure of the VSG C-terminal domains. Thus it is conceivable that the α-galactosylation machinery of African trypanosomes may have evolved specifically to optimise the packing of different VSG coats, and therefore increase viability in the presence of host serum complement.

b. The LPGs and GIPLs of *Leishmania*

(i) The LPGs

The structures of *Leishmania* LPGs are described in the chapter by McConville. These molecules are expressed in high copy number on the surface of the promastigote (insect dwelling) stage. The LPGs are either undetectable (McConville and Blackwell, 1991) or present in very low levels in the amastigote (mammalian) stage (Glaser et al, 1991; Moody et al, unpublished data), suggesting that they have evolved primarily for survival in the insect vector. Recent studies on infected sandflies suggest that surface LPG may be important in mediating the attachment of promastigotes to epithelial cells in the sandfly midgut (Davies et al, 1990). In the case of *L.major*, attachment to the midgut appears to involve the ß-galactose side chains of LPG and receptors on the epithelial cells (Sacks et al, unpublished results). As the promastigotes become metacyclic (infective stage) these side chains are no longer expressed on the LPG or become cryptic (McConville et al, unpublished results), which may allow the promastigotes to detach from the gut wall and migrate to the foregut or pharynx of the sandfly (Davies et al, 1990). Similar side chains are not present in the LPGs of *L.donovani* or *L.Mexicana* (see chapter by McConville), suggesting that this type of interaction may be specific to *L.major*. However, other parts of the LPG are highly conserved in the different species (see the chapter by McConville). In particular, all the LPGs contain a novel lyso alkyl-Pl lipid moiety, where the alkyl chains are mainly C24:0 or C26:0. This type of anchor may ensure that the LPG is only weakly associated with the cell surface; a significant proportion of the LPG, which still contains a lipid anchor, is lost from promastigotes in culture and in the sandfly gut (Davies et al 1990, Ilg et al 1991). The function of this extracellular LPG is unknown, although it appears to insert into the membranes of cells lining the gut wall (Davies et al, 1990) and may further modulate the interaction of the promastigotes with these cells.

LPG also appears to be essential for metacyclic promastigote infectivity in the mammalian host; variant strains which lack LPG are unable to survive in normally susceptible macrophages (Handman et al, 1985; Elhay et al, 1989; McNeely and Turco 1990). As discussed in the chapter by McConville, LPG is thought to be involved in:

(i) Preventing complement-mediated lysis (Puentes et al, 1990).

(ii) Mediating the attachment of promastigotes to macrophages and their uptake into the phagolysosome (Handman and Goding, 1985; Da Silva et al, 1989).

(iii) Protecting the parasite from hydrolytic enzymes and the effects of the oxidative burst (Eilam et al, 1985; Chan et al, 1989; McNeely and Turco 1990).

The structural features of LPG which are associated with these different functions are beginning to be determined, particularly with regard to the dual role of LPG in complement activation (which appears to be essential for mediating promastigote uptake via the complement receptors on the macrophage) and complement evasion. The role of LPG in complement evasion has been studied in log phase (complement susceptible) and metacyclic (complement resistant) promastigotes. In both forms of parasite, LPG is the main activator of complement, leading to the deposition of similar amounts of C3 on the parasite surface (Puentes et al, 1988). The lytic components, C5b-9, are also deposited on the cell surface (Puentes et al, 1990). This complex inserts into the membrane of log phase cells, causing cell lysis, but is shed from the surface of metacyclic promastigotes without causing cell death. The failure of C5b-9 to insert into the membrane of metacylic promastigotes is associated with an increase in the size of the LPG (Sacks et al, 1990), suggesting that LPG may sterically hinder access of the C5b-9 complex to the plasma membrane. This is consistent with molecular modelling studies which suggest that LPG will extend a considerable distance from the cell surface and have a large cross-sectional area (see the chapter by McConville and Homans et al, 1991).

Although there is no information on the site of complement deposition on LPG it may be significant that all the LPG chains are capped by an α-Man oligosaccharide (McConville et al, 1990b). We have proposed that these cap structures are recognized by serum mannose-binding proteins, which are known to activate complement by both the alternative and classical pathways. Activation of complement at the end of the LPG chains would further reduce the probability of the lytic components being inserted into the membrane and would ensure that earlier components of the complement cascade are accessible to macrophage receptors. Finally, shedding of the C5b-9 complex from the cell surface may be facilitated by the loose binding of LPG to the plasma membrane. Thus it is likely that LPG chains containing a large covalently bound protein complex would be selectively shed from the cell surface. These macrophage infectivity functions of LPG, associated with its structural changes upon meta-cyclogenesis, may represent one of the parasites evolutionary steps towards colonising mammals.

(ii) The GIPLs

The GIPLs are also major components on the parasite cell surface and, unlike the LPGs, are expressed in high copy number on both the promastigote and amastigote stages. While the function of these surface glycolipids is unknown, they may be analogous to the surface glycosphingolipids of higher eukaryotes, and have a role in protecting the outer bilayer of the plasma membrane and in mediating cell-cell interactions. This latter function may be particularly important for the amastigote stage where these glycolipids are likely to constitute the major surface components (McConville and Blackwell 1991).It is of interest that structurally distinct GIPL species are produced by different species. In *L.major*, the major GIPL species are structurally similar to the LPG anchor (series 1) while in *L. donovani* the major GIPL species are more analogous to the protein anchors (series 2) (McConville et al 1990a, McConville and Blackwell 1991, see the chapter by McConville for structures). These results suggest that *Leishmania* have adapted the biosynthetic machinery for making the LPG and protein anchors to make their major glycolipid species. Whether the expansion of different enzyme systems in different species is significant in terms of their function is unknown.

CONCLUSIONS

The GPI anchors of plasma membrane proteins are ubiquitous throughout the eukaryotes, and higher eukaryotes appear to have adopted them for certain specific tasks related to multicellular existence. The protozoa appear to have selected GPI anchors as the most useful form of protein anchorage, and the kinetoplastid parasites have evolved radical variations on the conserved GPI biosynthetic pathway to produce a bewildering array of exotic and unique structures. There is evidence that several of these parasite-specific adaptations are intimately involved in their ability to infect both their insect vectors and mammalian hosts. We postulate that the evolution of these novel GPI metabolic pathways may have occurred in parallel with the evolution of these organisms to occupy new ecological niches as mono- and di-genetic parasites.

ACKNOWLEDGEMENT
We thank The Wellcome Trust for support. MJM is a C.J. Martin research fellow.

REFERENCES

Bülow, R., Nonnengasser, C. and Overath, P. (1989). Mol. Biochem. Parasit. **32:** 85-92.

Chan, J., Fujiwara, T., Brennan, P., McNeil, M., Turco, S. J., Sibille, J. C., Snapper, M., Aisen, P. and Bloom, B. R. (1989). Proc. Natl. Acad. Sci. USA **86:** 2453-2457.

Clayton, C. E. and Mowatt, M. R. (1989). J.Biol.Chem. **264(25):** 15088-15093.

Conzelmann, A., Frankhauser, C. and Desponds, C. (1990). EMBO J. **9:** 653-661.

Cross, G. A. M. (1990). Ann. Rev. Cell Biol. **6:** 1-39.

Da Silva, R. P., Hall, B. F., Joiner, K. A. and Sacks, D. L. (1989). J.Immunol. **143:** 617-622.

Davies, C. R., Cooper, A. M., Peacock, C., Lane, R. P. and Blackwell, J. M. (1990). Parasitol, **101:** 337-343.

Degasperi, R., Thomas, L. J., Sugiyama, E., Chang, H. M., Beck, P. J., Orlean, P., Albright, C., Waneck, G., Sambrook, J. F., Warren, C. D. and Yeh, E. T. H. (1990). Science **250:** 988-991.

Dotti, C. G., Parton, R. G. and Simons. K. (1991). Nature **349:** 158-161.

Eilam, Y., El-On, J. and Spira, D. T. (1985). Exp. Parasitol. **59:** 161-168.

Elhay, M. J., Kelleher, M., Bacic, A., McConville, M. J., Tolsen, D. L., Pearson, T. W. and Handman, E. (1989). Mol. Biochem. Parasitol. **40:** 167-174.

Ferguson, M. A. J., Homans, S. W., Dwek, R. A. and Rademacher, T. W. (1988). Science **239:** 753-759.

Ferguson, M. A. J. and Homans, S. W. (1989). in New Strategies in Parasitology, McAdam,K.P.W. ed., Churchill Livingstone Press, U.K. pp 120-139.

Ferguson, M. A. J. (1991). Current Opinion in Structural Biology **1(4):** in press.

Field, M. C., Menon, A. K. and Cross, G. A. M. (1991). J.Biol. Chem **266:** in press.

Glaser, T. A., Moody, S. F., Handman, E., Bacic, A. and Spithill, T. W. (1991). Mol. Biochem. Parasitol. **45:** 337-345.

Guther, M. L. S., Cardoso de Almeida, M. L., Yoshida, N. and Ferguson, M. A. J. (1991). Unpublished data.

Handman, E. and Goding, J. W. (1985). EMBO J. **4:** 329-336.

Handman, E., Schnur, L. F., Spithill, T. W. and Mitchell, G. F. (1986). J.Immunol. **137:** 3608-3613.

Homans, S. W., Edge, C. J., Ferguson, M. A. J., Dwek, R. A. and Rademacher, T. W. (1989). Biochemistry **28:** 2881-2887.

Homans, S. W., Mehlert, A. M. and Turco, S. W. (1991). Unpublished data.

Ilg, T., Etges, R., Overath, P., Thomas-Oates, M. J., McConville, M. J. Romans, S. W. and Ferguson, M. A. J. (1990). Unpublished data.

Lamont. G. S., Fox, J. A. and Cross, G. A. M. (1987). Mol. Biochem. Parasitol. **24:** 131-136.

Lederkremer, R. M. de., Lima, C., Ramirez, M. I. and Casal, O. L. (1990). Eur. J.Biochem. **197:** 337-345.

Lemansky, P., Fatemi, S. H., Gorican, B., Meyale, S., Rossero, R. and Tartakoff, A. M. (1990). J.Biol.Chem. **110:** 1525-1531.

Lisanti, M. P., Caras, I. W., Gilbert, T., Hanzel, D. and Rodríguez-Boulan, E. (1990). Proc.Natl.Acad.Sci. USA **87:** 7419-7423.

Low, M. G. (1989). Biochim.Biophys. Acta. **988:** 427-454.

Maudlin, I. and Welburn, S. C. (1987). Trop. Med. Parasit. **38:** 172-175.

Masterson, W. J., Raper, J., Doering, T. L., Hart, G. W. and Englund, P. T. (1990). Cell **62:** 73-80.

Metcalf, P., Blum, M., Freymann, D., Turner, M. and Wiley, D. C. (1987). Nature **325:** 84-86.

McConville, M. J., Homans, S. W., Thomas-Oates, J. E., Dell, A. and Bacic, A. (1990a). J. Biol. Chem. **265:** 7385-7394.

McConville, M. J., Thomas-Oates, J. E., Ferguson, M. A. J. and Homans, S. W. (1990b). J. Biol. Chem. **265:** 19611-19623.

McNeely, T. B., and Turco, S. J. (1990). J. Immunol. 3099-3106.

Previato, J.O., Gorin, P. A. J., Mazurek, M., Xavier, M. T., Fournet, B., Wieruszesk, J. M. and Mendonca-Previato, L. (1990). J. Biol. Chem. **265:** 2518-2526.

Puentes, S. M., Sacks, D. L., Da Silva, R. P. and Joiner, K. (1988). J. Exp. Med. **167:** 887-902.

Puentes, S. M., Da Silva, R. P., Sacks, D. L., Hammer, C. H. and Joiner, K. A. (1990). Immunol. **145:** 4311-4316.

Robinson, P. J. (1991). Immunol. Today. **12:** 35-41.

Roditi, I., Carrington, M. and Turner, M. (1987). Nature **325:** 272-274.

Sacks, D. L., Brodin, T. N. and Turco, S. J. (1990). Mol. Biochem. Parasitol. **42:** 225-234.

Schneider, P., Ferguson, M. A. J., McConville, M. J., Mehlert, A., Homans, S. W. and Bordier, C. (1990). J. Biol. Chem. **265:** 16955-16964.

Thomas, J. R., Dwek, R. A. and Rademacher, T. W. (1990). Biochemistry **29:** 5413-5422.

Zaretskaia, M. S. H., Khorokhorina, V. A. and Sukhareva, N. N. (1989). Isv. Akad. Nauk. SSSR. Biol. **2:** 267-273.

BIOSYNTHESIS OF GLYCOSYL-PHOSPHATIDYLINOSITOL

***[1]Menon, Anant K.**

**Laboratory of Molecular Parasitology,
The Rockefeller University, 1230 York Avenue,
New York, N.Y. 10021, USA.**

* Corresponding author.

INTRODUCTION

Numerous proteins from eukaryotic organisms are anchored to cell membranes by glycosyl-phosphatidylinositol (GPI) (see reviews by Ferguson and Williams, 1988, Low, 1989, Thomas et al., 1990, Doering et al., 1990, Cross, 1990). Structural features of GPIs are discussed in detail in other articles in this volume.

Current models suggest that GPI anchors are synthesized as precursor glycolipids which are then covalently coupled to target proteins soon (1-5 min) after completion of protein synthesis, presumably after translocation of the polypeptide into the lumen of the endoplasmic reticulum (ER). In all cases the glycolipid anchor replaces a short run of mainly hydrophobic amino acids at the carboxy terminus of the protein (see article by Caras in this volume). Candidate GPI anchor precursors have been identified in African trypanosomes (Krakow et al., 1986, 1989, Menon et al., 1988a, Mayor et al., 1990a and b, Field et al., 1991a) and in mammalian cells (DeGasperi et al., 1990), and the transfer of structurally characterized precursors to protein has been demonstrated in vitro (Mayor et al., 1991). This article is concerned with the biochemical and membrane topological sequence of reactions involved in the assembly of the anchor precursors. In addition to providing a summary of published work on GPI biosynthesis, the article aims to describe some new results and introduce a model of the anchor assembly pathway.

GPI ANCHOR PRECURSORS IN TRYPANOSOMES

African trypanosomes have unique advantages as a model system for studying glycolipid anchor biosynthesis. A considerable proportion of the organism's biosynthetic program is dedicated to the synthesis of unusually abundant (10^7 molecules/cell) glycolipid-anchored surface coat proteins, VSG and PARP[1], in two life-cycle stages (bloodstream and procyclic (insect-stage)). Structurally distinct GPI anchor precursors have been isolated from both stages of the trypanosome life cycle (Krakow et al., 1986, 1989, Menon et al., 1988a, Mayor et al., 1990a and b, Field et al., 1991). The lipids (P2 and P3[2] in bloodstream forms, and PPI in procyclic trypanosomes) are GPI species containing a single N-unsubstituted ethanolamine residue, a Man_3GlcN core indistinguishable from the conserved core glycan seen in a variety of GPI-anchored proteins, and a phosphatidylinositol (PI) moiety. As shown in Figure 1, the PI moiety is different in each of the lipids. Unlike P2, both P3 and PP1 contain a fatty acid (palmitic acid, as determined in radiolabelling studies) esterified to the inositol ring, a structural feature that renders these lipids insusceptible to phosphatidylinositol-specific phospholipase C (PI-PLC).

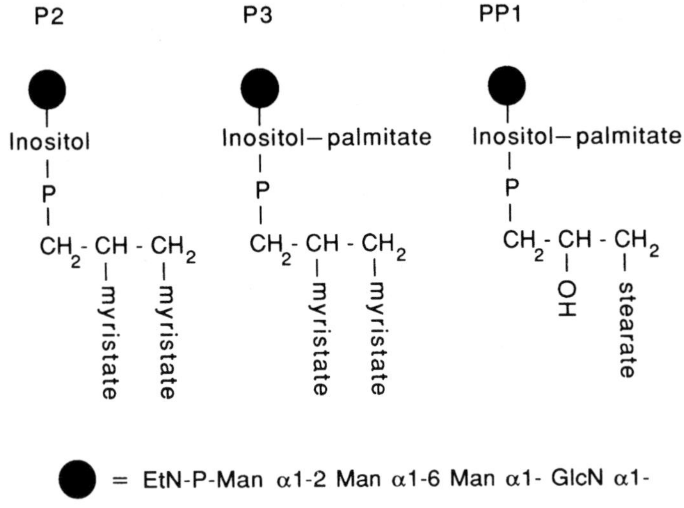

Fig.1. Structures of glycolipid anchor precursors in trypanosomes. The structures are taken from Menon et al. (1988a), Mayor et al. (1990a, b), and Field et al. (1991). P2 and P3 are found in bloodstream trypanosomes, and PP1 is the major [^{3}H]ethanolamine-labelled GPI in procyclic (insect-stage) trypanosomes. All fatty acids shown are in ester linkage. The identification of stearate in PP1, and palmitate in P3 and PP1 is based on metabolic labelling studies with [^{3}H]fatty acids. All three lipids contain the same glycan moiety (represented by the filled circle); the linkages not explicitly described as likely to be EtN-P-6Man, Manα1-4GlcN, and GlcNα-6Inositol (in analogy to the anchors of several proteins where data are available (Thomas et al., 1990)), and the amino group of the ethanolamine residue is unsubstituted. EtN, ethanolamine, P, phosphate, Man, mannose, GlcN, glucosamine.

OUTLINE OF THE BIOSYNTHETIC ROUTE

On inspection of the structures shown in Figure 1, the simplest model for the assembly of GPI precursors would appear to involve sequential addition of the various components (glucosamine, mannose, ethanolamine-phosphate) to appropriate PI molecules. In support of this model, partially glycosylated GPI intermediates containing glucosamine and O-3 mannose residues have been identified in radiolabelling experiments using lysates of bloodstream trypanosomes (Masterson et al., 1989, Menon et al., 1990a). The data indicate that the first committed step in the assembly process is the addition of GlcNAc (from UDP-GlcNAc) to PI, followed by the deacetylation of GlcNAc-PI to give GlcN-PI (Doering et al., 1990). Each of the three mannose residues is derived from dolichol-P-mannose although it is not entirely clear whether these residues are added sequentially, or via a "trimming and addition" pathway (Menon et al., 1990b, Field and Menon, 1991). The terminal ethanolamine residue is derived from phosphatidylethanolamine[3,4].

The essential ingredients of this model have been described in studies of the biosynthesis of the lipid(dolichol)-linked oligosaccharide precursors of N-linked sugars. It is generally believed that the assembly of these structures occurs in the ER and involves the stepwise construction of the oligosaccharide chain on dolichyl-phosphate (dol-P). The first seven sugars (two GlcNAc and five Man residues) are derived from the corresponding nucleotide sugars, UDP-GlcNAc and GDP-Man, while the additional mannose residues and the terminal glucose residues are derived from the lipids dol-P-Man and dol-P-Glc (Kornfeld and Kornfeld, 1985). However, unlike the dolichol pathway, GPI assembly is complicated by the presence of two classes of lipids (PI-PLC sensitive and resistant, depending on inositol acylation). It is not clear whether PI-PLC-sensitive PI-based GPI's and PI-PLC-resistant acyl PI-based GPI's are synthesized via parallel, non-interacting pathways or whether there is a more complex relationship between the two types of lipid. In this regard, GPI assembly may resemble the biosynthesis of the polymannophosphoinositides of *Mycobacteria*, a family of hyperacylated PI-based lipids. Some speculations are given below (see section on inositol acylation).

SELECTION OF THE PI SUBSTRATE FOR GPI ASSEMBLY

The parameters that determine which PI species will be glycosylated are unclear. Data described by Butikofer et al. (1990) indicate that, of the spectrum of PI molecular species available in cells of the electric organ of *Torpedo marmorata*, only a minor subset appears in the GPI anchor of *Torpedo* acetylcholinesterase. There are at least two plausible explanations. One is that only a particular subset of PI molecular species is available to the first GPI glycosyltransferase, the enzyme that transfers GlcNAc from UDP-GlcNAc to PI. This can occur by transporting or confining the relevant PI species to a particular ER membrane domain (such as the inner leaflet of the membrane bilayer) containing the enzyme activity. Circumstantial data support this line of reasoning. The inability to label GPI's via [^{3}H]inositol in trypanosome lysates, despite efficient labelling of PI, suggests that the labelled PI is unable to reach the site of GlcNAc addition - possibly the inner leaflet of the ER - within the time frame (30-60 minutes) of the experiment (Menon et al., 1990a). Curiously, when cultures of procyclic trypanosomes are incubated overnight with [^{3}H]inositol it is not possible to generate [^{3}H]inositol-labelled PPI or PARP, even though PI is labelled extremely efficiently (M.C. field, unpublished observations), indicating that the PI for GPI synthesis cannot rapidly equilibrate with newly synthesized PI.

The second explanation involves fatty acid re-modelling, a feature of the bloodstream trypanosome GPI biosynthetic pathway (Masterson et al., 1989, 1990, Doering et al., 1990) that may be relevant in other systems as well. The glycerol-linked fatty acids in both P2 and P3 (Figure 1) are replaced by myristic acid in the final stage of GPI assembly; the original PI species is thought to contain stearic acid in the 1-position and an as yet undefined fatty acid in the 2-position of the glycerol moiety (Masterson et al., 1990). Similar fatty acid re-modelling reactions might occur in other systems, giving rise to unique molecular species (glycerol fatty acids) in GPI anchors, not representative of the molecular species found in total cellular PI. However, it is difficult to see how this process could occur with ether-linked fatty acids.

ORIGIN OF THE THANOLAMINE-PHOSPHATE BRIDGE BETWEEN GLYCOLIPID AND PROTEIN

As mentioned above, crude preparations of bloodstream trypanosome membranes are capable of synthesizing P2 and P3 (Masterson et al., 1989, Menon et al., 1990a). Not surprisingly, the membrane preparations require supplementation with sugar nucleotides (UDP-GlcNAc or GDP-mannose) for efficient GPI synthesis. However, GPI synthesis in these preparations proceeds efficiently in the absence of exogenously added sources of the terminal ethanolamine-phosphate group such as CDP-ethanolamine (Masterson et al., 1989). Since it is unlikely that significant quantities of CDP-EtN are present in the membrane preparation, the inability of exogenously added

CDP-EtN to influence GPI assembly suggests that the immediate donor of the GPI phosphoethanolamine group is not CDP-EtN.

Three lines of evidence suggest that the terminal ethanolamine residue is derived from a phospholipid donor, phosphatidylethanolamine (PE). The data do not address the issue of whether phosphoethanolamine is transferred as a group, although this is likely to be the case.

[1] Incubation of lysates with CDP-[^{3}H]EtN results in labelling of PE, lyso PE, P2 and P3. However, pulse-chase labelling with CDP-[^{3}H]EtN shows that although P2 is not detectably labelled during the pulse (1 minute), labelling steadily increases over a 15 minute chase period clearly indicating that the immediate precursor of the P2 ethanolamine residue is not CDP-EtN[3]. The data also suggest that the ethanolamine residue is not derived from a small, rapidly metabolized donor pool but rather from a large, relatively stable pool such as that of PE.

[2] Metabolic labelling with [^{3}H]serine can be used to test the hypothesis that PE functions as the ethanolamine donor in GPI biosynthesis. This labelling strategy permits labelling of phosphatidylserine (PS), and therefore PE (via decarboxylation of PS, Kennedy, 1961), without simultaneously generating any other labelled ethanolamine metabolites such as CDP-EtN. When lipids extracts from [^{3}H]serine-labelled trypanosomes were analyzed by thin layer chromatography, they were found to contain labelled PS, PE, P2, and P3; the radioactivity in PE, P2, and P3 was confirmed to be ethanolamine after acid hydrolysis[3]. Since no other radiolabelled ethanolamine metabolites are generated in the experiment the results indicate the involvement of PE in GPI biosynthesis.

[3] Yeast *Saccharomyces cerevisiae* are known to synthesize GPI-anchored proteins (Conzelmann et al., 1988, Conzelmann et al., 1990); the yeast GPI anchors contain the conserved core glycan sequence with an extra mannose residue branching of the phosphoethanolamine-linked terminal mannose. A yeast mutant generated by disruption of the non-essential ethanolaminephosphotransferase *(EPTI)* gene (Hjelmstad and Bell, 1988) is defective in transferring phosphoethanolamine from CDP-ethanolamine to diacylglycerol to form PE. The mutant displays some activity, however, because the same reaction can be inefficiently catalyzed by the cholinephosphotransferase gene *(CPTI)* product in vivo (Hjelmstad and Bell, 1987, 1990)[4]. A double mutant *(eptl, cptl)*, with both genes disrupted, has enormously reduced activity (1%) compared to wild-type, and cannot synthesize radiolabelled PE from [^{3}H]EtN. Nevertheless, both single *(eptl)* and double *(eptl,*

cpt1) mutants have a normal phospholipid composition (since they generate PE from PS, and PC from PE, both are capable of synthesizing phosphoethanolamine and CDP-EtN as efficiently as wild-type (determined by [^{3}H]EtN labelling), and both synthesize GPI-anchored proteins (as judged by [^{3}H]inositol labelling)[4]. Coincident with their inability to synthesize labelled PE from [^{3}H]EtN, neither mutant is capable of efficiently incorporating [^{3}H]EtN into protein (protein labelling could be seen faintly in the *ept1* strain and not at all in the double mutant)[4].

INOSITOL ACYLATION

P3 and PP1 (Figure 1), as well as the GPI anchors of PARP (Clayton and Mowatt, 1989)[5], human erythrocyte acetycholinesterase (Roberts et al., 1988) and DAF (Walter et al., 1990), contain acylated inositol. Partially glycosylated GPI intermediates containing acyl inositol have also been identified, e.g. Man_{2-3}GlcN-acyl PI (Masterson et al., 1989, Menon et al., 1990a), and indicates that inositol acylation occurs early in the GPI assembly process. Available data suggest that inositol acylation occurs after the formation of GlcN-PI, before mannose addition.

Both PPI and PARP contain one glycerol-linked fatty acid, stearic acid esterified to the glycerol 1-position, and one inositol-linked fatty acid, palmitic acid, esterified to an inositol hydroxyl (Figure 1). Since procyclic trypanosomes are capable of elongating fatty acids, metabolic labelling with [^{3}H]palmitic acid results in labelling of both inositol and glycerol-linked fatty acids in PPI and PARP. The two fatty acids can be separated by GPI-specific phospholipase D (GPI-PLD) cleavage of the GPI structure, and their identities verified after conversion to the corresponding methyl esters (Field et al., 1991)[5].

The obvious candidate for a donor of the inositol-linked fatty acid in GPI's is fatty acyl Coenzyme A; palmitoyl CoA is the donor of the inositol and mannose-linked fatty acids in related glycosylated PI's, the polymannosphosphoinositides of *Mycobacterium phlei* (Ballou, 1972). However, it is possible that the inositol-linked fatty acid in GPI's is derived from a more complex reaction such as a transacylation reaction involving a phospholipid donor (Menon et al., 1990a). The two classes of donor can be distinguished in pulse-chase experiments with [^{3}H]palmitic acid. Analyses of protein in procyclic trypanosomes pulse-labelled for 2-4 hours and "chased" for 0-15 hours clearly show continued incorporation of radioactivity into PARP through the chase period[5]. This is not entirely surprising since the PI labelled during the pulse period would continue to be incorporated into PARP during the chase with a very gradual decrease in specific activity. However, fatty acid analyses clearly show that both [^{3}H]stearate and [^{3}H]palmitate in PARP increase during the chase[5], indicating that both the fatty acids are derived from a precursor pool labelled during the pulse and not drastically affected by the chase. In particular, the data indicate that the inositol-linked fatty acid is derived from a large, relatively stable precursor pool such as phospholipid, rather than from fatty acyl CoA.

It has been proposed that inositol acylation is chemically important in GPI biosynthesis (Menon et al., 1990a). The stereochemical constraint provided by inositol acylation may make inositol acylated lipids more appropriate substrates for glycosyltransferases than their non-inositol-acylated counterparts. Subsequent interconversion between the two forms may occur via inositol acylating and deacylating enzymes. Formally similar structures and interconversion reactions have been described in mycobacteria. *Mycobacterium phlei* contains a family of phosphatidylinositol dimannosides that occur in multiply acylated forms, with two fatty acids contained in the diacylglycerol moiety of the phospholipid and additional fatty acids esterified to available hydroxyls on mannose and myoinositol (Ballou, 1972, Brennan and Ballou, 1967, 1968, Prottey and Ballou, 1968). The additional fatty acids are added only after the formation of a dimannophosphoinositide from GDP-mannose and phosphatidylinositol. The various acylated forms of the phosphatidylinositol dimannosides can be enzymatically interconverted and this may be an important feature of the assembly pathway. These complex structures are precursors of the higher mycobacterial polymannophosphoinositides, including the pentamannoside and hexamannoside. It is possible that hyperacylation is an essential stereochemical prerequisite for continued mannosylation.

MEMBRANE TOPOLOGY OF GPI ASSEMBLY

The proposal that GPI anchors are assembled via the sequential glycosylation of PI, or hyperacylated PI, or via an interactive process involving both PI and acyl PI species, is complicated by membrane topological issues. The end products of the assembly process are presumed to be associated with the luminal leaflet of the ER in order to be transferred to a newly synthesized, glycosylated polypeptide. However, it is highly probable that the starting material, PI, is synthesized from CDP-diacylglycerol and inositol on the cytoplasmic face of the ER (Bell et al., 1981) suggesting that at some stage of the assembly process a cytosolically disposed lipid intermediate flips to the luminal face of the ER. There are several other indications that GPI assembly is likely to involve flipping of intermediates between the cytoplasmic and luminal leaflets of the ER accompanied by the transport of cytosolically synthesized precursors into the lumen of the ER. The three components required for the synthesis of GPI anchors from PI are glucosamine, mannose, and phosphoethanolamine. Activated forms of all three, UDP-GlcNAc, GDP-Man, and CDP-EtN, are synthesized in the cytoplasm and of these apparently only UDP-GlcNAc can be transported into the ER (Hirschberg and Snider, 1987). The synthesis of dolichol-P-mannose from cytosolic GDP-Man followed by (protein-mediated) partitioning of the lipid into the luminal leaflet (Haselbeck and Tanner, 1982) provides a source of mannose on both sides of the ER. It may be appropriate to postulate a similar translocation and distribution scheme for ethanolamine since the GPI ethanolamine donor, PE, is available in both leaflets of the ER (Higgins and Pigott, 1982).

TRANSFER OF GPI PRECURSORS TO PROTEIN

Membranes prepared from bloodstream trypanosomes contain a small amount of unprocessed (non-glycolipid-anchored) VSG (approximately 3600 molecules/cell equivalent, or 60 fmol per assay (10^7 cell equivalents) which can accept glycolipid precursors synthesized in situ or added exogenously (Mayor et al., 1991). Thus, in the course of a standard labelling incubation with GDP-[^{3}H]mannose, in the presence of tunicamycin to suppress the dolichol pathway of N-linked glycosylation, P2, P3 and VSG are labelled; the label incorporated into VSG is exclusively associated with the Man_3GlcN core GPI glycan structure (Mayor et al., 1991). Although both radiolabelled P2 and P3 are synthesized by the membranes, the VSG that is labelled is exclusively PI-PLC-sensitive, indicating that only P2 is transferred to the protein. However, when radiochemically pure glycolipids are added to membranes, a spectrum of lipids including P2, P3, and PP1 can be transferred to VSG acceptors indicating that all the lipids are bona fide GPI anchor precursors. The reaction does not require ongoing protein translation and its efficiency is undiminished in the presence or absence of non-hydrolyzable ATP analogues or an ATP regenerating system (Mayor et al., 1991), consistent with a transpeptidation reaction mechanism (see article by Udenfriend in this volume). Curiously, in the experiments with P3, the labelled VSG was found to be PI-PLC-resistant (Mayor et al., 1991) suggesting a lack of specificity under the assay conditions that is not reflected in living cells or when the precursors were synthesized in situ in membranes. In these situations P2 may have a competitive advantage over P3 resulting in exclusively PI-PLC-sensitive VSG. Alternatively, both lipids may be transferred to VSG acceptors but P3-anchored VSG may be rapidly inositol-deacylated to give PI-PLC-sensitive VSG.

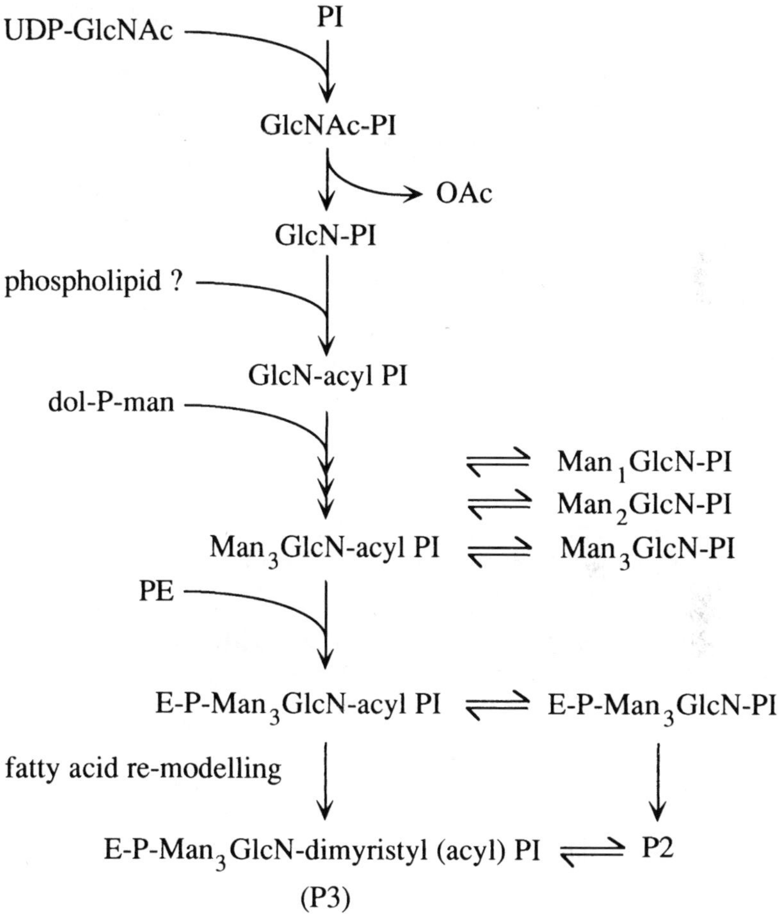

Fig. 2. A model for GPI biosynthesis. The figure shows a scheme for the biosynthesis of P2 and P3 (see Figure 1) in bloodstream trypanosomes. PI is synthesized on the cytoplasmic face of the ER and translocated into the luminal leaflet. It is proposed that all subsequent steps occur in the luminal leaflet, that GPI assembly occurs via inositol-acylated species after the synthesis of GlcN-PI, and that all inositol acylated species can interconvert with their non-inositol acylated counterparts.. The donors of the GlcN, Man, and ethanolamine residues in P2 and P3 are UDP-GlcNAc, Dol-P-Man, and PE respectively. It is likely that ethanolamine-phosphate is transferred as a group. The donor of the inositol-linked fatty acid is not fatty acyl CoA, and may be a phospholipid.

A MODEL FOR GPI ANCHOR BIOSYNTHESIS

A model biosynthetic scheme is illustrated in Figure 2. As indicated in the section on membrane topology, all the precursors for the different GPI components are available on both faces of the ER membrane bilayer. The simplest topological model that can be envisaged involves PI synthesis on the cytoplasmic face of the ER, followed by protein-mediated translocation of the phospholipid to the luminal face (Higgins et al., 1989, Bell et al., 1981). All subsequent assembly steps can then occur in the luminal leaflet to generate an anchor precursor appropriately located for transfer to newly translated and translocated target proteins. Enlarging on material presented in the section on inositol acylation, it can be proposed that inositol acylation occurs immediately following the synthesis of GlcN-PI to give GlcN-acyl PI, the substrate for mannose addition. At every stage of mannosylation, addition occurs to an inositol acylated lipid substrate which can then interconvert/equilibrate via inositol deacylase and acylase activities with its non-inositol-acylated counterpart. Phosphoethanolamine is then transferred from PE to complete the structure. The addition of GPI to protein is likely to be tightly coupled to the translocation of polypeptide across the ER membrane and may involve aspects of the translocation machinery.

Many aspects of this model remain to be tested: no data are currently available on the transbilayer distribution of individual biosynthetic steps, the precise role of inositol acylation and the relationship between PI- and acyl PI-linked lipids, the enzymology and control of GPI biosynthesis, and topological and mechanistic details of the glycolipid:protein addition reaction. Future experimentation is likely to depend on both biochemical and genetic approaches. The currently available spectrum of GPI biosynthesis mutants in a variety of mammalian cell types (Yeh et al., 1988, Hyman, 1988, Hollander et al., 1988, Schubert et al., 1990) offer numerous opportunities for experimentation.

ACKNOWLEDGEMENTS

This work was supported by NIH grant A121531 to George A M Cross, and by an award from the Irma T Hirschl Trust to the author. The author is pleased to acknowledge the contributions of M Eppinger and R T Schwarz (Philipps-Universitaet Marburg), G A M Cross, M C Field, S Mayor, and M R Rifkin (The Rockefeller University), and V Stevens (Emory University) to the work described in this article, and would like to thank Robin Cooper for comments on the manuscript, and B Dylan, I Richardson, and I Tomar for stimulation.

Footnotes

[1] The following abbreviations are used in the text: ATP, adenosine 5' -triphosphate, CDP-EtN, cytidine 5' -diphosphoethanolamine, CoA, coenzyme A, CPT, cholinephosphotransferase, DAF, decay accelerating factor, dol-P, dolichyl phosphate, EPT, ethanolaminephosphotransferase, ER, endoplasmic reticulum, EtN, ethanolamine, GDP-Man, guanosine 5' -diphosphomannose, Glc, glucose, GlcN, glucosamine, GlcNAc, N-acetylglucosamine, GPI, glycosylphosphatidylinositol, GPI-PLD, GPI-specific phospholipase D, Man, mannose, PARP, procyclic acidic repetitive protein (also known as procyclin), PC, phosphatidyl-choline, PE, phosphatidylethanolamine, PI, phosphatidylinositol, PIPLC, phosphatidylinositol-specific phospholipase C, PS, phosphatidylserine, VSG, variant surface glycoprotein, UDP-GlcNAc, uridine 5' -diphospho-N-acetylglucosamine.

[2] P2 and P3 are also known as lipids A and C respectively (Krakow, et al., 1986).

[3] A. K. Menon, S. Mayor, M. Eppinger, R. T. Schwarz, manuscript in preparation.

[4] A. K. Menon, V. L. Stevens, manuscript in preparation.

[5] M. C. Field, A. K. Menon, G. A. M. Cross, manuscript submitted.

REFERENCES

Ballou, C. E. (1972). Biosynthesis of mannosphosphoinositides in *Mycobacterium pheli*. Methods Enzymol. **28:** 493-500.

Bell, R. M., Ballas, L. M. and Coleman, R. A. (1981). Lipid topogenesis. J. Lipid Res. **22:** 391-403.

Brennan, P. and Ballou, C. E. (1967). Biosynthesis of mannosphosphoinositides by *Mycobacterium phlei*: the family of dimannosphosphoinositides. J. Biol. Chem. **242:** 3046-3056.

Brennan, P. and Ballou, C. E. (1968) Biosynthesis of mannophosphoinositides by *Mycobacterium phlei*: enzymatic acylation of the dimannophosphoinositides. J. Biol. Chem. **243:** 2975-2984.

Butikofer, P., Kuypers, F. A., Shackleton, C., Brodbeck, U. and Stieger, S. (1990) Molecular species analysis of the glycosylphosphatidylinositol anchor of *Torpedo marmorata* acetylcholinesterase. J. Biol. Chem. **265:** 18983-18987.

Conzelmann, A., Riezman, H., Desponds, C. and Bron, C. (1988) A major 125-kD membrane glycoprotein of *Saccharomyces cerevisiae* is attached to the lipid bilayer through an inositol-containing phospholipid. EMBO J. **7:** 2233-2240.

Conzelmann, A., Fankhauser, C. and Desponds, C. (1990). Myoinositol gets incorporated into numerous membrane glycoproteins of *Saccharomyces cerevisiae*; incorporation is dependent on phosphomannomutase (SEC53). EMBO J. **9:** 653-661.

Cross G. A. M. (1990). Glycolipid anchoring of plasma membrane proteins. Ann. Rev. Cell Biol. **6:** 1-39.

Clayton, C. E. and Mowatt, M. R. (1989). The procyclic acidic repetitive proteins of *Trypanosoma brucei*: purification and post-translational modification. J. Biol. Chem. **264:** 15088-15093.

Dahl, C. E. and Dahl, J. S. (1984). Phospholipids as acyl donors to membrane proteins of *Mycoplasma capricolum*. J. Biol. Chem. **259:** 10771-10776.

DeGasperi, R., Thomas. L. J., Sugiyama, E., Chang, H. M., Beck, P. J., Orlean, P., Albright, C., Waneck, G., Sambrook, J. F., Warren, C. D. and Yeh, E. T. H. (1990). Correction of defect in mammalian GPI anchor biosynthesis by a transfected yeast gene. Science **250:** 988-991.

Doering, T. L., Masterson, W. J., Englund, P. T., and Hart, G. W. (1989). Biosynthesis of the glycosyl-phosphatidylinositol membrane anchor of the trypanosome variant surface glycoprotein: origin of the non-acetylated glucosamine. J. Biol. Chem. **264:** 11168-11173.

Doering, T. L., Masterson, W. J., Hart, G. W. and Englund, P. T. (1990). Biosynthesis of glycosyl-phosphatidylinositol membrane anchors. J. Biol. Chem. **265:** 611-614.

Ferguson, M. A. J. and Williams, A. F. (1988). Cell-surface anchoring of proteins via glycosyl-phosphatidylinositol structures. Ann. Rev. Biochem. **57:** 285-320.

Field, M. C. and Menon, A. K. (1991). Biosynthesis of glycolipid anchors in *Trypanosoma brucei*. Trends Glycosci. Glycotech. **3:** 107-115.

Field, M. C., Menon, A. K. and Cross, G. A. M. (1991). Developmental variation of glycosyl-phosphatidylinositol membrane anchors in *Trypanosoma brucei*: identification of candidate biosynthetic precursor for the GPI-anchor of the major procyclic stage surface glycoprotein. J. Biol. Chem. **266:** in press.

Haselbeck, A. and Tanner, W. (1982). Dolichyl phosphate-mediated mannosyl transfer through liposomal membranes. Proc. Natl. Acad. Sci. U.S.A. **79:** 1520-1524.

Higgins, J. A., Hitchin, B.W. and Low, M. G. (1989). Phosphatidylinositol-specific phospholipase C of *Bacillus thuringiensis* as a probe for the distribution of phosphatidylinositol in hepatocyte membranes. Biochem. J. **259:** 913-916.

Higgins, J. A. and Pigott, C. A. (1982). Asymmetric distribution of phosphatidyl-ethanolamine in the endoplasmic reticulum demonstrated using trinitrobenzene-sulphonic acid as a probe. Biochim. Biophys. Acta **693:** 151-158.

Hirschberg, C.B. and Snider, M. (1987). Topography of glycosylation in the rough endoplasmic reticulum and Golgi apparatus. Ann. Rev. Biochem. **56:** 63-87.

Hjelmstad, R. H. and Bell, R. M. (1987). Mutants of *Saccharomyces cerevisiae* defective in *sn*-1,2-diacylglycerol cholinephosphotransferase: isolation, characterization, and cloning of the *CPT1* gene. J. Biol. Chem. **262:** 3909-3917.

Hjelmstad, R. H. and Bell, R. M. (1988). The *sn*-1,2-diacylglycerol ethanolaminephosphotransferase activity of *Saccharomyces cerevisiae*: isolation of mutants and cloning of the *EPT1* gene. J. Biol. Chem. **263:** 19748-19757.

Hjelmstad, R. H. and Bell, R. M. (1990). The *sn*-1,2-diacylglycerol cholinephosphotransferase of *Saccharomyces cerevisiae*: nucleotide sequence, transcriptional mapping and gene product analysis of the *CPT1* gene. J. Biol. Chem. **265:** 1755-1764.

Hollander, N., Selvaraj, P. and Springer, T. A. (1988). Biosynthesis and function of LFA-3 in human mutant cells deficient in phosphatidylinositol-anchored proteins. J. Immunol. **141:** 4283-4290.

Hyman, R. (1988). Somatic analysis of the expression of cell surface molecules. Trends in Genetics **4:** 5-8.

Kennedy, E. P. (1961). Biosynthesis of complex lipids. Fed. Proc. Fed. Am. Soc. Exp. Biol. **20:** 934-960.

Kornfeld, R. and Kornfeld, S. (1985). Assembly of asparagine-linked oligosaccharides. Ann. Rev. Biochem. **54:** 631-664.

Krakow, J. L., Hereld, D., Bangs, J. D., Hart, G. W. and Englund, P. T. (1986). Identification of glycolipid precursor of the *Trypanosoma brucei* variant surface glycoprotein. J. Biol. Chem. **261:** 12147-12153.

Krakow, J. L., Doering, T. L. Masterson, W. J., Hart, G. W. and Englund, P. T. (1989). A glycolipid from *Trypanosoma brucei* related to the variant surface glycoprotein membrane anchor. Mol. Biochem. Parasitol. **36:**263-270.

Low, M.G. (1989). The glycosyl-phosphatidylinositol anchor membrane proteins. Biochim. Biophys. Acta **988:** 427-454.

Masterson, W. J., Doering, T. L., Hart, G. W. and Englund, P. T. (1989). A novel pathway for glycan assembly: biosynthesis of the glycosyl-phosphatidylinositol anchor of the trypanosome variant surface glycoprotein. Cell **56:** 73-80.

Masterson, W. J., Raper, J., Doering, T. L., Hart, G. W. and Englund, P. T. (1990). Fatty acid remodelling: a novel reaction sequence in the biosynthesis of trypanosome glycosyl-phosphatidylinositol membrane anchors. Cell **62:** 73-80.

Mayor, S., Menon, A. K., Cross, G. A. M., Ferguson, M. A. J., Dwek, R. A. and Rademacher, T. W. (1990a). Glycolipid precursors for the membrane anchor of *Trypanosoma brucei* variant surface glycoproteins. I. Glycan structure of the phosphatidylinositol-specific phospholipase C sensitive and resistant glycolipids. J. Biol. Chem. **265:** 6164-6173.

Mayor, S., Menon, A. K., Cross, G. A. M. (1991). Transfer of glycosyl-phosphatidylinositol membrane anchors to polypeptide acceptors in a cell-free system. J. Cell Biol.: in press.

Menon, A. K., Mayor, S., Ferguson, M. A. J., Duszenko, M. and Cross, G. A. M. (1988a). Candidate glycophospholipid precursor for the glycosyl-phosphatidylinositol membrane anchor of *Trypanosoma brucei* variant surface glycoproteins. J. Biol. Chem. **263:** 1970-1977.

Menon, A. K., Schwarz, R. T., Mayor, S. and Cross, G. A. M. (1988b). Glycolipid precursor of *Trypanosoma brucei* variant surface glycoproteins: incorporation of radiolabelled mannose and myristic acid in a cell-free system. Biochem. Soc. Trans. Lond. **16:** 996-997.

Menon, A. K., Schwarz, R. T., Mayor, S. and Cross, G. A. M. (1990a). Cell-free synthesis of glycosyl-phosphatidylinositol precursors for the glycolipid membrane anchor of *Trypanosoma brucei* variant surface glycoproteins: structural characterization of putative biosynthetic intermediates. J. Biol. Chem. **265:** 9033-9042.

Menon, A. K., Mayor, S. and Schwarz, R. T. (1990b). Biosynthesis of glycosyl-phosphatidylinositol lipids in *Trypanosoma brucei*: involvement of mannosphosphoryldolichol as the mannose donor. EMBO J. **9:** 4249-4258.

Orlean, P. (1990). Dolichol phosphate mannose synthase is required in vivo for glycosyl-phosphatidylinositol membrane anchoring, O-mannosylation and N-glycosylation of protein in *Saccharomyces cerevisiae*. Mol. Cell Biol. **10:** 5796-5805.

Prottey, C. and Ballou, C. E. (1968). Diacyl myoinositol monomannoside from *Propionibacterium shermanii*. J. Biol. Chem. **243:** 6196-6201.

Roberts, W. L., Myher, J. J., Kuksis, A., Low, M. G. and Rosenberry, T. L. (1988). Lipid analysis of the glycoinositol phospholipid membrane anchor of human erythrocyte acetycholinesterase: palmitoylation of inositol results in resistance to phosphatidylinositol-specific phospholipase C. J. Biol. Chem. **263:** 18766-18775.

Schubert, J., Uciechowski, P., Delany, P., Tischler, H-J., Kolanus, W. and Schmidt, R. E. (1990). The PIG-anchoring defect in NK lymphocytes of PNH patients. Blood **76:** 1181-1187.

Thomas, J. T., Dwek, R. A. and Rademacher, T. W. (1990). Structure, biosynthesis and function of glycosyl-phosphatidylinositols. Biochemistry **29:** 5413-5422.

Walter, E. I., Roberts, W. L., Rosenberry, T. L., Ratnoff, W. D. and Medoff, M. E. (1990). Structural basis for variations in the sensitivity of human decay accelerating factor to phosphatidylinositol-specific phospholipase C cleavage. J. Immunol. **144:** 1030-1036.

Yeh, E. T. H., Reiser, H., Bamezai, A. and Rock, K. L. (1988). TAP transcription and phosphatidylinositol linkage mutants are defective in activation through the T cell receptor. Cell **52** 665-674.

POLARIZED SORTING OF GPI-LINKED PROTEINS IN EPITHELIA AND MEMBRANE MICRODOMAINS

***M P Lisanti and E Rodríguez-Boulan**

Department of Cell Biology and Anatomy, Cornell University Medical College, New York, NY 10021, USA.

* Corresponding author.

THE EPITHELIAL SORTING PARADIGM

Transport of membrane proteins to the cell surface is thought to occur by default, while retention at locations along the secretory pathway (i.e. ER, Golgi) is thought to be signal-mediated (Pfeiffer & Rothman, 1987; Weiland et al, 1987). However, in epithelial cells sorting of membrane proteins to either the apical or basolateral surface must occur, indicating that sorting of proteins to at least one domain must be signal-mediated (Simons and Fuller, 1985; Rodríguez-Boulan and Nelson, 1989; Wandinger-Ness and Simons, 1991). Transfection of chimeric viral glycoproteins into MDCK cells has indicated that protein ectodomains may contribute apical or basolateral sorting information (Caplan and Matlin, 1989). Additional sorting information appears to be conveyed by signals in the cytoplasmic domain (Casanova et al, 1990); Breitfeld et al, 1990).

We decided to explore the polarized sorting of a newly discovered class of membrane glycoproteins, i.e. proteins anchored via glycosyl-phosphatidylinositol (GPI; Cross, 1987; Low and Saltiel, 1988; Ferguson and Williams, 1988; Low, 1989; Lisanti et al., 1990). As these proteins contain only two topological domains, sorting information must be present either in the protein ectodomain or the glycolipid membrane anchor (or both).

APICAL SORTING OF ENDOGENOUS GPI-ANCHORED PROTEINS IN RENAL AND INTESTINAL EPITHELIA

As several apical markers (alkaline phosphatase, 5′ nucleotidase, trehalase, and renal dipeptidase) were anchored to the membrane via GPI, we suspected that this mechanism of anchoring might contain apical targeting information. To test this hypothesis we designed a method to survey the surface distribution of endogenous GPI-anchored proteins in a variety of available epithelial cells lines (Table 1). Cells were surface-labelled apically or basolaterally with sulfo-NHS-biotin (a tight-

Table 1. Characteristics of polarized epithelial cell lines.

line	orign	TER	phenotype	B/A ratio
MDCK I	dog kidney	1,000	distal tubule	7.6
MDCK II	dog kidney	100-200	distal tubule	2.8-4.0
LLC-PK_1	pig kidney	400	proximal tubule	1-1.5
Caco-2	human colonic carcinoma	200	enterocytic	?
SK-CO15	human colonic carcinoma	2,500	colonic	?

See Gstraunthaler (1988) and Le Bivic et al., 1989 for specific references. TER= transmonolayer epithelial resistance, expressed in ohms.cm^2. B/A ratio= ratio between the surface areas of the basolateral and the apical domains.

junction impermeant probe) and extracted with the detergent Triton X-114 (Lisanti et al, 1988; Sargiacomo et al, 1989). Detergent rich phases were collected and treated with phosphatidylinositol-specific phospholipase C (PI-PLC) to release GPI-anchored proteins to the detergent-poor phase. Biotinylated proteins were detected after transfer to nitrocellulose by blotting with iodinated streptavidin. In MDCK cells (dog kidney/distal tubule), six GPI-anchored proteins were detected and were all apically polarized (Lisanti et al, 1988). The same striking correlation was observed with several other epithelial cells lines: LLC-PK_1 (pig kidney/proximal tubule phenotype); Caco-2 (human adenocarcinoma/enterocytic phenotype); SK-C015 (human adenocarcinoma/colonic phenotype). At least 13 different proteins were detected based on distinct MWs (Lisanti et al, 1990b). The polarized apical distribution of GPI-anchored proteins in MDCK cells was confirmed by Wilson et al, 1990. Furthermore, a recent report indicates that GPI-anchored proteins are restricted to the apical domain (bile canilicular surface) of hepatocytes in vivo (Ali and Evans, 1990). To date, a total of 12 known GPI-anchored proteins have been localized to the apical surfaces of polarized epithelia (summarized in Table 2 see over page).

Table 2. Known GPI-anchored proteins which are apically distributed in polarized epithelia.

Proteins	MW (kD)	References
Alkaline phosphatase[u]	70	Jung and Pergande, 1983; Brown et al., 1989; Le Bivic et al., 1990
5' Nucleotidase	73	Simons and Fuller, 1985
Trehalase	96	Takesue et al., 1986
Renal dipeptidase	47	Littlewood, Hooper, Turner, 1989
Aminopeptidase P	91	Hooper and Turner, 1988; Hooper et al., 1990
Carcinoembryonic antigen	180	Lisanti et al., 1990b
Decay-accelerating factor[p,t,d,u]	70	Medof et al., 1987; Lisanti et al., 1989a
Folate receptor[p,u]	28-40	Hansen et al., 1980; Selhub and Franklin, 1984; Rothberg et al., 1990a
Carboxypeptidase M[d,u]	54	Deddish et al., 1990
Carbonic anhydrase IV[p,u]	35	Brown et al., 1990
GP-2	78	LeBel & Beattie, 1988; Rindler and Hoops, 1990
Tamm-Horsfall urinary[t,d,u] glycoprotein	85	Rindler et al., 1990

Of the above 12 proteins, only two are not expressed in renal epithelial cells (CEA and GP-2). Their precise location in kidney [proximal tubule (p); thick ascending limb (t); distal tubule/collecting duct (d)] and their presence is urine (u) is indicated where known.

RECOMBINANT ADDITION OF GPI TO FUSION PROTEINS LEADS TO APICAL LOCALIZATION IN MDCK CELLS

To explore further the correlation between GPI-anchoring and apical localization in epithelia, we used the cut and paste approach (summarized in Table 3). We wondered whether GPI-anchoring of a basolateral antigen (Herpes

Table 3. Polarity of chimeric, native and anchorless GPI-anchored proteins in transfected MDCK cells.

Protein ectodomain	Membrane anchoring mechanism	Polarity
Herpes simplex gD-1 or VSV G	-native, transmembrane	basolateral
	-anchorless, secreted (gD-1 only)	basolateral
	-fusion protein, GPI-anchored	apical
hGH	-native anchorless, secreted	unsorted
	-fusion protein, GPI-anchored	apical
DAF, PLAP, or Thy-1	-native, GPI-anchored	apical
	-anchorless, secreted	apical
	-fusion protein, with G protein transmembrane and cytoplasmic domains (PLAP only)	basolateral
N-CAM	-transmembrane form (ld;180K)	basolateral
	-transmembrane form (sd;140K)	basolateral
	-GPI-anchored form (ssd;120K)	apical

See text for specific references.

simplex gD-1) or of an unsorted secretory protein (human growth hormone) might target these fusion proteins to the apical cell surface. For these experiments the DAF signal for GPI-attachment (37 C-terminal amino acids) was transferred to the ectodomains of these proteins (Caras et al, 1987, 1989). DAF and both fusion proteins were GPI-anchored and targeted to the apical surface of transfected MDCK cells (Lisanti et al, 1989a; Lisanti, et al., 1991). Similar results were obtained by J Rose's group using the VSV G protein (a basolateral antigen) and the Thy-1 signal for GPI-attachment (Brown et al, 1989). Opposite asymmetric segregation of transmembrane and GPI-anchored forms of VSV G protein is also observed after expression in cleavage stage Xenopus embryos (Roberts et al, 1990). These results provide direct evidence that GPI-anchoring may function as an apical sorting signal for post-Golgi targeting of membrane proteins in epithelial cells (Lisanti and Rodríguez-Boulan, 1990). This is the first function assigned to membrane attachment via GPI.

In accordance with the above fusion protein experiments, GPI-anchored and transmembrane forms of N-CAM are targeted to the apical and basolateral surfaces of transfected MDCK cells, respectively (Powell et al, 1990). These results provide a functional basis for generating alternatively spliced isoforms of a given protein that differ in their mechanism of membrane anchorage. Although the functional consequences of the apical expression of a cell-adhesion molecule (CAM) are unknown, this unusual localization may play a role in the formation of multiple cell layers as in the generation of a stratified epithelium.

We next explored the possibility that the ectodomains of GPI-anchored proteins may contain redundant or contradictory information. Anchor minus forms of DAF (Lisanti et al, 1989a) and Thy-1 (Powell et al.,1991) were expressed in MDCK cells and secreted in a predominantly apical fashion. Similar results were obtained for an anchor minus form of PLAP (Brown et al, 1989). Taken together, these results indicate that the ectodomain of a GPI-anchored protein may possess redundant or contradictory sorting information, or may lack sorting information. Thus, a hierarchy of sorting signals seems to exist with GPI-anchoring acting as a dominant apical sorting signal (Lisanti and Rodríguez-Boulan, 1990).

EXOCYTIC AND ENDOCYTIC SORTING OF A GPI-ANCHORED PROTEIN IN MDCK CELLS

To determine the mechanisms by which the apical polarity of GPI-anchored proteins is generated and maintained, we used an established MDCK cell line expressing a transfected GPI-anchored fusion protein (gD-1-DAF). To determine whether GPI-anchored proteins are sorted intracellularly or at the level of the basolateral membrane (via a transcytotic route as for the polymeric Ig receptor), we developed a biotin-based targeting assay (Lisanti et al, 1990a). Briefly, cells were pre-incubated at 20°C to accumulate the protein in the trans-Golgi network (TGN; Griffiths and Simons, 1986), surface amino groups were quenched with a sulfo-NHS analog that lacks a biotin moiety at 4°C, cells were recultured for various times at 37°C, and surface delivery monitored by domain-selective biotinylation. Using this assay system, apical surface delivery was direct (without transient basolateral expression) and was observed as early as 5 min after transfer to 37°C. Such apical delivery was 10-11 times the rate of basolateral delivery and was dependent on continued protein synthesis and independent of N-glycosylation, as determined by preincubations with cycloheximide and tunicamycin, respectively. Furthermore, tunicamycin treatment resulted in increased surface delivery, suggesting that GPI-synthesis or attachment was enhanced by inhibition of N-glycosylation.In accordance with these observations, inhibition of N-glycosylation by tunicamycin treatment increases the pool of Dol-P-Man in mutant cells defective in Dol-P-Man synthetase, allowing the synthesis of GPI-precursors and their transfer to protein (Singh and Tartokoff, 1990). Similarly, N-glycosylation is not required for the apical routing of a GPI-anchored hGH fusion protein, as hGH does not contain a consensus sequence for N-glycosylation (Lisanti et al, 1989a).

To study the polarized endocytosis and transcytosis of a GPI-anchored protein we used MDCK cells expressing gD-1-DAF, as a detectable amount (5-15%) is missorted to the basolateral membrane. To detect endocytosis, cells were apically or basolaterally labelled with a reducible analog of sulfo-NHS-biotin and after various times subjected to domain-selective reduction with glutathione (also tight junction impermeant) (Bretscher, 1989; Graeve et al, 1989; Le Bivic et al, 1989). Protection from reduction indicates that endocytosis occurs. Endocytosis of GPI-anchored gD-1 occurred slowly and preferentially from the apical side, while transcytosis did not occur at a detectable rate (Lisanti et al, 1990a). Apical and basolateral fractions of GPI-anchored gD-1 were endocytosed at rates 2.5 and 33.50 times slower than bulk endocytosis, consistent with the observations that several GPI-anchored proteins are excluded from coated pits and that a signal in the cytoplasmic domain is required for rapid endocytosis via coated pits (reviewed in Lisanti et al., 1990). (For comparison, 80% of the transferrin receptor is endocytosed in 5 min, while only 40% of apical GPI-anchored gD-1 is endocytosed in 120 min). Our results indicate that the basolateral fraction does not behave as a precursor to the apical pool (as in the transcytotic route to the apical surface). Similarly, direct targeting was observed for GPI-anchored alkaline phosphatase endogenously expressed in Caco-2 cells, while other apical markers (such as aminopeptidase N) followed the transcytotic route to varied extents (Le Bivic et al, 1990; Matter et al, 1990). Direct targeting of alkaline phosphatase was dependent on an intact microtubular network, as is the case for apical delivery of other antigens (Gilbert et al, 1991.

MANNOSAMINE, A NOVEL GPI-SYNTHESIS INHIBITOR, CONVERTS AN APICAL GPI-ANCHORED PROTEIN TO A NON-POLARIZED SECRETORY PRODUCT

We wished to study the effect of depleting GPI-linked proteins in polarized cells. For this purpose, we characterized a new GPI synthesis inhibitor, as few are known and none allow evaluation of their effect on the processing of the protein product. The amino-sugar mannosamine biosynthetically prevents the formation of alpha 1,2 mannose linkages in N-linked glycans, but does not affect their transfer to protein and processing from high mannose to hybrid and complex structures (Pan and Elbein, 1985). As a single alpha 1,2 mannose linkage is present in the conserved core glycan of the GPI-anchor

(Ferguson and Williams, 1988; Low and Altiel, 1988; Low, 1989), we suspected that mannosamine might act as an inhibitor of GPI-attachment. To test this hypothesis, we utilized MDCK cells permanently transfected with gD-1-DAF, a recombinant molecule containing the ectodomain of a basolateral viral glycoprotein (gD-1) fused to the DAF signal for GPI-attachment (Caras et al, 1987). We have previously shown that this fusion protein is GPI-anchored and expressed on the apical surface of polarized MDCK cells (Lisanti et al, 1989).

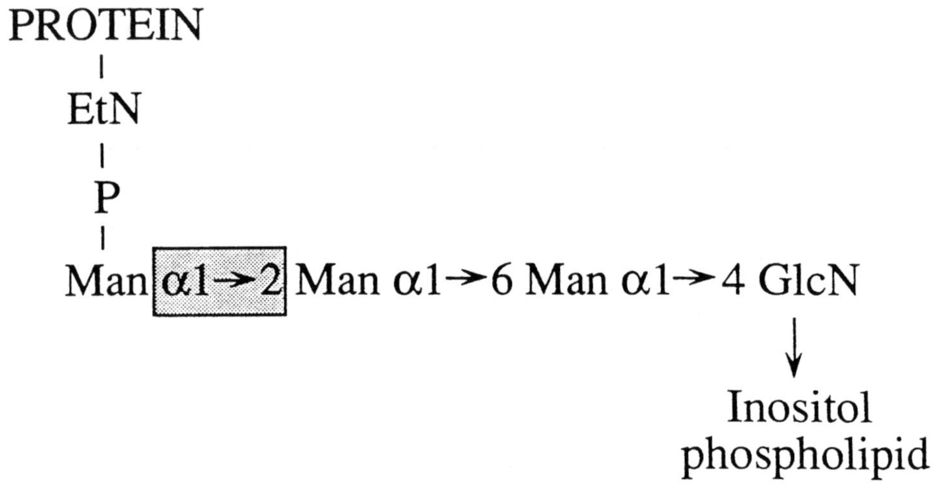

Fig 1. Structure of the conserved core of the GPI-glycan. The alpha 1,2 linkage is highlighted to illustrate the expected site of mannosamine action. Adapted from Lisanti et al., 1991.

Mannosamine specifically inhibited the synthesis of complete GPI-precursor lipids and their attachment to protein, probably by preventing the addition of the third mannose (an alpha 1,2 linked residue) to the conserved core of the GPI-glycan (Figure 1; Lisanti et al, 1991). In MDCK cells, the inhibitor converted apical gD-1-DAF to a non-polarized secretory product. This mannosamine-induced secretory product behaved similarly to another form of the fusion protein (truncated gD-1) that lacks the C-terminal signal for GPI-attachment. These data argue further that GPI was responsible for the apical sorting of this recombinant GPI-linked protein. In addition, mannosamine treatment caused a generalized reduction in the expression of endogenous GpI-anchored proteins. However, the mannosamine treated MDCK cells formed tight monolayers and remained polarized, suggesting that GPI-anchoring is not required for the maintenance of cell polarity by MDCK cells.

It is not known how mannosamine inhibits the formation of alpha 1,2 mannose linkages. As it is ineffective in a cell-free system that reconstitutes N-linked glycosylation, mannosamine may require conversion to an active metabolite (Pan and Elbein, 1985). One possibility is that it is transferred to glycans from a mannose donor and subsequent sugars cannot be added to the 2 positions, since a primary amino group replaces the hydroxyl group at this position. Alternatively, it may act as an enzymatic inhibitor of alpha-1,2-specific-mannosyl transferases. Both scenarios would explain why mannosamine inhibits specifically the incorporation of ethanolamine into GPI-precursor lipids, since ethanolamine is normally attached to a mannose residue of the conserved GPI-core structure that is alpha 1,2 linked.

A GPI-ANCHORED PROTEIN IS POLARIZED TO THE FREE SURFACE IN THE ABSENCE OF CELL-CELL CONTACT

To examine the role of the tight junctional fence in maintaining the asymmetric distribution of a GPI-anchored protein, we studied the polarity of GPI-anchored gD-1 under conditions 1-5 uM calcium) which prevent cell-cell contact and, therefore, tight junction formation. Using confocal microscopy, we observed that GPI-anchored gD-1 was localized to the free surface and excluded from the attached surface of the cell (Lisanti et al, 1990a). In addition, a fraction was retained intracellularly in an apical post-Golgi compartment, i.e. VAC (vacuolar apical compartment). Similar results have been obtained previously for other apical antigens (non-GPI-anchored), while basolateral antigens are randomly distributed on the cell surface and excluded from VACs (reviewed in Rodríguez-Boulan and Nelson, 1989).

As tight junctions are required to maintain the asymmetry of an apical endogenous glycosphingolipid (the Forsman antigen) or artificial lipid probes (Simons and van Meer, 1988), our results indicate that a GPI-anchored protein behaves more like an apical antigen than as a free lipid. Thus, a GPI-recognition mechanism may exist at the cell surface to maintain their polarized distribution. This notion is further supported by the polarized distribution of GPI-anchored proteins in non-epithelial cell types. In these cases, tight junctions are absent. For example, a GPI-anchored protein (PH-20) is asymmetrically distributed to the head regions of guinea pig sperm (Phelps et al, 1988). In addition, several GPI-anchored proteins are localized to the anterior pole (leading edge) of locomotor lymphocytes (Haston & Wilkinson, 1988; Ploug et al, 1991) and to neurite outgrowths (axons) and nerve terminals of neurons (Rotundo and Carbonetto, 1987; Gennarini et al, 1989; Grondal and Zimmerman, 1987; Dotti and Simons, 1991). The polarized localization of GPI-anchored proteins suggests that these specialized domains in non-epithelial cells may be analogous to the apical domain of epithelial cells and that the sorting of GPI- anchored proteins may occur by a common molecular mechanism.

GPI-ANCHORED PROTEINS ARE APICALLY SORTED BY A MECHANISM THAT IS INDEPENDENT OF THE APICAL ROUTING OF TRANSMEMBRANE PROTEINS

The core structure of the glycan of the GPI-anchor is highly conserved, but variations in side chain modifications have been observed (Ferguson et al, 1988; Homans et al, 1988; Roberts et al, 1988; Schneider et al, 1990). More specifically, the VSG anchor contains 0-8 galactose residues, while Thy-1 contains an extra ethanolamine phosphate, a mannose residue, and a beta-N-acetyl glucosamine residue. To investigate the possible effect of side chain modifications on the apical targeting of GPI-anchored proteins, we examined their distribution in lectin-resistant mutant MDCK cell lines. GPI-anchored proteins assumed a polarized apical distribution in a RCA-resistant MDCK cell line (Lisanti et al, 1990b). As these cells are unable to incorporate galactose due to a defect in uptake of UDP-galactose at the level of the Golgi (Meiss et al, 1982; Brandli et al, 1988), galactosylation of either O- or N-linked sugars, or the GPI-anchor is not required for apical targeting.

In Con A-resistant MDCK cells, two of five GPI-anchored proteins were relatively unpolarized (50-60% apical), while three were as apically polarized as in wild type cells (80-99% apical) (Lisanti et al, 1990b). As the correct polarized viral budding of influenza and VSV is preserved in these cells (Green et al, 1981; Meiss et al, 1982), this defect may only affect a subset of glyco-proteins indicating that the sorting of GPI-anchored proteins may be independent of the polarized sorting of transmembrane glycoproteins. Because GPI-anchored proteins may contain redundant apical information in their protein ectodomain, the proteins that were still very polarized may be sorted based on this information. In an attempt to test this hypothesis, we expressed gD-1-DAF in these mutant cells as it contains either no information or basolateral information in its ectodomain. gD-1-DAF was expressed with an apical polarity of 50% in Con A mutants (Lisanti, Caras and Rodríguez-Boulan, unpublished observations), while wild-type cells demonstrate an apical polarity of 85-95% (Lisanti et al, 1989). These results are consistent with the conclusion that the recognition of the GPI-anchor as an apical targeting signal in these cells is partially disrupted. Elucidation of the glycosylation defect should provide a more definitive conclusion as to the effect on the GPI-anchor structure. As the purified GPI-anchor binds Con A (Rosenberry et al, 1986) and mannose may be a constituent of the conserved core glycan as well as a side chain (Ferguson & Williams, 1988), the mutation may result in changes in mannose linkages or side chain addition.

Further evidence that the sorting of GPI-anchored proteins may be receptor-mediated and independent of the sorting of other apical antigens comes from the study of a thyroid epithelial cell line (FRT, Fisher rat thyroid). Although FRT cells retain the polarized characteristics of thyroid cells in vivo (Nitsch et al, 1985), they lack their differentiated phenotype (ie synthesis of thyroglobulin and responsiveness to TSH) and, therefore, possess a mutant phenotype in comparison with the native epithelium. In FRT cells, GPI-anchored proteins are not confined to the apical surface, while other apical (dipeptidyl peptidase IV) and basolateral antigens (Na/K ATPase, transferrin receptor, uvomorulin) demonstrate the same polarity as observed in other epithelia (Zurzolo et al, 1990). This lack of correlation between GPI-anchoring and apical localization was confirmed by transfection studies with GPI-anchored proteins which assume a polarized apical distribution in MDCK cells. These results argue that GPI-anchored proteins are apically sorted by a mechanism that is independent of the apical routing of transmembrane proteins. Thus, FRT cells sort GPI-anchored proteins as would be expected of a cell line that lacks a putative apical GPI-sorting receptor.

CLUES TO THE MOLECULAR MECHANISM OF INTRACELLULAR SORTING OF GPI-ANCHORED PROTEINS

Little is known about the molecular machinery responsible for segregation of apical from basolateral proteins in epithelia. It has been proposed that sorting receptors may mediate their segregation by analogy with the mannose-6-phosphate receptor that targets tagged hydrolases to lysosomes (Sly, 1982; Kornfeld, 1990). Alternatively, as exocytic carrier vesicles are presumed to resemble the lipid composition of the membrane to which they are directed, the preferential apical localization of glycosphingolipids (and the apical depletion of phosphatidyl-choline) may provide the mechanism for sorting (Simons and van Meer, 1988; Rodríguez-Boulan and Nelson, 1989). Either of these possibilities for sorting (receptor-mediated or glycolipid-mediated) are compatible with the apical sorting of GPI-linked proteins.

Further clues come from the cell surface localization of GPI-anchored proteins. GPI-anchored folate receptor (Rothberg et al, 1990a) and glycosphingolipids (Montesano et al, 1982) cluster in cell surface microinvaginations termed caveolae. This localization is dependent on the presence of cholesterol as inhibitors of cholesterol synthesis or cholesterol binding antibiotics disrupt such clustering (Rothberg et al, 1990b). The beta hydroxyl group of cholesterol may interact via hydrogen bonding with the ether linked fatty acid in the Sn-2 position in the GPI-anchor, thereby facilitating clustering. Increasing evidence also indicates that the endocytic behaviour of a molecule correlates with its cell surface localization in polarized cells. Membrane glycoproteins that lack endocytic signals for clathrin-coated pit localization (eg influenza HA and GPI-anchored proteins) are sorted to the apical surface, while receptors that are rapidly endocytosed (eg the transferrin, mannose-6-phosphate, and asialoglycoprotein receptors) are basolaterally localized. In addition, isoforms of the FcR II that differ in their cytoplasmic tails are either excluded from coated pits and apically localized (B1 isoform) or efficiently cluster in coated pits and assume a basolateral distribution (B2 isoform) (Hunziker and Mellman, 1989). Similarly, addition of the transmembrane and cytoplasmic domains of the VSV G protein (containing an endocytic sorting signal) to a GPI-anchored protein, converts this fusion protein to a basolateral antigen (Brown et al, 1989). Thus, apical versus basolateral sorting in the TGN may be achieved by exclusion from or inclusion into Golgi-associated clathrin-coated pits, respectively. As GPI-anchored proteins ae excluded from coated pits, but included in caveolae (Rothberg et al, 1990a), a similar clustering mechanism may exist in the TGN to provide the molecular means for apically routing GPI-anchored proteins.

At least two GPI-anchored proteins are embedded in the membrane of the erythrocyte (acetylcholinesterase and decay accelerating factor). When erythrocytes are subjected to a variety of stresses (aging, CA^{2+} exchange, or ATP depletion) they shed vesicular membranes from their cell surface. These vesicles are enriched in the above two GPI-anchored proteins, while the remaining red cell membrane is depleted (Shukla et al, 1977; Butikofer et al, 1989). Furthermore, these vesicles have a lower phospholipid content and are not enriched in other red cell proteins (either transmembrane proteins or cytoskeletal elements), indicating that this event is specific for GPI-anchored proteins. Similar segregation of GPI-anchored proteins from their transmembrane counterparts occurs during viral budding. In VSV-infected leukemic cells, several transmembrane glycoproteins are excluded from budding virions, while GPI-anchored Thy-1 was specifically incorporated (Calafat et al, 1983). In addition, as the scrapie prion protein (PrP) is GPI-anchored (Stahl et al, 1990), cell surface shedding of membrane vesicles enriched in GPI-anchored proteins may be the underlying mechanism in the transmission of infectious prion particles. Further investigation of these shedding phenomena may shed light on how the apical sorting of GPI-anchored proteins takes place intracellularly within the lumen of the TGN.

Several GPI-anchored proteins (Thy-1, Ly-6, Qa-2) are present on the cell surfaces of T-lymphocytes where they may mediate T cell activation via antibody crosslinking. Replacement of the GPI-anchor of these antigens (Ly-6 or Qa-2) with transmembrane domains prevents the mitogenic response (Robinson et al, 1989; Su et al, 1991), whereas conversion of H-2 from a transmembrane to a GPI-anchored form confers antibody-dependent activation (Robinson et al, 1989). These results suggest that antibody cross-linking of GPI-anchored proteins is sufficient to induce T-cell activation, implying that the GPI-anchor itself is crucial for the signal transduction event. As the GPI-anchor does not span the lipid bilayer, this poses a topological problem for conveying signals to the cytoplasm. To resolve this apparent paradox, it has been proposed that either 1) the degradation products of the GPI-anchor may act as second messengers themselves, or 2) that a signal transducing molecule (a transmembrane glycoprotein with a cytoplasmic tail) may recognize the GPI-anchor at the cell surface. In addition, as GPI-anchored proteins are excluded from coated pits and concentrated in caveolae, this pattern of cell surface localization may also play a role in the signal transduction event. Further study of GPI-dependent T cell activation may lead to the discovery of cell surface molecules that recognize the GPI-anchor.

Although membrane lipids may provide the medium in which the GPI-anchor is embedded, the importance of the glycan portion as a potential signal for apical sorting should not be overlooked. As the GPI-glycan contains several novel features not detected in other forms of glycosylation, it can be easily distinguished from N-linked glycans. These differences are: 1) glcN with a free amino group that is linked alpha 1,4 (glcN is N-acetylated and beta 1,4 in N-glycans); and 2) the three mannoses in the conserved core structure are linked 1,2-1,6-1,4, a linkage sequence not detected in N-glycans (see Table 4). In fact, several GPI-specific phospholipases (C or D) have been isolated that only hydrolyze the GPI-anchor, but have little or no activity for phosphatidylinositol (Fox et al, 1986; Davitz et al, 1987; Lower and Prasad, 1988; Cardoso de Almeida et al, 1988; Fouchier et al, 1990) or for GPI-precursors lacking mannose (GlcN-PI, GlcNAc-PI; discussed in Menon et al, 1990). The observation that specific GPI-binding proteins can exist lends support to the idea that a protein that recognizes the glycan portion of GPI (an endogenous lectin) may function as an apical sorting receptor. The availability of the cDNAs of these phospholipases (Hereld et al, 1988) may help in the search for non-enzymatic GPI-binding proteins.

Table 4. Comparison of the core structure and linkages of GPI and N-glycans.

Glycosyl-phosphatidylinositol

α1,2	α1,6	α1,4	free N	inos	p	diacyl-, alkylacyl-glycerol, or ceramide
Man ---	**Man ---**	**Man ---**	**Glc N ---**	**sugar ---**	**p ---**	**lipid moiety**
α1,2	α1,3	β1,4	NAc	GlcNAc	pp	dolichol

N-linked glycan

The common compostional features of GPI and N-linked glycans are shown in the center (in bold), while the specific linkages and other variations are listed above and below.

There are precedents that endogenous lectins can act as receptors for sorting events, i.e the mannose-6-phosphate receptors (Sly, 1982; Kornfeld, 1990) and the asialoglycoprotein receptor (Drickamer, 1988). Further support for this idea comes from results with a lectin resistant MDCK cell line (Con A selected) that missorts certain GPI-anchored proteins. Although the defect is this cell line is unknown, it seems to involve mannosylation, consistent with the mannose binding capability of Con A (see above).

The precursor of mannose-6-phosphate also bears structural similarities to the terminal end of the completed GPI precursor lipid. Both structures contain an amino group located two carbons from a phosphate group, linked to the 6 position of a terminal alpha 1,2 linked mannose residue

(Table 5). The chemical identities of these substrates imply that the enzymes that govern their synthesis and processing may rely on similar recognition mechanisms. These may include: 1) the enzymes that mediate the transfer of GlcNAc-P or EthN-P to the terminal alpha 1,2 linked mannose residues of N-linked glycans or GPI precursors, respectively; and 2) the uncovering enzyme (for transfer of GPI to protein).

Table 5. Comparison of the terminal end of the completed GPI precursor with an intermediate in the generation of a lysosomal targeting signal (Mannose-6-phosphate).

Glycosyl-phosphatidylinositol

EthN

...Man a1,2 Man 6-P-O-CH_2-CH_2-NH_2

...Man a1,2 Man 6-P-O-CH-CH-NH-Ac

GlcNAc

N-linked glycan with Mannose-6-P intermediate

To generate the mannose-6-phosphate group on N-linked glycans, GlcNAc-P is transferred from UDP-GlcNAc to terminal alpha 1,2 linked mannose residues (Madiyalakan et al., 1987). Then GlcNAc is removed by the uncovering enzyme to reveal the mannose-6-phosphate group (Sly, 1982; Kornfeld, 1990). The intermediate before the action of the uncovering enzyme is shown and bears striking resemblence to the terminal end of the completed GPI precuror lipid.

APICAL CELL SURFACE RELEASE OF GPI-ANCHORED PROTEINS

Several GPI-anchored proteins are found in a soluble form in addition to the membrane-anchored form. These soluble forms may be generated by alternative mRNA splicing or by degradation of the site of membrane attachment (reviewed in Low, 1989; Lisanti et al., 1990). It has been suggested that the latter mechanism may occur via a cell surface protease, glycosidase or phospholipase, although no convincing evidence has been presented to support or refute any of these possibilities (Chan et al, 1988; Lisanti et al, 1989b). As these soluble forms are often found in bodily secretions which face the apical surfaces

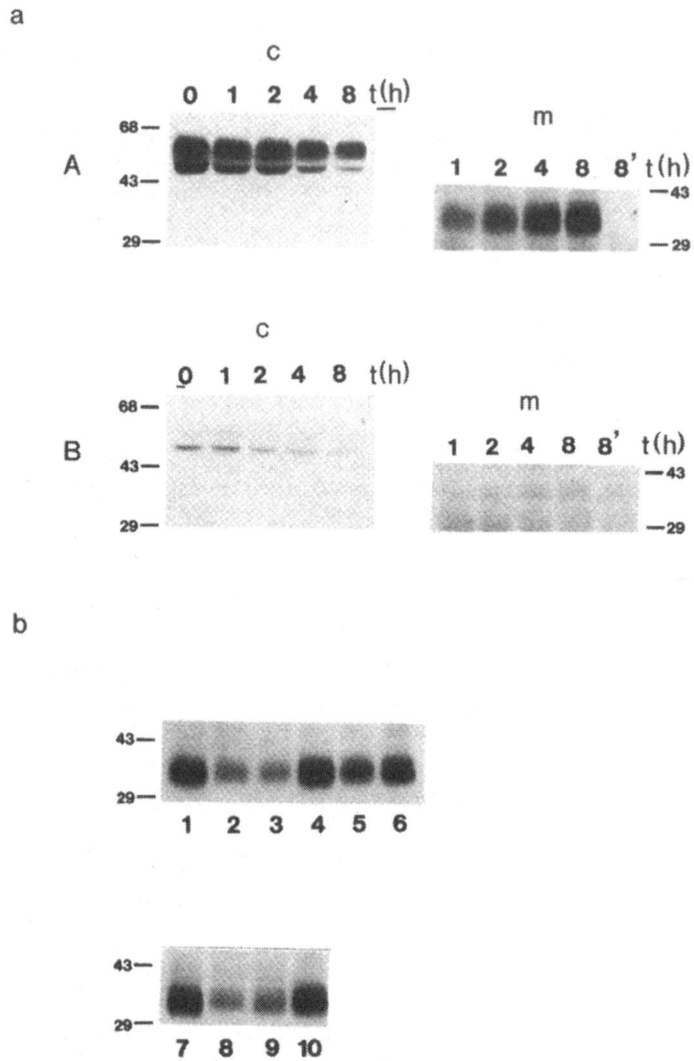

Figure 2. Apical release of gD-1-DAF is mediated by a trypsin-like protease.

a) Filter-grown monolayers were apically (A) or basolaterally (B) biotinylated, an recultured at 37°C for various times (0, 1, 2, 4, and 8 hours). Cells (c) and apical and basolateral media (m) were collected and immunoprecipitated. Apical media was analyzed for apically-biotinylated samples and visa versa. Lanes 8'represent media collected after 8 hours from the side opposite biotinylation.

b) Filter-grown monolayers were metabolically-labeled overnight and incubated in media containing various protease inhibitors for 8 hours. Apical media was collected and immunoprecipitated. Lanes 1-6: untreated control; leupeptin; antipain; pepstatin; chymostatin; and E-64. Lanes 7-10: untreated control; aprotinin; soybean trypsin inhibitor; and chloroquine. Similar results were obtained when cell-surface biotinylation was employed to evalute the sensitivity of release to protease inhibitors.

of epithelia (tears, saliva, milk, CSF, and urine), we decided to investigate the mechanism of release in a polarized system.

The degradation of gD-1-DAF was followed in MDCK cells via domain-selective biotinylation and analysis of the resulting media. gD-1-DAF was converted from a membrane bound, cell surface form to a soluble form with a t1\2 of 8 hours. The soluble form was present mainly in the apical media (>90%) and was 10 K smaller than the membrane associated form, suggesting that release occurred via a proteolytic event. To investigate this possibility, we evaluated the effect of a battery of protease inhibitors (either peptides or proteins) on apical cell surface release. Release was inhibited by various serine protease inhibitors, suggesting that the protease was a trypsin-like enzyme (Figure 2; Lisanti et al, 1991). The existence of a cell surface trypsin-like protease has been demonstrated in hepatocytes (Tanaka et al, 1986) and this enzyme possesses the same differential sensity to protease inhibitors as we have observed for the release of gD-1-DAF. In fact, only a single cleavage site that conforms to the specificity of the hepatic enzyme (leu-arg) is present in the ectodomain of gD-1 and cleavage at this site would yield a fragment of the predicted size that we observe experimentally (residues 1-221). We are currently investigating the sensitivity of release of other GPI-anchored proteins to protease inhibitors.

The half-time of degradation and release of gD-1-DAF is similar to that observed for other GPI-anchored proteins ((scrapie prion protein (5.2 h; Borchelt et al, 1990) N-CAM 120 (11 h; He et al, 1987), CEA (>24 h; Sack et al, 1988)). Several other GPI-anchored proteins undergo release and the soluble form is significantly smaller than the membrane-associated form or does not efficiently incorporate ethanolamine ((for example DAF (Seya et al, 1987; Tausk et al, 1989), FcR III (Huizinga et al, 1990), N-CAM 120 (He et al, 1987), heparan sulfate proteoglycan (David et al, 1990)), suggesting that proteolytic release may be a general mechanism for shedding a GpI-anchored proteins.

The cell surface release of GPI-anchored proteins may have clinical implications for the detection of epithelial cancers and renal disease. For example, CEA (a GPI-anchored protein) is normally apically located in intestinal cells and is undetectable in the serum. However, in adenocarcinoma of the colon the loss of the apical polarity of CEA (its random distribution on the cell surface) correlates with the increasing levels of CEA observed in the blood (Hamada et al, 1985).Thus, it may be envisioned that loss of polarity in epithelial cancers results in expression of CEA on the basolateral membrane and subsequent release into the blood. Normal release of CEA from the apical surface would go undetected in the blood. As certain GPI-anchored proteins are specific for a given epithelial cell type, the use of GPI-anchored proteins to follow the development of epithelial cancers (which represent >80-90% of all cancers) should not be overlooked.

In the kidney, the major enzyme responsible for reabsorption of bicarbonate is GPI-anchored carbonic anhydrase IV and is located at the apical surface of proximal tubular cells (Brown et al, 1990). Similarly, GPI-anchored carboxypeptidase M is involved in the metabolism of kinins and is apically located in distal tubular cells (Deddish et al, 1990). Finally, the major urinary glycoprotein (Tamm-Horsfall; uromodulin) is a GPI-anchored protein of unknown function and localized to the apical surfaces of the thick ascending limb of Henle and distal tubular cells (Rindler et al, 1990). Soluble forms of these proteins are present in elevated levels in the urine in pathogenic states (Iimura, 1990; Hoyer and Seiler, 1979).

Table 6. Summary of the experimental observations on the polarized sorting of GPI-anchored proteins.

Major

1- GPI-anchoring correlates with apical localization (conserved across species and tissue type).

2- Reachoring of basolateral or unsorted proteins via GPI leads to apical expression.

3- GPI-anchored proteins are sorted intracellulary and delivered directly to the apical surface, without first reaching the basolateral surface.

4- Inhibition of GPI-synthesis by mannosamine converts an apical GPI-anchored protein to a non-polarized secretory product.

5- Endocytosis of a GPI-anchored protein is slow and occurs preferentially from the apical side; the basolateral fraction behaves as a stable missorted population and does not transcytose.

6- The apical polarity of a GPI-anchored protein is independent of tight junctions and cell-cell contact.

7- The apical sorting of GPI-anchored proteins is independent of the polarized sorting of transmembrane glycoproteins.

8- A GPI-anchored proteins is released from the apical cell surface by a trypsin-like protease.

Minor

1- Redundant apical targeting information may be present within the ectodomains of GPI-anchored proteins.

2- A GPI-anchored protein may be accumulated intracellularly in the TGN (at 20 oC) or VAC's (an apical post-Golgi compartment).

3- The apical polarity of GPI-anchored proteins is independent of O-glycosylation (RCA-resistant mutants with severely truncated N-and O-linked glycans).

4- Tunicamycin treatment increases the amount of a GPI-anchored protein delivered to the cell surface (possibly by increasing GPI-biosynthesis), but does not affect GPI-anchoring or apical sorting.

5- GPI-anchoring is not required to maintain the polarized phenotype of MDCK cells.

SUMMARY

Taken together, our results with endogenous and transfected GPI-anchored proteins (summarized in Table 6) suggest that covalent attachment to GPI functions as an apical transport signal in polarized epithelial cells. This is the first example of a well-defined targeting signal for the post-Golgi sorting of plasma membrane proteins in polarized epithelia; the only other known post-Golgi targeting signal is mannose-6-phosphate, which functions in the recognition of lysosomal hydrolases and directs them to lysosomes.

ACKNOWLEDGEMENT

We thank Dr K S Perrego for critically reading the manuscript. MPL is supported by a Medical Scientist Training grant from the Cornell University MD/PhD program and ERB by grants from the NIH (GM-34107 and GM-41771), the American Cancer Society and the New York Heart Association.

REFERENCES

Ali, N. and Evans, W. H. (1990). Priority targeting of GPI-anchored proteins to the bile-canicular (apical) plasma membrane of hepatocytes. Biochem. J. **271:** 193-199.

Borchelt, D. R., Scott, M., Taraboulos, A., Stahl, N. and Prusiner, S. B. (1990). Scrapie and cellular prion proteins differ in their kinetics of synthesis and topology in cultured cells. J. Cell Biol. **110:** 743-752.

Brandli, A. W., Hansson, G. C., Rodríguez-Boulan, E. and Simons, K. (1988). A polarized epithelial cell mutant deficient in translocation of UDP-galactose into the Golgi complex. J. Biol. Chem. **263:** 16283-16290.

Breitfeld, P. P., Casanova, J. E., McKinnon, W. C. and Mostov, K. (1990). Deletions in the cytoplasmic domain of the polymeric Ig receptor differentially affect endocytic rate and post-endocytic traffic. J. Biol. Chem. **265:** 13750-13757.

Bretscher, M. S. (1989). Endocytosis and recycling of the fibronectin receptor in CHO cells. EMBO J. **8:** 1341-1348.

Brown, D. A., Crise, B. and Rose, J. K. (1989). Mechanism of membrane anchoring affects polarized expression of two proteins in MDCK cells. Science **245:** 1499-1501.

Brown, D., Zhu, X. L. and Sly, W. S. (1990). Localization of membrane-associated carbonic anhydrase type IV in kidney epithelial cells. Proc. Natl. Acad. Sci USA **87:** 7457-7461.

Butikofer, P., Kuypers, F. A., Xu, C. M., Chi, D. T. Y. and Lubin, B. (1989). Enrichment of two GPI-anchored proteins, AChE and DAF, in vesicles released from human red blood cells. Blood **74:** 1481-1485.

Calafat, J., Janssen, H., Demant, P., Hilgers, J. and Zavada, J. (1983). Specific selection of host cell glycoproteins during assembly of murine leukaemia virus and vesicular stomatitis virus : Presence of Thy-1 glycoprotein and absence of H-2, Pgp-1 and T-200 glycoproteins on the envelopes of these glycoproteins. J. Gen. Virol. **64:** 1241-1253.

Caplan M. and Matlin, K. S. (1989). Sorting of membrane and secretory proteins in polarized secretory cells. In Functional epithelial cells culture, Modern Cell Biology, vol 8 (Matlin, K. S. and Valentich, J. D. eds). pp. 71-127. Alan R. Liss NY.

Cardoso de Almeida, M. L., Turner, M. J., Stambuk, B. B. and Shenkman, S. (1988). Identification of an acid-lipase in human serum which is capable of solubilising GPI-anchored proteins. Biochem. Biophys. Res. Comm. **150:** 476-482.

Caras, I. W., Weddell. G. N., Davitz, M. A., Nussenweig, V. and Martin, D. W. (1987). Signal for attachment of a phospholipid membrane anchor in decay accelerating factor, Science **238:** 1280-1283.

Caras, I. W., Weddell, G. N. and Williams, S. R. (1989). Analysis of the signal for attachment of a glycophospholipid membrane anchor. J. Cell Biol. **108:** 1387-1396.

Casanova, J. E., Breitfeld, P. P., Ross, S. A. and Mostov, K. E. (1990). Phosphorylation of the polymeric Ig receptor required for efficient transcytosis. Science. **248:** 742-745.

Chan, B. L., Lisanti, M. P., Rodríguez-Boulan, E. and Saltiel, A. R. (1988). Insulin-stimulated release of lipoprotein lipase by metabolism of its phosphatidyl anchor. Science. **241:** 1670-1672.

Cross, G. A. M. (1987). Eukaryotic protein modification and membrane attachment via phosphatidylinositol. Cell **48:** 179-181.

David, G., Lories, V., Decock, B., Marynen, P., Cassiman, J.-J. and Van den Berghe, H. (1990). Molecular cloning of a PI-anchored membrane heparan sulphate proteoglycan from human lung fibroblasts. J. Cell Biol. **111:** 3165-3176.

Davitz, M. A., Hereld, D., Skak, S., Krakow, J., Englund, P. T. and Nussenweig, V. (1987). A glycan-phosphatidylinositol-specific phospholipase D in human serum. Science **238:** 81-84.

Davitz, M. A., Hom, J. and Schenkman, S. (1989). Purification of a GPI-specific phospholipase D from human plasma. J. Biol. Chem. **264:** 13760-13764.

Deddish, P. A., Skidgel R. A., Kriho, V. B., Li, X.-Y., Becker, R. P. and Erdos, E. G. (1990). Carboxypeptidase M in MDCK cells: Evidence that carboxypeptidase M has a PI-glycan anchor. J. Biol. Chem. **265:** 15083-15089.

Dotti, C. G., Parton, R. G. and Simons, K (1991) Polarized sorting of glypiated proteins in hippocampal neurons. Nature **349:** 158-161.

Drickamer, K. (1988). Two distinct classes of carbohydrate recognition domains in animal lectins. J. Biol. Chem. **263:** 9557-9560.

Ferguson, M. A. J., Homans, S. W., Dwek, R. A. and Rademacher, T. W. (1988). Glycosyl-phosphatidylinositol moiety that anchors *Trypanosoma brucei* variant surface glycoprotein to the membrane. Science. **239:** 753-759.

Ferguson, M. A. J. and Williams, A. F. (1988). Cell-surface anchoring of proteins via glycosyl-phosphatidylinositol structures. Ann. Rev. Biochem. **57:** 285-320.

Fouchier, F., Baltz, T. and Rougon, G. (1990). Identification of GPI-specific phospholipase C in mouse brain membranes. Biochem. J. **269:** 321-327.

Fox, J. A., Duszenko, M., Ferguson, M. A. J., Low, M. G. and Cross, G. A. M. (1986). Purification and characterization of a novel glycan-PI specific phospholipase C from *T. Brucei*. J. Biol. Chem. **261:** 15767-15771.

Fox, J. A., Soliz, W. J. and Saltiel, A. R. (1987) Purification of a PI-glycan specific phospholipase C from liver plasma membranes; a possible target of insulin action. Proc. Natl. Acad. Sci. USA. **84:** 2663-2667.

Gennarini, G., Cibelli, G., Rougon, G., Mattei, M.-G. and Gordis, C. (1989). The mouse neuronal cell surface protein F3: A phosphatidylinositol-anchored member of the immunoglobulin superfamily related to chicken contactin. J. Cell. Biol. **109:** 775-788.

Gilbert, T., Le Bivic, A., Quaroni, A. and Rodríguez-Boulan, E. (1991). Microtubular organization and its involvement in the biogenetic pathways of plasma membrane proteins in Caco-2 intestinal epithelial cells. J. Cell Biol. in press.

Graeve, L., Drikamer, K. and Rodríguez-Boulan, E. (1989). Polarized endocytosis by MDCK cells transfected with functional chicken liver glycoprotein receptor. J. Cell Biol. **109:** 2809-2816.

Green, R., Meiss, H. K. and Rodríguez-Boulan, E. (1981). Glycosylation does not determine the segregation of viral envelope glycoproteins in the plasma membrane of epithelial cells. J. Cell Biol. **89:** 23-239.

Griffiths, G. and Simons, K. (1986). The trans Golgi network: Sorting at the exit site of the Golgi complex. Science **234:** 438-443.

Grondal, E. J. M. and Zimmerman, H. (1987). Purification, characterization and cellular localization of 5'nucleotidase from *Torpedo* electric organ. Biochem. J. **245:** 805-810.

Gstraunthaler, G. J. A. (1988). Epithelial cells in Tissue Culture. Renal Physiol. Biochem. **11:** 1-42.

Hamada, Y., Yamamura, M., Koshiro, H., Yamamato, M., Nagura, H. and Watanabe, K. (1985). Immunohistochemical study of CEA in patients with colorectal cancer. Cancer **55:** 136-141.

Hansen, S. I., Holm, J. and Lyngbye, J. (1980). High affinity folate binding protein in urine. ICRS Med. Sci. **8:** 846-847.

Haston, W. S. and Wilkinson, P. C. (1988). Locomotion and chemotaxis of leukocytes: gradient perception and locomotion capacity. Curr. Opinion Immunol. **1:** 5-9.

He, H.-T., Finne, J. and Goridis, C. (1987). Biosynthesis, Membrane association, and release of N-CAM-120, a phosphatidylinositol-linked form of the neural cell adhesion molecule. J. Cell Biol. **105:** 2489-2500.

Hereld, D., Hart, G. W. and Englund, P. T. (1988). cDNA encoding the glycosyl-PI-specific phospholipase C of T. brucei. Proc. Natl. Acad. Sci. USA **85:** 8914-8918.

Homans, S. W., Ferguson, M. A. J., Dwek, R. A., Rademacher, T. W., Anand, R. and Williams, A. F. (1988). Complete structure of the glycosyl-phosphatidylinositol membrane anchor of rat brain Thy-1 glycoprotein. Nature **333:** 269-272.

Hooper, N. M., Hryszko, J. and Turner, A. J. (1990). Purification and characterization of pig kidney aminopeptidase P. Biochem. J. **267:** 509-515.

Hooper, N. M. and Turner, A. J. (1988). Ectoenzymes of the kidney microvillar membrane. FEBS Letts. **229:** 340-344.

Hoyer, J. R. and Seiler, M. W. (1979). Pathophysiology of Tamm-Horsfall protein. Kidney Int. **16:** 279-289.

Huizinga, T. W., de Hass, M., Kleijer, M., Nuijens, J. H., Roos, D. and von dem Borne, A. E. G. Kr. (1990). Soluble FcR III in human plasma originates from release by neutrophils. J. Clin. Invest. **86:** 416-423.

Hunziker, W. and Mellman, I. (1989). Expression of macrophage-lymphocyte Fc receptors in MDCK cells: Polarity and transcytosis differ for isoforms with or without coated pit localization domains. J. Cell Biol. **109:** 3291-3302.

Iimura, O. (1989). in Renal Function, Hypertension and Kallikrein-Kinin System (Iimura, O. and Margolius, H. S., eds) pp. 3-18, Hokusen-sha, Sapporo, Japan.

Jung, K. and Pergande, M. (1983). Particulate and free enzyme activity in urine as a result of the shedding of brush border membranes from kidney. Clin. Chem. **29:** 392-393.

Kornfeld, S. (1990). Lysosomal enzyme targeting. Biochem. Soc. Trans. **18:** 367-374.

Le Bel, D. and Beattie, M. (1988). The major protein of pancreatic zymogen granule membranes (GP-2) is anchored via covalent bonds to phophatidylinositol. Biochem. Biophys. Res. Comm. **154:** 818-823.

Le Bivic, A., Quaroni, A., Nichols, B. and Rodríguez-Boulan, E. (1990). Biogenetic pathways of plasma membrane proteins in Caco-2, a human intestinal epithelial cell line. J. Cell Biol. **111:** 1351-1361.

Le Bivic, A., Real, F. X. and Rodríguez-Boulan, E. (1989). Vectorial targeting of apical and basolateral plasma membrane proteins in a human adenocarcinoma epithelial cell line. Proc. Natl. Acad. Sci. USA **86:** 9313-9317.

Lisanti, M. P., Caras, I. W., Davitz, M. A. and Rodríguez-Boulan, E. (1989a) A glycophospholipid membrane anchor acts as an apical targeting signal in polarized epithelial cells. J. Cell Biol. **109:** 2145-2156.

Lisanti, M. P., Caras, I. W., Gilbert, T., Hanzel, D. and Rodríguez-Boulan, E. (1990a). Vectorial apical delivery and slow endocytosis of a glycolipid-anchored fusion protein in transfected MDCK cells. Proc. Natl. Acad. Sci. USA **87:** 7419-7423.

Lisanti, M. P., Caras, I. W. and Rodríguez-Boulan, E. (1991). Fusion proteins containing a minimal GPI-attachment signal are apically expressed in transfected MDCK cells. J. Cell Sci. in press.

Lisanti, M. P., Darnell, J. C., Chan, B. L., Rodríguez-Boulan, E. and Saltiel, A. R. (1989b). The distribution of glycosyl-phosphatidylinositol anchored proteins is differentially regulated by serum and insulin. Biochem. Biophys. Res. Comm. **164:** 824-832.

Lisanti, M. P., Field, M., Caras, I. W., Menon, A. and Rodríguez-Boulan, E. (1991). Mannosamine, a novel inhibitor of glycosyl-phosphatidylinositol incorporation into proteins. EMBO J. in press.

Lisanti, M. P., Le Bivic, A., Saltiel, A. R. and Rodríguez-Boulan, E. (1990b). Preferred apical distribution of glycosyl-phosphatidylinositol (GPI) anchored proteins: A highly conserved feature of the polarized epithelial cell phenotype. J. Membr.Biol. **113:** 155-167.

Lisanti, M. P., Rodríguez-Boulan, E. and Saltiel, A. R. (1990) Emerging functional roles for the GPI membrane protein anchor. J. Membr. Biol. **117:** 1-10.

Lisanti, M. P. and Rodríguez-Boulan, E, (1990). Glycophospholipid membrane anchoring provides clues to the mechanism of protein sorting in polarized epithelial cells. Trends Biochem. Sci. **15:** 113-118.

Lisanti, M. P., Sargiacomo, M., Graeve, L., Saltiel, A. R. and Rodríguez-Boulan, E. (1988). Polarized apical distribution of glycosyl-phosphatidylinositol anchored proteins in a renal epithelial cell line. Proc. Natl. Acad. Sci. USA **85:** 9557-9561.

Littlewood, G. M., Hooper, N. M. and Turner, A. J. (1989). Ectoenzymes of the kidney microvillar membrane. Biochem. J. **257:** 361-367.

Low, M. G. (1989). Glycosyl-phosphatidylinositol: A versatile anchor for cell surface proteins. FASEB J. **3:** 1600-1608.

Low, M. G. and Prasad, A. R. S. (1988). A phospholipase D for the PI-anchor of cell surface proteins is abundant in plasma. Proc. Natl. Acad. Sci. USA **85:** 980-984.

Low, M. G. and Saltiel, A. R. (1988). Structural and functional roles of glycosyl-phosphatidylinositol in membranes. Science **239:** 268-275.

Madiyalakan, R., Jain, R. K., and Matta, K. L. (1987). Phosphorylation of the alpha 1,2 linked mannosylsaccharide by N-acetylglucosamine-1-phosphotransferase from fibroblasts occurs at the terminal mannose. Biochem. Biophys. Res. Comm. **142:** 354-348.

Matter, K., Brauchbar, M., Bucher, K. and Hauri, H-P. (1990). Sorting of endogenous plasma membrane proteins occurs from two sites in cultured human intestinal cells (Caco-2). Cell **60:** 429-437.

Medof, M. E., Walter, E. I., Rutgers, J. L., Knowles, D. M. and Nussenweig, V. (1987). Identification of DAF on epithelium and glandular body fluids. J. Exp. Med. **165:** 848-864.

Meiss, H. K., Green, R. F. and Rodríguez-Boulan, E. (1982). Lectin resistant mutants of polarized epithelial cells. Mol. Cell Biol. **2:** 1287-1294.

Menon, A., Schwartz, R. T., Mayor, A. and Cross, G. A. M. (1990). Cell-free synthesis of GPI-precursors for the glycolipid membrane anchor of *T. Brucei* variant surface glycoproteins: Structural characterization of putative biosynthetic intermediates. J. Biol. Chem. **265:** 9033-9042.

Montesano, R., Roth, J., Robert, A. and Orci, L. (1982). Non-coated membrane invaginations are involved in binding and internalization of cholera and tetanus toxins. Nature **296:** 651-653.

Nitsch, L., Tramontano, D. and Ambesi-Impiombato, F. S. (1985). Morphological and functional polarity of an epithelial thyroid cell line. Eur. J. Cell Biol. **38:** 57-66.

Pan, Y. T. and Elbein, A. D. (1985). The effect of mannosamine on the formation of lipid-linked oligosaccharides and glycoproteins in canine kidney cells. Arch. Biochem. Biophys. **242:** 447-456.

Pfeiffer, S. R. and Rothman, J. E. (1987). Biosynthetic protein transport and sorting by the endoplasmic reticulum and Golgi. Ann. Rev. Biochem. **56:** 829-852.

Phelps, B. M., Primakoff, P., Koppel, D. E., Low, M. G. and Myles, D. G. (1988). Restricted lateral diffusion of PH-20, a PI-anchored sperm membrane protein. Science **240:** 1780-1782.

Ploug, M., Ronne, E., Behrent, N., Jensen, A. L., Blasi, F. and Dano, K. (1991). Cellular receptor for urokinase plasminogen activator: C-terminal processing and membrane anchoring by GPI. J. Biol. Chem. **266:** 1926-1933.

Powell, S. K., Lisanti, M. P. and Rodríguez-Boulan, E. (1991). Thy-1 expresses two signals for apical localisation in epithelial cells. Am. J. Physiol. **260:** C715-C720.

Powell, S. K., Cunningham, B. C., Edelman, G. M. and Rodríguez-Boulan, E. (1990). Opposite polarity of GPI and transmembrane-anchored forms of N-CAM in MDCK cells. J. Cell Biol. **111:** 327a.

Rindler, M. J. and Hoops, T. C. (1990). The pancreatic membrane protein GP-2 localizes specifically to secretory granules and is shed into the pancreatic juice as a protein aggregate. Eur. J. Cell Biol. **53:** 154-163.

Rindler, M. J., Naik, S. S., Li, N., Hoops, T. C. and Peraldi, M.-N. (1990). Uromodulin (Tamm-Horsfall glycoprotein) is a PI-linked membrane protein. J. Biol. Chem. **265:** 20784-20789.

Roberts, S. J., Leaf, D. S., Moore, H.-P. and Gerhart, J. C. (1990). Polarity of membrane traffic in Xenopus embryos. J. Cell Biol. **111:** 489a.

Roberts, W. L., Santikarn, S., Reinhold, V. N. and Rosenberry, T. L. (1988). Structural characterization of the glycoinositol phospholipid membrane anchor of human erythrocyte acetylcholinesterase by fast atom bombardment mass spectrometry. J. Biol. Chem. **263:** 18776-18784.

Robinson, P. J., Millrain, M., Antoniou, J., Simpson, E. and Mellor, A. L. (1989). A glycophospholipid anchor is required for Qa-2-mediated T cell activation. Nature **342:** 85-87.

Rodríguez-Boulan, E. and Nelson, W. J. (1989). Morphogenesis of the polarized epithelial cell phenotype. Science. **245:** 718-725.

Rosenberry, T. L., Roberts, W. L. and Haas, R. (1986). Glycolipid membrane-binding domain of human erythrocyte acetylcholinesterase. FASEB J. **45:** 2970-2975.

Rothberg, K. G., Ying, Y., Kolhouse, J. F., Kamen, B. A. and Anderson, R. G. W. (1990a). The glycophospholipid-linked folate receptor internalizes folate without entering the clathrin-coated pit endocytic pathway. J. Cell Biol. **110:** 637-649.

Rothberg, K. G., Ying, Y., Kamen, B. A. and Anderson, R. G. W. (1990b). Cholesterol controls the clustering of the glycophospholipid-anchored membrane receptor for 5-methyltetrahydrofolate. J. Cell Biol. **111:** 2931-2938.

Rotundo, R. L. and Carbonetto, S. T. (1987). Neurons segregate clusters of membrane-bound acetylcholinesterase along their neurites. Proc. Natl. Acad. Sci. USA **84:** 2063-2067.

Sack, T. L., Gum, J. R., Low, M. G. and Kim, Y. S. (1988). Release of CEA from human colon cancer cells by PI-PLC. J. Clin. Invest. **82:** 586-593.

Sargiacomo, M., Lisanti, M. P., Graeve, L., Le Bivic, A. and Rodríguez-Boulan, E. (1989). Integral and peripheral protein composition of the apical and basolateral membrane domains in MDCK cells. J. Membr. Biol. **107:** 277-286.

Schneider, P., Ferguson, M. A. J., McConville, M. J., Mehlert, A., Homans, S. W. and Bordier, C. (1990). Structure of the GPI-membrane anchor of the *Leishmania major* promastigote surface protease. J. Biol. Chem. **265:** 16955-16964.

Selhub, J. and Frankin, W. A. (1984). The folate-binding protein of rat kidney: Purification, properties and cellular distribution. J. Biol. Chem. **259:** 6601-6606.

Seya, T., Farries, T., Nickells, M. and Atkinson, J. P. (1987). Additional forms of human DAF. J. Immunol. **139:** 1260-1267.

Shukla, S. D., Berriman, J., Coleman, R., Finean, J. B. and Michell, R. H. (1978). Membrane protein segregation during release of microvesicles from human erythrocytes. FEBS Lett. **90:** 289-292.

Simons, K. and Fuller, S. D. (1985). Cell surface polarity in epithelia. Ann. Rev. Cell Biol. **1:** 243-288.

Simons, K. and van Meer, G. (1988). Lipid sorting in epithelial cells. Biochemistry **27:** 6197-6202.

Singh, N. and Tartakoff, A. (1990). The importance of dolichol-p-mannose for the synthesis of glycolipid anchors of membrane proteins. J. Cell Biol. **111:** 197a.

Sly, W. (1982). The uptake and transport of lysosomal enzymes. In the Glycoconjugates (M. I. Horowitz, ed) **vol IV,** pp.3-25. Academic Press. N.Y.

Stahl, N. S., Borchelt, D. R. and Prusiner, S. B. (1990). Differential release of cellular and scrapie prion proteins from cellular membranes by PI-specific phospholipase C. Biochemistry **29:** 5405-5412.

Su, B., Waneck, G. L., Flavell, R. A. and Bothwell, A. L. (1991). The GPI-anchor is critical for Ly-6A/E-mediated T cell activation. J. Cell Biol. **112:** 377-384.

Takesue, Y., Yokota, K., Nishi, Y., Taguchi, R. and Ikezawa, H. (1986). Solubilisation of trehalase from rabbit renal and intestinal brush-border membranes by a phosphatidylinositol-specific phospholipase C. FEBS Lett. **201:** 5-8.

Tanaka, K., Nakamura, T. and Ichihara, A. (1986). A unique trypsin-like protease associated with the plasma membranes of rat liver: Purification and characterization. J. Biol. Chem. **261:** 2610-2615.

Tausk, F., Fey, M. and Gigli, I. (1989). Endocytosis and shedding of DAF on human polymorphonuclear cells. J. Immunol. **143:** 3295-3302.

Wandinger-Ness, A. and Simons, K. (1991). The polarized transport of surface proteins and lipids in epithelial cells. In: Intracellular Trafficking of Proteins. J. Hanover and C. Steer, eds. Cambridge University Press (in press).

Wieland, F. T., Gleason, M. L., Serafini, T. A. and Rothman, J. E. (1987). The rate of bulk flow from the endoplasmic reticulum to the cell surface. Cell **50:** 289-300.

Wilson, J. M., Fasel, N. and Kraehenbuhl, J.-P. (1990). Polarity of endogenous and exogenous GPI-anchored membrane proteins in MDCK cells. J. Cell Sci. **96:** 143-149.

Zurzolo, C., Nitsch, L., Rodríguez-Boulan, E. and Lisanti, M. P. (1990). Reversed polarity of GPI-anchored proteins in a polarized thyroid cell line. J. Cell Biol. **111:** 327a.

EVALUATION OF SOMATIC CELL VARIANTS DEFICIENT IN GLYCOSYLPHOSPHATIDYL-INOSITOL ANCHORING AS CANDIDATES FOR GENETIC CORRECTION

***[1]Fasel, N., Deglon, N, Rousseaux M., Beghdadi-Rais, C., Bernard, M and Bron, C.**

Institute of Biochemistry, University of Lausanne, CH-1066 EPALINES, Switzerland.

* Corresponding author.

ABSTRACT

Many cell surface glycoproteins are anchored in the lipid bilayer by a glycosylphosphatidyl-inositol (GPI) structure. Recently, a number of cell lines which are deficient in the biosynthesis and/or addition of this anchor have been described. In this report, we summarize the current knowledge on these lines and discuss their potential use to isolate the genes involved in the GPI anchor biosynthetic pathway with a specific emphasis on L cell fibroblasts.

INTRODUCTION

Attachment of numerous glycoproteins to the cell surface via a glycosylphosphatidyl-inositol anchor has been described in various organisms (Cross, 1990; Thomas et al., 1990). This complex structure is synthesized as a precursor glycolipid (Menon et al., 1988; Mayor et al., 1990a; Mayor et al., 1990b) which is transferred onto glycoproteins in the endoplasmic reticulum (Bangs et al., 1985; Ferguson et al., 1986; Conzelmann et al., 1987).

Somatic cell variants leading to changes in expression of cell surface antigens provide useful tools to define genes important for post-translational modifications such as GPI anchor addition. Inability to place or maintain GPI-anchored molecules on the cell surface has been described for paroxysmal nocturnal hemoglobinuria (PNH), a complex acquired stem cell disorder characterised by an unusual susceptibility of red cells, platelets and granulocytes to complement lysis. Attempts to derive stable lines from PNH patients cells have been thus far unsuccessful. Cell lines unable to express properly GPI-anchored molecules were obtained by specific selection. T lymphoma GPI anchor mutants have been isolated using Thy-1 antigen as selective cell surface market (Hyman, 1988). Other somatic variants have been described with a general defect in the glycosylphosphatidyl-inositol synthesis or addition affecting cell surface expression of a number of GPI-linked antigens. Biochemical characterization of these mutants and structural definition of some of the GPI anchors suggested that GPI biosynthesis and addition is an unexpectedly complex process.

Isolation of the genes involved in the biosynthesis and addition of the GPI is essential in the understanding of the GPI anchor biosynthetic pathway. In this chapter, we evaluate different GPI anchor deficient lines as potential candidates for complementation.

I. GPI DEFICIENT T LYMPHOMAS

A number of T lymphoma variant cell lines have been obtained by mutagenesis followed by cytotoxic immunoselection with anti-Thy-1 antibody and complement (reviewed in Hyman, 1988; Hyman, 1973; Hyman and Stallings, 1974; Hyman and Cunningham, 1986; Evans et al., 1987; Sneller and Gunter, 1987). These Thy-1 lymphoma lines were classified into nine complementation classes (A to I) by somatic cell fusion. Six classes (A, B, C, E, F and H) showed a recessive behaviour in hybrids and defined genes that act in the post-translational processing of probably all GPI-anchored molecules. The three other classes (D, G and I) had deficiencies affecting specifically Thy-1 gene expression and will not be discussed further.

MUTANT CELL LINE OF COMPLEMENTATION CLASSES A, C, F AND H

In these four complementation classes, detailed analysis revealed that Thy-1 was not correctly processed (Conzelmann et al., 1986; Conzelmann et al., 1988; Fatemi and Tartakoff, 1988). The C-terminal hydrophobic domain present on the precursor polypeptide was neither cleaved off nor was the GPI anchor added (Conzelmann et al., 1986; Conzelmann et al., 1988; Fatemi and Tartakoff, 1988) as it normally occurs in wild type cells. Such molecules were therefore resistant to phosphatidylinositol-specific phospholipase C (PI-PLC) treatment (Conzelmann et al., 1986; Conzelmann et al., 1988) and accumulated in the endoplasmic reticulum as shown by the sensitivity of the glycan moiety to endoglycosidase H (Conzelmann et al., 1988). In addition, the half-life of the Thy-1 antigen was shown to be drastically reduced (Trowbridge et al., 1978; Fatemi and Tartakoff, 1988).

MUTANT CELL LINE OF COMPLEMENTATION CLASS B

Release in the extracellular fluid of Thy-1 molecules has been shown to occur in the lymphoma of the complementation class B (Fatemi and Tartakoff, 1988; Conzelmann et al., 1988). When characterized biochemically, it was demonstrated that the Thy-1 molecules present in the culture medium had a molecular mass of 2 to 3 kDa smaller than the form present at the surface of wild type cells (Conzelmann et al., 1988). Even if conflicting evidence concerning the presence of an ethanolamine group was reported, the presence of the C-terminal hydrophobic peptidic region can be ruled out. In this mutant, removal of the C-terminal hydrophobic segment important as the GPI anchor recognition signal occurred but the GPI anchor was not added. It could mean that the enzyme(s) important for the cleavage of the C-terminus and the transfer of GPI anchor could be defective. Detailed analysis of Thy-1 gene expression (N. Fasel and N. Deglon, unpublished data) showed that release of Thy-1 in the medium could not be explained by an aberrant or alternative RNA processing modifying the coding potential of Thy-1 mRNA. Furthermore, Thy-1 transcription level is reduced by approximately 8 fold suggesting an additional defect in the Thy-1 gene transcription machinery. It is also conceivable that the defect in the GPI anchor biosynthesis affects Thy-1 gene expression via a mechanism which could involve structures like PI-glycans which have been shown to be involved in IL-2, insulin and NGF triggering pathways (Eardley and Koshland, 1991; Merida et al., 1990; Saltiel and Cuatrecasas, 1986; Saltiel et al., 1986; Romero et al., 1988; Gaulton et al., 1988; Chan et al., 1989) or alternatively via a GPI-anchored cell surface receptor influencing Thy-1 gene transcription.

COMPLEMENTATION CLASS E

The deficiency of class E mutant cell lines was shown to be due to a defect in the dolichol-phosphate-mannose synthase (DPM1; E.C. 2.4.83), an enzyme which controls the transfer of GDP-mannose onto dolichol-phosphate (Chapman et al., 1980). These cells have therefore a pleiotropic defect in N-linked oligosaccharides biosynthesis (Trowbridge and Hyman, 1979; Chapman et al., 1979). The finding that Thy-1 and other GPI-anchored proteins are not expressed on the surface of class E mutant raised the possibility that Dol-P-Man may also function as a mannosyl donor during the biosynthesis of the glycophospholipid structure, which is known to contain mannose. In fact it has been shown recently that the three mannose residues in the core glycan of trypanosome GPI lipids are derived from Dol-P-Man (Menon et al., 1990). The yeast cDNA coding for the Dol-P-Man synthase has been isolated (Orlean et al.,

1988; Deglon et al., 1991). Its expression was reported to correct glycosylation and GPI anchor deficiency in yeast (Orlean, 1990) and in different mammalian cell lines (Beck et al., 1990; De Gasperi et al., 1990; Deglon et al., 1991).

The involvement of Dol-P-Man in GPI synthesis and N-linked glycosylation raised a topological question (Hirschberg and Snider, 1987; Lennarz, 1987). Synthesis and utilisation of the Dol-P-Man occur on opposite sides of the rough endoplasmic reticulum membrane. Isolation of the Dol-P-Man synthase gene and its *in vitro* expression allowed analysis of Dol-P-Man synthase accessibility to protease digestion. It was demonstrated (Deglon et al., 1991) that this enzyme was translocated into microsomes suggesting a luminal orientation of this essential enzyme in the GPI anchor biosynthesis and in N- and O-linked glycosylation.

II. OTHER GPI-ANCHOR DEFICIENT T LYMPHOMA LINES

Thy-1$^-$ mutant lines:
Cells obtained from a radiation-induced T lymphoma, RL-1 were adapted to long-term tissue culture (Buxbaum et al., 1977) and Thy-1$^-$ clones were obtained. Evidence for a defect in the GPI anchor biosynthesis was obtained (Teng et al., 1989). Interestingly, these Thy-1$^-$ lymphoma cells have a reduced capacity to induce tumors in nude mice compared to a Thy-1$^+$ revertant or Thy-1$^+$ clones (Buxbaum and Basch, 1980; Basch et al., 1984). It is however not known whether reduced tumorigenicity of the GPI deficient line is linked to the absence of Thy-1 or another GPI-anchored molecule. Determination of the linkage to one of the complementation classes has not been shown yet, but release of Thy-1 in the medium suggests that it could belong to the class B.

TAP$^-$ mutant lines:
Another series of murine T lymphoma mutant lines have been described. These variants have been selected using antibodies to Ly-6.2(TAP),H9/25 or Thy-1 and in all cases cell lines with a pleiotropic defect in cell surface expression were obtained. In fact, selection against any of these GPI-anchored antigens led to the loss of all three determinants (Horton and Hyman, 1983).

A T-T hybridoma was mutagenized and negatively selected with anti-TAP antibodies. A mutant line with a defect in the GPI anchor biosynthetic pathway was obtained (Yeh et al., 1988) but comparison of this line with the six complementation classes has not been reported yet.

III. GPI DEFICIENT B LYMPHOMA

A GPI-deficient line has been derived from the human B lymphoblastoid JY cell line (Hollander et al., 1988). This cell line has been shown to be negative for two GPI-anchored molecules, decay accelerating factor (DAF) and Blast-1. The defect in JY cell line has not been characterised, yet this line is not deficient in conversion of GDP-mannose to Dol-P-Man, thus we can conclude that its genetic lesion is different from class E mutant line.

IV. MOUSE L CELL FIBROBLASTS

Mouse L cell fibroblasts have been reported to be GPI anchor deficient (Ferguson and Williams, 1988; Stroynowski et al., 1987; Edelman et al., 1987; Camerini et al., 1989). This was suggested by the release in the culture medium of three GPI-anchored proteins: the neural-cell adhesion molecule (NCAM-120), the Leu-8 antigen, the human equivalent of the lymph node homing receptor (Mel-14) and the Qa-2, the MHC class IQ-7 gene product (Edelman et al., 1987; Camerini et al., 1989; Stroynowski et al., 1987). Furthermore transfection of chimaeric constructs showed that sequences leading to deficient surface expression in L cells contained the amino acid signal for GPI anchoring (Stroynowski et al., 1987). Biochemical analysis of GPI-anchored molecules in mouse L cell fibroblasts could demonstrate that this cell line is a fibroblast equivalent line of one of the T lymphoma mutant lines and be used as recipient in DNA transfer complementation experiments. To test this hypothesis, we expressed Thy-1 antigen in L cells and analyzed its processing.

Biochemical characterization of Thy-1 molecules expressed in L cells: L cells were co-transfected with a Thy-1 cDNA under the control of an SV40 promoter and pSV2 hygromycin selective plasmid (Wilson et al., 1990). Northern blot analysis of cytoplasmic RNA isolated from different hygromycin resistant L cell clones allowed identification of Thy-1 expressing cells (data not shown). The clone T.9.6 expressing the highest level of Thy-1 mRNA was used in all subsequent experiments.

As a preliminary approach, Thy-1^{+}T.9.6 cells and non transfected L cells were pulse labelled for two hours in the presence of ^{35}S-methionine. The culture medium was collected and immunoprecipitation using a rabbit anti-Thy-1 antiserum (MacDonald et al., 1985) was performed. As shown in Figure 1, a protein with an apparent molecular mass of 21 to 23 kDa as determined by SDS poly-acrylamide gel could be detected in the culture medium of T.9.6 L cells (Fig.1, lane b) but not in untransfected L cells

(Fig.1, lane a). Such a result provides additional evidence that GPI-anchored molecules are secreted in L or NIH 3T3 fibroblastic cells (Ferguson and Williams, 1988; Stroynowski et al., 1987; Edelman et al., 1987; Camerini et al., 1989). Since we introduced a Thy-1 cDNA and not a genomic Thy-1, we can definitively exclude that alternative RNA processing, as it was demonstrated for Qa-2 (Ulker et al., 1990), is responsible for the presence of Thy-1 in the extracellular fluid of L cells. Interestingly the form present in the medium had a smaller molecular mass than the Thy-1 molecule found at the surface of T lymphomas (data not shown).

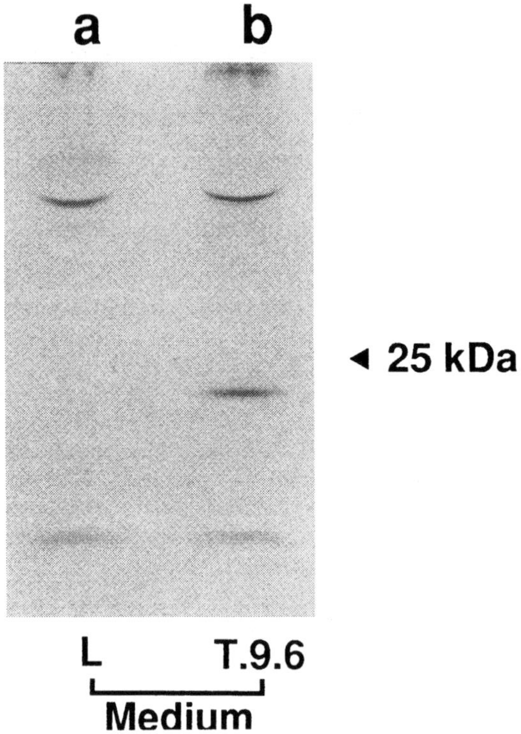

Figure 1. Biosynthetic labelling of Thy-1 released by L cells: Non transfected (line a) and transfected T.9.6 cells were pulse labelled for 2 hrs. After the labelling period, Thy-1 was immuno-precipitated from the culture medium using a polyclonal antibody against Thy-1 antigen (MacDonald et al., 1985). Immunoprecipitates were analysed by electrophoresis on SDS-PAGE and subsequent autoradiography. Position of the wild type Thy-1 is shown by an arrow at 25 kDa.

It was of interest to determine the intracellular nature of Thy-1 molecules in T.9.6L cells. For this purpose, the cells were labelled for two hours with ^{35}S-methionine followed by lysis in the presence of TX-114 detergent

(Bordier, 1981) to obtain a separation between hydrophilic and hydrophobic proteins. Under these conditions, it has been established that Thy-1 partitions solely in the detergent phase (Conzelmann et al., 1986). Immunoprecipitation of either the detergent or the aqueous phase was done with the rabbit anti-Thy-1 polyclonal antiserum and the samples were analysed by SDS polyacrylamide gel electrophoresis. A species of 25 kDa molecular mass is observed in the TX-114 detergent phase (Fig. 2, lane 1). This corresponds to the size of Thy-1 polypeptide described in T lymphoma lines. Its hydrophobic nature can be explained by the presence of a GPI anchor or of the C-terminal hydrophobic peptidic segment present on the polypeptide precursor. Since its hydrophobic nature is not altered by PI-PLC treatment (data not shown), we think that the peptidic C-terminal drh,rmy pg Yjy-1 has not been removed. ^{3}H - tryptophan labelling of Thy-1 molecules should resolve this issue, as demonstrated for Thy-1 in some of the T lymphoma mutant lines (Conzelmann et al., 1988). An intracellular form, with a molecular mass similar to the secreted Thy-1 antigen, is present in the aqueous phase (Fig. 2, lane 3). It is reasonable to think that the intracellular form partitioning? in the aqueous phase (Fig.2, lane 3) is the precursor of the form present in the extracellular fluid (Fig. 2, lane 2) and that this latter is not generated by a specific cleavage occurring at the cell surface.

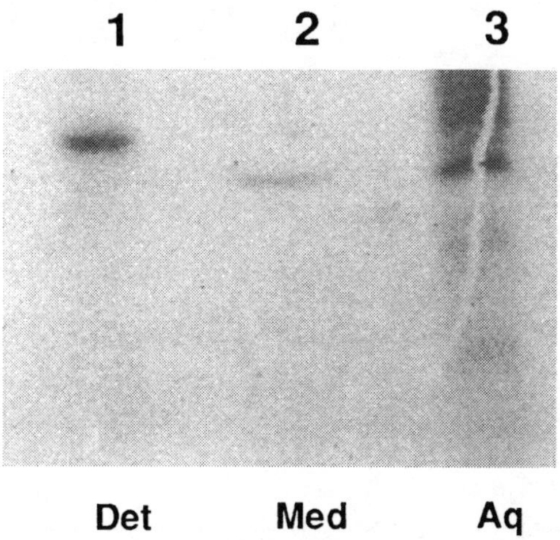

Figure 2. Analysis of Thy-1 molecules expressed in L cells by biosynthetic labelling: T.9.6 cells were pulse labelled for 2 hrs. After the labelling period and cell lysis in presence of TX-114, the culture medium (Med), the aqueous (Aq) and the detergent (Det) phases were subjected to anti-Thy-1 immunoprecipitation. Analysis was done on a 12.5% SDS-PAGE and subsequent autoradiography.

We subsequently analysed the glycosylation status of Thy-1 molecules by endoglycosidase F treatment. The intracellular hydrophilic and the secreted Thy-1 molecules (Fig. 3, lanes a and b) had an intermediate migration pattern after incubation with endo F. The hydrophobic form showed a molecular mass corresponding to a totally deglycosylated protein (Luescher and Bron, 1985; Fasel et al., 1989). Comparison of the migration pattern of the endo F treated (Fig. 3, lanes a, b and c) and untreated forms (Fig. 3, lanes d, e and f) leads to the conclusion that the intracellular hydrophilic and the secreted Thy-1 molecules have endo F resistant glycans in contrast to the TX-114 binding forms where the three N-linked sugars can be removed (Fig. 3, lane c). The reason for this endo F resistance is still unknown. This type of aberrant secondary modification could be a consequence of the defect in GPI anchor biosynthesis: aberrant N-glycosylation has never been reported in this cell line which is often used as recipient in transfection experiments. If this is the case, genetic correction of the GPI defect should also restore the glycosylation pattern expected for Thy-1 molecules. Recent experiments using endoglycosidase H showed that glycosylation is different in L cells and in S1A^{-b} cells, arguing in favour of different phenotypes in these cell lines.

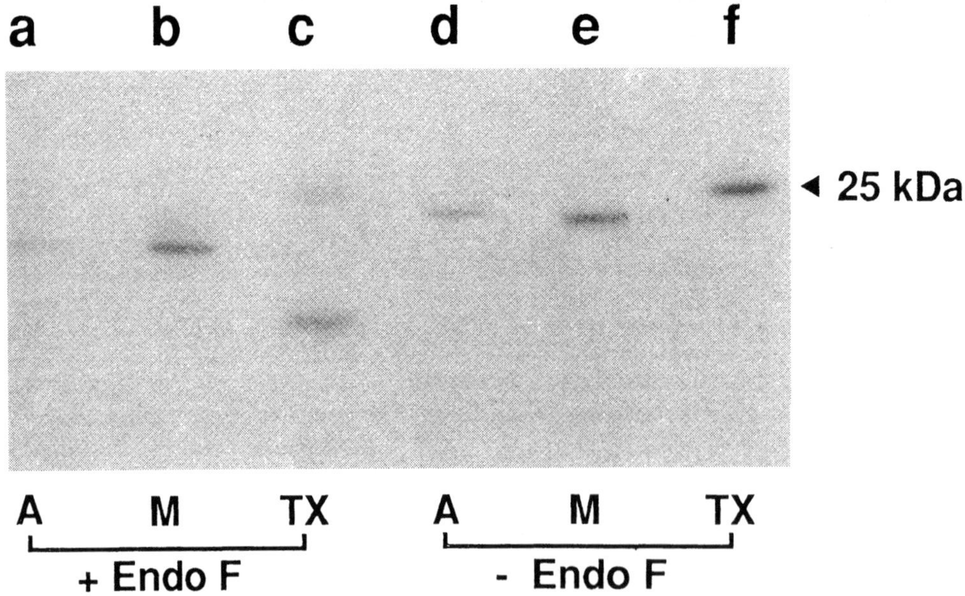

Figure 3. Endoglycosydase F analysis of Thy-1 expressed in L cells: Thy-1 molecules present in the TX-114 detergent (lanes c and f) or aqueous (lanes a and d) phases and the culture medium (lanes b and e) were prepared as described in figure 2. Endo F treated samples were loaded in lanes a, b and c; untreated samples are in the lanes d, e and f.

Finally, our effort to label the Thy-1 form released in the L cell culture medium with ^{3}H-ethanolamine has been so far unsuccessful. This leads us to propose that no GPI anchor or part of the anchor has been added to Thy-1 molecules present in L cells.

In conclusion, our data not only confirm that L fibroblast cells are deficient in GPI anchor addition but also suggest that N-glycosylation is aberrant. Since this cell line can be transfected to a high efficiency, it is one of the most adequate line for genetic complementation. Recently, it has been shown that L cells could have the same genetic lesion than T lymphoma of class H (Singh et al., 1991). If the L cell complementing gene can be isolated, genetic correction of class H cell will definitively prove it.

V. COMPLEMENTATION BY DNA MEDIATED GENE TRANSFER

The cell lines described in this chapter are potential hosts for DNA mediated complementation using cDNA expression libraries. Corrections of the defect would lead to re-expression of GPI-anchored molecules at the cell surface. Enrichment for complemented cells can be obtained by successive rounds of antibody panning (Seed and Aruffo, 1987) and transient expression. This approach was, until now, hampered by the difficulty to transfect efficiently the T lymphoma lines. Despite intensive efforts, complementation of the S1A^{-b} line using cDNA libraries has failed so far (N Deglon, unpublished results). For the other T lymphoma lines expression of reporter genes was even lower when compared to S1A^{-b} line. Difficulty to transfect these lines was observed using different techniques. These results lead us to the conclusion that mouse L cells offer a good alternative to isolate at least one of the genes involved in this complex pathway.

PERSPECTIVES

To circumvent the problem of low efficiency transfection of T lymphomas variant lines, new lines should be generated by mutagenesis and compared, by somatic cell fusion, with previously defined GPI deficient lines. With such an approach we should be able to isolate GPI deficiencies in cells where transient expression of heterologous DNA is more efficient. Complementation of these lines will allow us to isolate genes essential in the GPI anchor biosynthesis and help in the characterization of this pathway and in the understanding a complex acquired stem cell disorder such as paroxysmal nocturnal hemoglobinuria.

ACKNOWLEDGEMENTS

We are grateful to Dr P Shaw for critical reading and S Gardiol and Z Freiwald for preparation of the manuscript. This work was supported by grants from the Swiss National Fund for Scientific Rsearch No. 31-26320.89 and the Swiss Cancer League FOR-383.90.2.

REFERENCES

Bangs, J. D., Hereld, D., Krakow, J. L., Hart, G. W. and Englund, P. T.(1985). Rapid processing of the carboxyl terminus of a *Trypanosome* variant surface glycoprotein. Proc. Natl. Acad. Sci. USA **82:** 3207-3211.

Basch, R. S., Lakow, E., Teng, M.-H. and Buxbaum, J. N. (1984). Isolation of highly malignant Thy-1 positive revertants from cultured cloned Thy-1 negative lymphoma cells of low tumorigenicity. Som. Cell Mol. Genet. **10:** 37-44.

Beck, P. J., Orlean, P., Albright, C., Robbins, P. W., Gething, M.-J. and Sambrook, J. F. (1990). The *Saccharomyces cerevisiae* DPM1 gene encoding dolichol-phoshate-mannose synthase is able to complement a glycosylation-defective mammalian cell line. Mol. Cell Biol. **10:** 4612-4622.

Bordier, C. (1981). Phase separation of integral proteins in Triton-X114 solution. J. Biol. Chem. **256:** 1604-1607.

Buxbaum, J. N., Basch, R. S. and Szabadi, R. R. (1977). Analysis of Thy-1 variants of murine lymphoma cells. Som. Cell Genet. **3:** 1-15.

Buxbaum, J. N. and Basch, R. S. (1980). Immunologic basis of resistance to RL 1 induced by immunoselected Thy-1.2 negative variants. J. Immunol. **125:** 673-678.

Camerini, D., James, S. P., Stamenkovik, L. and Seed, B. (1989). Leu-8/TQ1 is the human equivalent of the Mel-14 lymph node homing receptor. Nature **342:** 78-82.

Chan, B. L., Chao, M. V. and Saltiel, A. R. (1989). Nerve growth factor stimulates the hydrolysis of glycosyl-phosphatidylinositol in PC-12 cells: A mechanism of protein kinase C regulation. Proc. Natl. Acad. Sci. USA. **86:** 1756-1760.

Chapman, A., Fujimoto, K. and Kornfeld, S. (1980). The primary glycosylation defect in class E Thy-1-negative mutant mouse lymphoma cells is an inability to synthesize dolichol-P-mannose. J. Biol. Chem. **255:** 4441-4446.

Chapman, A., Trowbridge, I. S., Hyman, R. and Kornfeld, S. (1979). Structure of the lipid-linked oligosaccharides that accumulate in class E Thy-1-negative mutant lymphomas. Cell **17:** 509-515.

Conzelmann, A., Spiazzi, A., Hyman, R. and Bron, C. (1986). Anchoring of membrane proteins via phosphatidyl inositol is deficient in two classes of Thy-1 -negative mutant lymphoma cells. EMBO J. **5:** 3291-3296.

Conzelmann, A., Spiazzi, A. and Bron, C. (1987). Glycolipid anchors are attached to Thy-1 glycoprotein rapidly after translation. Biochem. J. **246:** 605-610.

Conzelmann, A., Spiazzi, A., Bron, C. and Hyman, R. (1988). No glycolipid anchors are added to Thy-1 glycoprotein in Thy-1 negative mutant thymoma cells of four different complementation classes. Mol. Cell Biol. **8:** 674-678.

Cross, G. A. M. (1990). Glycolipid anchoring of plasma membrane proteins. Ann. Rev. Cell Biol. **6:** 1-39.

De Gasperi, R., Thomas, L. J., Sugiyama, E., Chang, H. M., Beck, P. J.,Orlean, P., Albright, C., Waneck, G. L., Sambrook, J. F., Warren, C. D. and Yeh, E. T. H. (1990). Correction of a defect in mammalian GPI anchor biosynthesis by a transfected yeast gene. Science **250:** 988-991.

Deglon, N., Krapp, A., Bron, C. and Fasel, N. (1991). Translocation of the yeast dolichol-phosphate-mannose synthase into microsomal membranes. Biochem. Biophys. Res. Comm. **174:** 1337-1342.

Eardley, D. D. and Koshland, M. E. (1991). Glycophosphatidylinositol: A candidate system for interleukin-2 signal transduction. Science **251:** 78-81.

Edelman, G. M., Murray, B.A., Mege, R.-M., Cunningham, B. A. and Gallin, W. J. (1987). Cellular expression of liver and neural cell adhesion molecules after transfection with their cDNAs results in specific cell-cell binding. Proc. Natl. Acad. Sci. USA **84:** 8502-8506.

Evans, G. A., Hyman, R. and Lewis, K. D. (1987). A mutant lymphoma cell with a defective Thy-1 glycoprotein gene. Immunogenetics **25:** 28-34.

Fasel, N., Rousseaux, M., Schaerer, E., Medof, M. E., Tykocinski, M. L. and Bron, C. (1989). In vitro attachment of glycosyl-phosphatidylinositol anchor structures to mouse Thy-1 antigen and human decay-accelerating factor. Proc. Nat. Acad. Sci. USA **86:** 6858-6862.

Fatemi, S. H. and Tartakoff, A. M. (1988). The phenotype of five classes of *T lymphoma* mutants. J. Biol. Chem. **263:** 1288-1294.

Ferguson, M. A. J., Duszenko, M., Lamont, G. S., Overath, P.and Cross, G. A. M. (1986). Biosynthesis of *Trypanosoma brucei* variant surface glycoproteins. J. Biol. Chem. **261:** 356-362.

Ferguson, M. A. J. and Williams, A. F. (1988). Cell surface anchoring of proteins via glycosyl-phosphatidylinositol structures. Ann. Rev. Biochem. **57:** 285-320.

Gaulton, G. N., Kelly, K. L., Pawlowski, J., Mato, J. M. and Jarett, L. (1988) Regulation and function of an insulin-sensitive glycosyl-phosphatidylinositol during T lymphocyte activation. Cell **53:** 963-970.

Hirschberg, C. B. and Snider, M. D. (1987). Topography of glycosylation in the rough endoplasmic reticulum and Golgi apparatus. Ann. Rev. Biochem. **56:** 63-87.

Hollander, N., Selvaraj, P. and Springer, T. A. (1988). Biosynthesis and function of LFA-3 in human mutant cells deficient in phosphatidylinositol-anchored proteins. J. Immunol. **141:** 4283-4290.

Horton, M. A. and Hyman, R. (1983). Genetic basis for Ly-6 defect: Complementation between Ly-6 and Thy-1- mutant cell lines. Immunogenetics **17:** 261-270.

Hyman, R. (1973). Studies on surface antigen variants. Isolation of two complementary variants for Thy-1.2. J. Natl. Cancer Inst. **50:** 415-422.

Hyman, R. (1988). Somatic genetic analysis of the expression of cell surface molecules. TIG **4:** 5-8.

Hyman, R. and Cunningham, K. (1986). A Thy-1- mutant defining a gene acting in trans position to regulate cell-surface Thy-1 glycoprotein expression and Thy-1 messenger RNA content. Immunogenetics. **23:** 312-321.

Hyman, R. and Stallings, V. (1974). Complementation patterns of Thy-1 variants and evidence that antigen loss variants "pre-exist" in the parental population. J. Natl. Cancer Inst. **52:** 429-436.

Lennarz, W. J. (1987). Protein glycosylation in the endoplasmic reticulum: Current topological issues. Biochem. **26: 7205-7210.**

Luescher, B. and Bron, C. (1985). Biosynthesis of mouse Thy-1 antigen. J. Immunol. **134:** 1084-1089.

MacDonald, H. R., Bron, C., Rousseaux, M., Horvath, C. and Cerottini, J.-C. (1985). Production and characterization of monoclonal anti-Thy-1 antibodies that stimulate lymphokine production by cytolytic T cell clones. Eur. J. Immunol. **15:** 495-501.

Mayor S., Menon, A. K., Cross, G. A. M., Ferguson, M. A. J., Dwek, R. A. and Rademacher, T. W. (1990a). Glycolipid precursors for the membrane anchor of *Trypanosoma brucei* variant surface glycoproteins I. Can structure of the phosphatidylinositol-specific phospholipase C sensitive and resistant glycolipids. J. Biol. Chem. **265:** 6164-6173.

Mayor, S., Menon, A. K. and Cross, G. A. M. (1990b) Glycolipid precursors for the membrane anchor of *Trypanosoma brucei* variant surface glycoproteins II. Lipid structures of phosphatidylinositol-specific phospholipase C sensitive and resistant glycolipids. J. Biol. Chem. **265:** 6174-6181.

Menon, A. K., Mayor, S., Ferguson, M. A. J., Duszenko, M. and Cross, G. A. M. (1988). Candidate glycophospholipid precursor for the glycosylphosphatidylinositol membrane anchor of *Trypanosoma brucei* variant surface glycoproteins. J. Biol. Chem. **263:** 1970-1977.

Menon, A. K., Mayor, S. and Schwarz, R. T. (1990). Biosynthesis of glycosyl-phosphatidylinositol lipids in *Trypanosoma brucei*: Involvement of mannosylphosphodolichol as the mannose donor. EMBO J. **9:** 4249-4258.

Merida, I., Pratt, J. C. and Gaulton, G. N. (1990). Regulation of interleukin-2 dependent growth responses glycosyl-phosphatidylinositol molecules. Proc. Natl. Acad. Sci. USA **87:** 9421-9425.

Orlean, P. (1990). Dolichol phosphate mannose synthase is required in vivo for glycosyl phosphatidylinositol membrane anchoring, O mannosylation, and N glycosylation of protein in *Saccharomyces cerevisiae*. Mol. Cell Biol. **10:** 5796-5805.

Orlean, P., Albright, C. and Robbins, P. W. (1988). Cloning and sequencing of the yeast gene for dolichol phosphate mannose synthase, an essential protein. J. Biol. Chem. **263:** 17499-17507.

Romero, G., Luttrell, L., Rogol, A., Zeller, K., Hewlett, K. and Larner, J. (1988). Phosphatidylinositol-glycan anchors of membrane proteins: Potential precursors of insulin mediators. Science **240:** 509-511.

Saltiel, A. R. and Cuatrecasas,P. (1986). Insulin stimulates the generation from hepatic plasma membranes of modulators derived from an inositol glycolipid. Proc. Natl. Acad. Sci. USA **83:** 5793-5797.

Saltiel, A. R., Fox, J. A., Sherline, P. and Cuatrecasas, P. (1986). Insulin stimulated hydrolysis of a novel glycolipid generates modulators of cAMP phosphodiesterase. Science. **233: 967-972.**

Seed, B. and Aruffo, A. (1987). Molecular cloning of the CD2 antigen, the T cell erythrocyte receptor, by a rapid immunoselection procedure. Proc. Natl. Acad. Sci. USA **84:** 3365-3369.

Singh, N., Singleton, D. and Tartakoff, A. M. (1991). Anchoring and degradation of glycolipid-anchored membrane proteins by L929 versus by LM-TK-mouse fibroblasts: Implications for anchor biosynthesis. Mol. Cell Biol. **11:** 2362-2374.

Sneller, M. C. and Gunter, K. C. (1987). DNA methylation alters chromatin structure and regulates Thy-1 expression in EL4 T cells. J. Immunol. **138:** 3505-3512.

Stroynowski, I., Soloski, M. J., Low, M. G. and Hood, L. E. (1987). A single gene encoded soluble and membrane-bound forms of the major histocompatibility Qa-2 antigen: Anchoring of the product by a phospholipid tail. Cell **50:** 759-768.

Teng, M.-H., Hedayati, S., Alexander, A. A., Barkin, R., Basch, R. S. and Buxbaum, J. N. (1989). A Thy-1 negative lymphoma cell variant defective in the formation of glycosyl-phosphatidylinositol membrane protein anchors. Mol. Immunol. **26:** 391-402.

Thomas, J. R., Dwek, R. A. and Rademacher, T. W. (1990). Structure, biosynthesis, and function of glycosyl-phosphatidylinositols. Biochem. **29:** 5413-5422.

Trowbridge, I. S., Hyman, R. and Mazauskas, C. (1978). The synthesis and properties of T25 glycoprotein in Thy-1 negative mutant lymphoma cells. Cell **14:** 21-32.

Trowbridge, I. S. and Hyman, R. (1979). Abnormal lipid-linked oligosaccharides in class E Thy-1 negative mutant lymphomas. Cell **17:** 503-508.

Ulker, N., Lewis, K. D., Hood, L. E. and Stroynowski, I. (1990). Activated T cells transcribe an alternatively spliced mRNA encoding a soluble form of Qa-2 antigen. EMBO J. **9:** 3839-3847.

Wilson, J. M., Fasel, N. and Kraehenbuhl, J.-P. (1990). Polarity of and exogenous glycosyl-phosphatidylinositol-anchored membrane proteins in Madin-Darby canine kidney cells. J. Cell Sci. **96:** 143-149.

Yeh, E. T. H., Reiser, H., Bemezai, A. and Rock, K. L. (1988). TAP transcription and phosphatidylinositol linkage mutants are defective in activation through the T cell receptor. Cell **52:** 665-674.

BIOCHEMICAL AND FUNCTIONAL CHARACTERIZATION OF A GLYCOLIPID ANCHORED CELL ADHESION MOLECULE IN *DICTYOSTELIUM DISCOIDEUM*

***Aline M da Silva and Claudette Klein.**

Departamento de Bioquímica, Instituto de Química, Universidade de São Paulo, Caixa Postal 20780 São Paulo, 01498 Brasil.

* Corresponding author.

Evidence has been accumulating over the last few years that a number of proteins from a diversity of eukaryotic organisms are anchored to the plasma membrane by a glycosyl-phosphatidylinositol (GPI) anchor (Ferguson and Williams, 1988; Low, 1989; Thomas et al, 1990). A characteristic of many, but not all, of the GPI-anchored proteins is that they can be released from the surface when cells are incubated with bacterial phosphatidylinositol-specific phospholipase C (PI-PLC) (Ferguson and Williams 1988; Low, 1989). Such findings suggest that this structure could allow for the release of anchored proteins in response to an equivalent, endogenous phospholipase activity.

A GPI-anchored protein has recently been reported for a cell adhesion molecule (gp80) in the slime mold *Dictyostelium discoideum* (Sadeghi et al, 1988b; Stadler et al, 1989). Also an enzymatic activity that resembles a PI-PLC seems to be involved in the release of this molecule from *the D discoideum* cell surface (da Silva and Klein, 1989). In this paper we have attempted to consolidate the current information in this system aiming to provide a general picture of the common and unique features of GPI-anchoring in *D. discoideum*.

BIOCHEMICAL CHARACTERIZATION OF GP80 AND OF ITS GPI-ANCHOR

The best characterized *D. discoideum* cell adhesion molecule is an 80 kDa glycoprotein (gp80) synthesized during the aggregation stage of the developmental program. This protein appears to mediate cell contact, via homophilic interactions, during this period (Muller and Gerisch, 1978, Siu et al, 1987). The ability of aggregating cells to form EDTA-resistant cell contacts (referred to as Contact Sites A or csA) during aggregation has been correlated with the presence of gp80 (Gerisch, 1980). A cell surface glycoprotein, referred to as antigen 117, has also been implicated in cell adhesion during

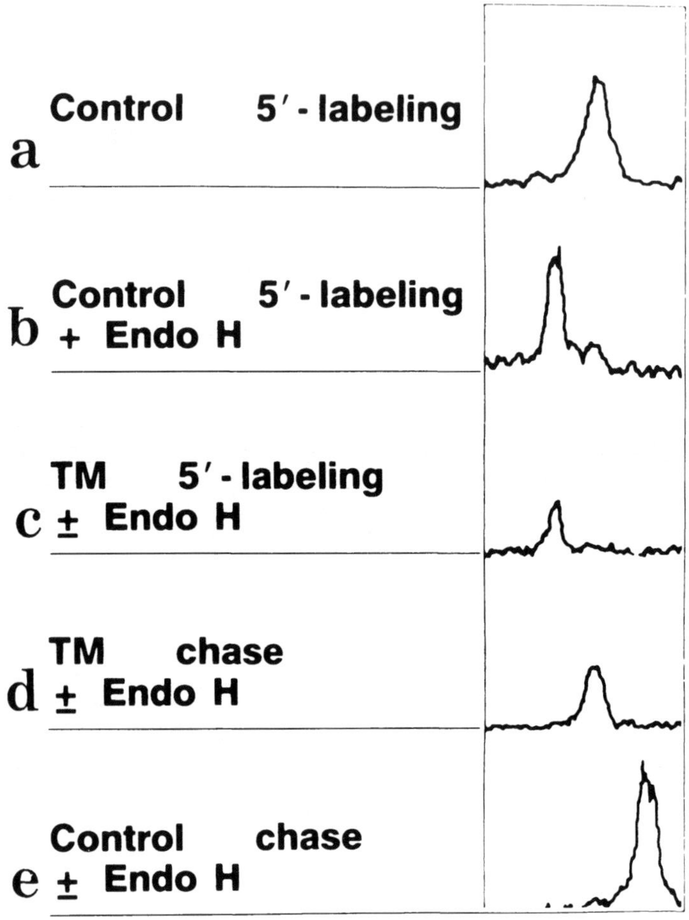

Figure 1. Pulse chase analysis of gp80. Cells were starved in the absence or presence of tunicamycin for 7 hr. The cells were then harvested, labelled with $[^{35}S]$methionine for 5 min or in the case of chase washed free of radioactive label and further incubated with 10 mM methionine for 1 hr. The protein was immunoprecipitated using the mono-clonal antibody 117 and one half of immunoprecipitated material was subjected to endo H digestion as described (Sadeghi et al, 1987). The products were analyzed by SDS-PAGE followed by fluorography and the figure shows the densitometer scans of the region of interest of the auto-radiogram. The scans show (a) the precursor labelled in control cells; (b) the precursor treated with endo H; (c) the precursor labelled in tunicamycin-treated cells ± endo H digestion; (d) the mature antigen formed in tunicamycin-treated cells ± endo H digestion; (e) the mature protein formed in control cells ± endo H digestion. Scans run from low (left) to high (right) molecular weight showing the position of 52, 60-62 and 69-72 kDa proteins. (From Sadeghi et al, 1988a).

aggregation and appears in SDS-PAGE as a doublet of approximately 69-72 kDa (Brodie et al, 1983). This protein has been the subject of investigation of our laboratory. A recent collaboration between our laboratory and that of Dr Gerisch has established that gp80 and antigen 117 are similar, if not identical and that reported differences most likely reflect differences in the processing of the protein is generated by extensive modification of 52/53 kDa peptide. Pulse-chase experiments have shown that the first detectable precursor after 5-min labelling with [^{35}S]methionine migrates on SDS-PAGE as a broad band with molecular weight between 60 and 62 kDa (Fig. 1a). This broad band can be resolved into two distinct bands. (In Dr Gerisch's laboratory, this precursor is a single band of approximately 68 kDa). This molecular weight reflects the cotranslational addition of N-linked oligosaccharides to the protein (Sadeghi et al, 1987; Hohmann et al, 1987). After endo H digestion, this precursor migrates on SDS-PAGE as a protein of 52 kDa (Fig. 1b). Cells incubated with tunicamycin and pulse-labelled for 5 min produce an endo H insensitive protein that comigrates with endo H-digested precursor of control cells (Fig. 1c). After a 1-hr chase period, the protein radiolabelled in tunicamycin-treated cells migrates on SDS-PAGE with a molecular weight of 62 kDa (Fig. 1d). This form of the protein is not endo H sensitive. As seen in Fig. 1e after a 1-hr chase, all the radiolabel in the control cells is present in the endo H-resistant 69-72 kDa bands of the mature protein. Thus, a post-translational modification which gives rise to the mature form of the protein involves an alteration of the carbohydrate structure such that it becomes endo-glycosidase H-resistant. The 60-62 kDa precursor (or 68 kDa precursor reported by Dr Gerisch) is modified by a post-translational addition of oligosaccharides referred to as type 2 carbohydrate (Hohmann et al, 1987). This modification, which accounts for the increase in molecular mass to produce the mature forms can occur independently of N-linked glycosylation and most likely reflects the addition of O-linked carbohydrates (Sadeghi et al, 1987; Hohmann et al, 1987; Sadeghi et al, 1988a).

Evidence for a GPI anchor on gp80 was first obtained by radiolabelling the protein with myristate, palmitate, myo-inositol, or ethanolamine (Sadeghi et al, 1988b; Stadler et al, 1989). Typical results are shown in Fig.2. In this experiment, cells were starved for 2 hr and then incubated with the appropriate radiolabelled compound for an additional 4 hr. The protein was then immunoprecipitated and analyzed by SDS-PAGE and autoradiography. Ethanolamine effectively labelled the protein and subsequent experiments verified that the radiolabel remained as ethanolamine. In our hands the protein appears to be equally labelled with either palmitate or myristate.

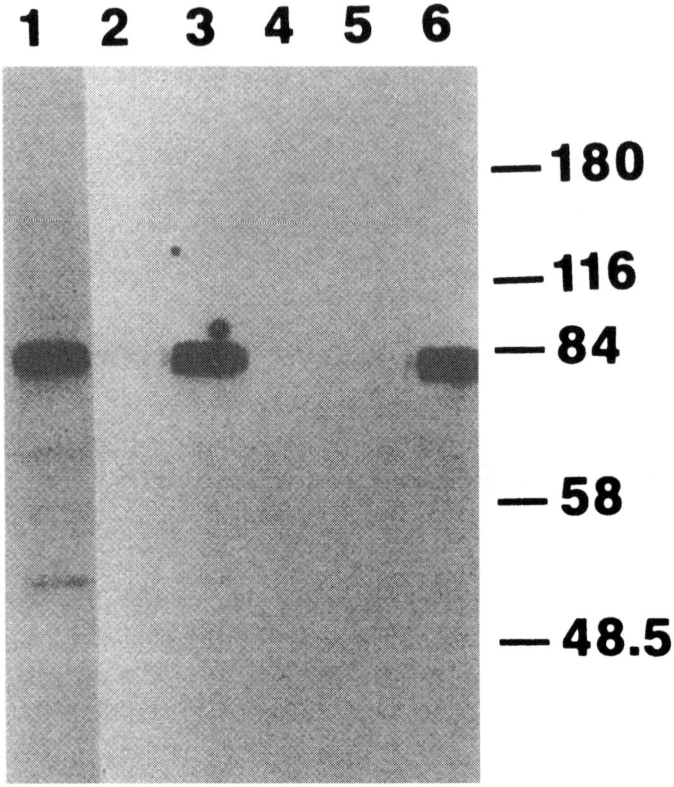

Figure 2. Radiolabelling of gp80. Cells were radiolabelled with [^{3}H]ethanolamine (lane 1), [^{3}H]myristate (lanes 2, 3, and 4) or [^{3}H]palmitate (lanes 5 and 6). After 6 hr of starvation, the protein was immunoprecipitated and the immunoprecipitate was treated with nitrous acid (lanes 2 and 5) or sodium periodate (lane 4). Controls for nitrous acid and periodate treatments involved incubating the immunoprecipitated protein in 50mM sodium citrate, pH 3,5 (lane 6) or adding ethylene glycol prior to sodium periodate (lane 3), respectively. The products were analyzed by SDS-PAGE and fluorography. The results of periodate treatment of gp80 isolated from cells incubated with [^{3}H]palmitate were the same as those obtained when [^{3}H]myristate was the source of radiolabel (not shown). The migration of molecular weight markers in kDa is on the right (from Sadeghi et al, 1988b).

Compositional analyses of the fatty acid label incorporated into the protein indicated that when cells were incubated with palmitate, 60 to 80% of the radiolabel incorporated could be released by saponification and chromatographed as palmitate. In contrast, only 10% of the radiolabel incorporated into the protein was released when cells were incubated with myristate and the released radiolabel chromatographed as palmitate. This may be a reflection of the unusual nature of the anchor synthesized by this organism. Data to be discussed below indicate that the anchor has a composition somewhat distinct from the majority of those previously described.

The ability to radiolabel gp80 with the above-mentioned compounds is consistent with the presence of GPI anchors on the protein. In the structures reported in other systems, the fatty acids are linked to the protein by a glycan, as opposed to being directly attached to the protein by an ester or amide bond (Ferguson and Williams, 1988; Low, 1989). Therefore, specific treatments that would destroy the glycan components or disrupt other linkages in the glycolipid structure would be expected to release the fatty acid from the protein moiety. As shown in Fig.2, periodate oxidation is able to release the fatty acid radiolabel from the protein, indicating that the structure that links the fatty acid to the protein contains carbohydrate. Characteristic of GPI anchors is a glucosamine residue that is not N-acetylated (Ferguson and Williams, 1988; Low, 1989). Deamination by nitrous acid can cleave the glucosamine glycosidic bond and release the phospholipid from the rest of the protein (Medof et al, 1986). As shown in Fig.2, lanes 2 and 5, such treatment also results in a loss of fatty acid label from gp80, suggesting that a non N-acetylated glucosamine is a component of the glycolipid tail of the protein.

Additional experiments have been performed to determine the nature of the fatty acid product released by nitrous acid treatment. Stadler et al (1989) observed that the released material migrated on TLC with a slower mobility in polar solvents than did phosphatidylinositol, the corresponding material released from myristate-labelled *T brucei* variant surface glycoprotein (VSG). Treatment of the nitrous acid released material with hydroxylamine did not liberate free fatty acids indicating that the lipid moieties were not present in the form of ester bonds but more likely present in amide linkages. In addition, when the product of nitrous acid hydrolysis was incubated with sphingomyelinase, the label produced migrated as a ceramide on TLC. From these data it would appear that GPI anchors in gp80 may contain, as opposed to diacylglycerol, a ceramide and may represent a prototype for a novel class of anchors (Stadler et al, 1989). Indeed, more recent data have identified a non N-acetylated glucosamine in the lipopeptido phosphoglycan (LPPG) from *Trypanosoma cruzi*, showing that LPPG which appears to contain a glycophospho-inositol-ceramide has structural features similar to the GPI anchors (Lederkremer et al, 1990).

Several glycolipid-tailed proteins have been shown to be sensitive to the action of bacterial PI-PLC (phosphatidylinositol-specific phospholipase C) (Ferguson and Williams, 1988; Low, 1989), and this sensitivity has been used as a criterion for GPI-anchor identification. However, several proteins with known GPI anchors are resistant to PI-PLC (Ferguson and Williams, 1988; Low, 1989; Thomas et al, 1990). It would appear that neither the purified PI-PLC from *B thuringensis* or *Staphylococcus aureus* (generous gifts of Dr M Low) nor that from *T brucei* (provided by Drs P T Englund or R Bülow) are capable of hydrolysing the anchor of gp80 (Sadeghi et al, 1988b; Stadler et al, 1989; da Silva and Klein, 1990). However, it should be pointed out that, if indeed a ceramide is part of the gp80 anchor, it is unlikely that this would be the cause of its resistance to PI-PLCs since the basic moiety recognized by this enzyme should still be readily exposed and accessible. In addition, it has recently been shown that *B thuringensis* PI-PLC can cleave the phosphoinositol ceramide moiety of *T cruzi* LPPG (Lederkremer et al, 1990). Thus, it appears that the actual nature of the GPI anchor in gp80 and the reasons for its insensitivity to exogenous PI-PLCs may only become clear upon further chemical analysis of the structure. It may be that such experimentation would reveal an additional modification of the anchor as has been shown in the case of the human erythrocyte acetylcholinesterase, where additional palmitylation of the inositol ring explains the resistance of this glycolipid anchor to the action of purified PI-PLCs (Roberts et al, 1988a; Roberts et al, 1988b).

It is not clear when, in the biogenesis of the protein, gp80 becomes anchored but in pulse chase experiments we have been able to radiolabel the 60-62 kDa precursor with either myristate or palmitate (unpublished observations). Assuming that only the anchor of the protein would contain fatty acid label, an assumption that seems reasonable in light of the experiments showing a glycan bridge between fatty acid and the protein, these data would argue that GPI anchoring in this system occurs co-translationally or immediately after translation. This is consistent with the current models for this process where generally it is believed that in an early event in biosynthesis the C-terminal amino acid residues are cleaved from the protein and replaced by the glycolipid anchor (Doering et al, 1990). Although the newly created amino acid C-terminal residue to which the GPI anchor is attached has not been identified in the case of gp80, a putative residue has been suggested based on the cDNA of the protein and the localization of the hydrophobic C-terminal domain that appear to be characteristic of such cleavages (Ferguson and Williams, 1988). As we have recently succeeded in expressing the cDNA for gp80 in cells that do not normally express the protein (see below), we are currently using such a transformation system to dissect the signal for GPI anchoring in *D. discoideum*.

PUTATIVE FUNCTIONS FOR GLYCOLIPID ANCHORING IN *D. DISCOIDEUM*

The reasons for which specific proteins are GPI-anchored remain somewhat of a mystery. In the case of gp80, this modification seems sufficient to localize the protein in the plasma membrane and orient it to the external domain. We and others have demonstrated that these events are undisturbed in cells treated with tunicamycin in which the gp80 that is made there is devoid of its N-linked oligosaccharides (Hohmann et al, 1987; Sadeghi et al, 1988s). Additionally, the protein synthesized in modB mutants, mutants that are defective in the posttranslational addition of carbohydrate (s) to a number of proteins (Murray et al, 1984), is accurately targeted to the cell surface (Sadeghi et al, 1988b). Thus the GPI anchor seems sufficient for targeting the protein to the plasma membrane. The GPI anchor may also provide the protein with an enhanced lateral mobility in the plane of the membrane compared to proteins that are inserted via transmembrane domains. This could be of particular advantage with respect to a cohesion protein which is functional during cell aggregation. During this time, cells migrate chemotactically toward central collection points, and in so doing, establish specific contacts with other cells. With this aim, the amoebae continuously emit and retract pseudopods to orient their migration and form the specific polar and side-side contacts that characterize their migration (Choi and Siu, 1987). It is possible, to ensure that contact be initially established, that gp80 should be distributed over the cell surface. Once initial contact between cells occurs, this contact could then be reinforced by increasing rapidly the localized concentration of the cohesion molecule. The capacity to do so may be imparted by the GPI anchor. Here we should emphasize the fact that in other organisms adhesion molecules, such as N-CAM and LFA-3, are also membrane associated by means of a GPI anchor (He et al, 1986; Cunningham et al, 1987; Seed et al, 1987; Dustin et al, 1987).

Additionally, it may be that the ability to hydrolyze an anchor by exogenous PI-PLCs is biologically significant. Should the endogenous equivalent be found to function under physiological conditions, disruption of the anchor would present a rapid means by which the corresponding protein could be shed from the cell surface. In *D. discoideum* it appears that such endogenous activity may be present. When amoebae are starved in spinner cultures to develop aggregation competence, approximately 20-25% of the protein was found to be present in the medium in a soluble form. The mobility of the released protein on SDS-PAGE was similar to that of the membrane bound form and thus unlikely reflect proteolysis (Sadeghi et al, 1988b; da Silva and Klein, 1989). The plasma membrane-enriched fraction retains the activity necessary to release the protein as a soluble form (Sadeghi et al, 1988b; da Silva and Klein, 1989) and gp80 is the major released protein during incubation of salt-washed membranes (Fig.3a). The releasing activity was shown to be dependent upon the pH and temperature of incubation and was found to be inhibited by divalent cation, salt and thiol-blocking agents. Protease inhibitors and various phospholipids were ineffective in altering it. The enzyme is an integral membrane protein as determined by the inability of high salt washes to remove it from the membranes and several lines of evidence suggest that this enzyme is a PI-PLC like activity.

As mentioned earlier, the apparent molecular mass of the released material appears unaltered from that found in the membranes. When cells were radiolabelled with myristate, palmitate, ethanolamine or methionine, and both membrane bound and released forms of gp80 analyzed by immunoprecipitation, SDS-PAGE, and autoradiography, we observed that the released material retained the methionine and ethanolamine labels but not that of the fatty acids, results which confirms that the enzyme cleaves the anchor between the ethanolamine and fatty acid backbone (Sadeghi et al, 1988b). The product of the releasing enzyme was investigated using an anti-cross reacting determinant (CRD) antibody (da Silva and Klein, 1989). The CRD is an antigenic determinant(s) present on the soluble form of VSG of *T brucei* but not detectable on the membrane-bound form of the protein (Cardoso de Almeida and Turner, 1983).The major epitope recognized by the antibody appears to be the cyclic phosphate on the inositol ring that is generated upon hydrolysis of the GPI anchor with PI-PLC (Zamze et al, 1988; Shak et al, 1988), but is absent from the epitope when the anchor is hydrolyzed by PI-PLD (Davitz et al, 1987; Cardoso de Almeida et al, 1988). When equivalent amounts of gp80 present in released or membrane-bound forms were analyzed by immunoblots with the anti-CRD antibody (a gift from Dr N Andrews) no band corresponding to gp80 was observed in the membrane fraction (Fig.3). However, gp80 that had been released into soluble fraction contained the CRD epitope, an observation that provides the first evidence the CRD epitope can be revealed in gp80 by an endogenous PI-PLC like enzyme present in the membrane fraction (da Silva and Klein, 1989). Therefore, in analogy to *T cruzi* LPPG (Lederkremer et al, 1990), and despite the anticipated unusual structural features of gp80 GPI, the CRD epitope was detected in a ceramide containing anchor.

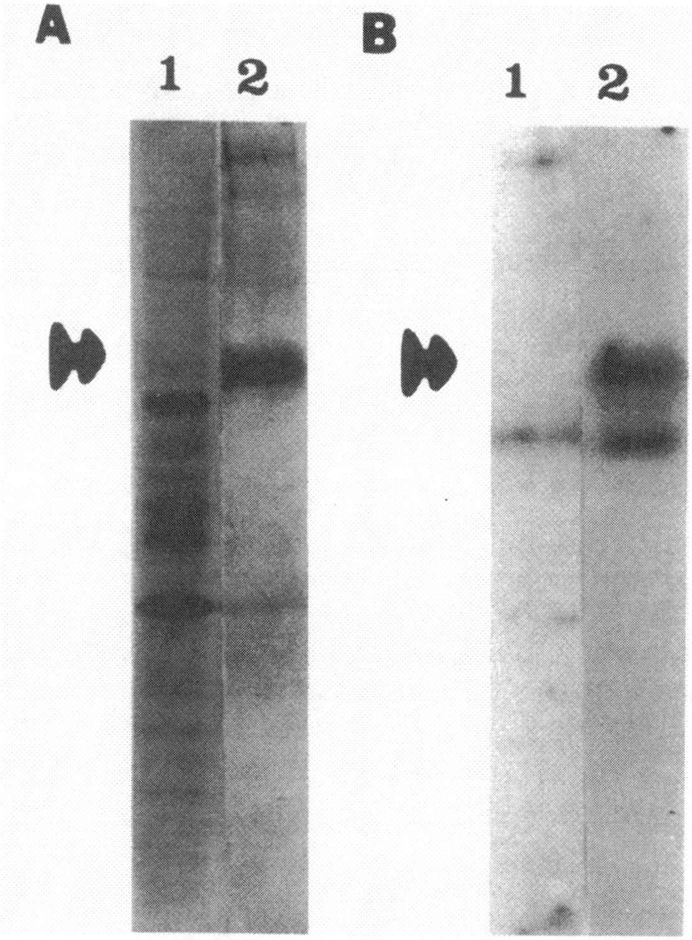

Figure 3. Specificity of gp 80 release and detection of CRD. (A) Membranes isolated from aggregating cells were incubated for 2 hr in the presence of 3 M NaCl (lane 1) and then for another 2 hr in absence of salt (lane 2). Silver-stained material released under the respective incubation conditions is shown. Lane 2 shows the protein released from a five fold greater quantity of membranes compared with that of lane 1. (B) The autoradiogram of immunoblots probed with anti-CRD antibody. Lane 1 contains total membrane proteins; lane 2, the equivalent released fraction. Release was performed in the absence of salt. The arrow indicates the position of gp80 (from da Silva and Klein, 1989).

The presence of such an enzyme that specifically releases gp80 from the cell surface under physiological conditions has interesting implications as to the role of this activity in development. As mentioned earlier, approximately 20-25% of the protein can be released from intact aggregation competent cells and the same was found to be true of the membranes prepared from such cells. Restricted release of the protein could occur if the majority of the protein is inaccessible to the releasing enzyme, for example due to different membrane distributions. Alternatively, limited release could be explained by the presence of two biochemically distinct populations of the protein, only a minority being sensitive to enzymatic hydrolysis and the majority being modified to prevent its release. Independent of the exact mechanism, the limited release of the protein during aggregation is necessary to assure that the protein is present on the cell surface in order to function at that developmental stage. However, since the protein rapidly disappears from the cell surface of cells that have entered into aggregates and begun to form slugs, it is conceivable that this loss reflects an activation of the releasing enzyme described above, or a conversion of the gp80 into a form more readily hydrolyzed by it. Unfortunately, several technical problems prevent us from determining the activity of the releasing enzyme at this later developmental stage.

These and many other questions concerning the processes involved in the addition and function of the GPI anchor we addressed using the transformation system since we have recently succeeded in expressing gp80 in vegetatively growing cells (da Silva and Klein, 1990). In collaboration with Drs T Graham and A Kaplan (Saint Louis University), we have constructed an expression vector, pDNeo67, in which we cloned the cDNA for gp80 downstream from the actin 6 promoter. The construct, pDNeo67.80 allows for the expression of the protein in growth phase cells, which do not normally express the protein. Fig 4 shows gp80 expression levels during growth and development monitored by immunoblots of cells transformed with the construct pDNeo67.80 (AX2.T80 cells) or with the vector alone (AX2-TC cells). AX2-T80 cells expressed the protein during growth as well as during the first few hours of starvation (lanes 1-3). After that time, the actin 6 promoter is minimally functional and the endogenous promoter for gp80 becomes functional. Thus we see the increase in the level of gp80 after 4 hours of starvation (lane 4) similar to the rise seen in control transformants (lane 10). Growing AX2-T80 cells expressed approximately one-fourth of the protein expressed by fully aggregation competent cells, a result consistent with the relative amounts of the mRNA present in each of the cells. The external cell surface localization of the protein synthesized by AX2-T80 cells was verified by radioimmune assay (da Silva and Klein, 1990).

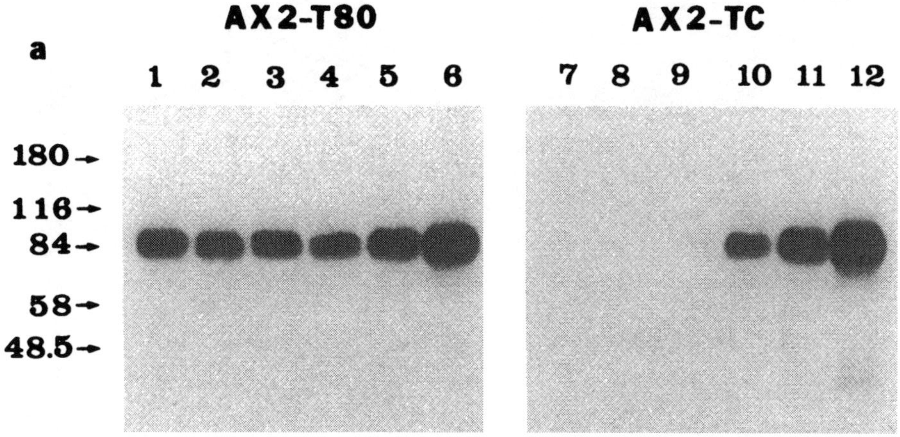

Figure 4. Accumulation of gp80 during growth and development of the transformants. Vegetative amoebae were washed free of growth medium and starved as a spinner suspension. At indicated times, cells were harvested and 100ug of total cell proteins was analyzed by immunoblotting. Lanes 1-6 and 7-12 show the results obtained using AX2-T80 and AX2-TC cells, respectively. Lanes 1 and 7, vegetative cells; lanes 2 and 8, 1-hr-starved cells; lanes 3 and 9, 2-hr-starved cells; lanes 4 and 10, 4-hr-starved cells; lanes 5 and 11, 6-hr-starved cells; lanes 6 and 12, 8-hr-starved cells. Cells had formed tight aggregates by 8 hr of starvation. Molecular weight markers (in kDa) are indicated on the left (adapted from da Silva and Klein, 1990).

The protein synthesized by growing AX2-T80 cells appears to be identical to that made by aggregation competent cells. The precursor is a doublet of approximately 60-62kDa which is processed to the mature product of 69-72kDa, (Fig 5a.) (We remind the reader that in the AX2 strain of Dr Gerisch's laboratory, the protein is not a doublet and at maturity migrates as 80kDa protein). Thus, it would appear that the protein in AX2-T80 cells is correctly modified by N- and O- linked oligosaccharides. The protein is also radiolabelled by sulfate, phosphate, ethanolamine, myristate and palmitate, (Fig.5b), lanes 1,2,3,4 and 5; lanes 6 and 7 show the retention of fatty acid label when gp80 is treated with PI-PLC. These results show that growing cells also modify the N-linked oligosaccharides with sulfate, as do developing cells, and that they modify the protein with a GPI-anchor similar to that of developing cells. These observations have been

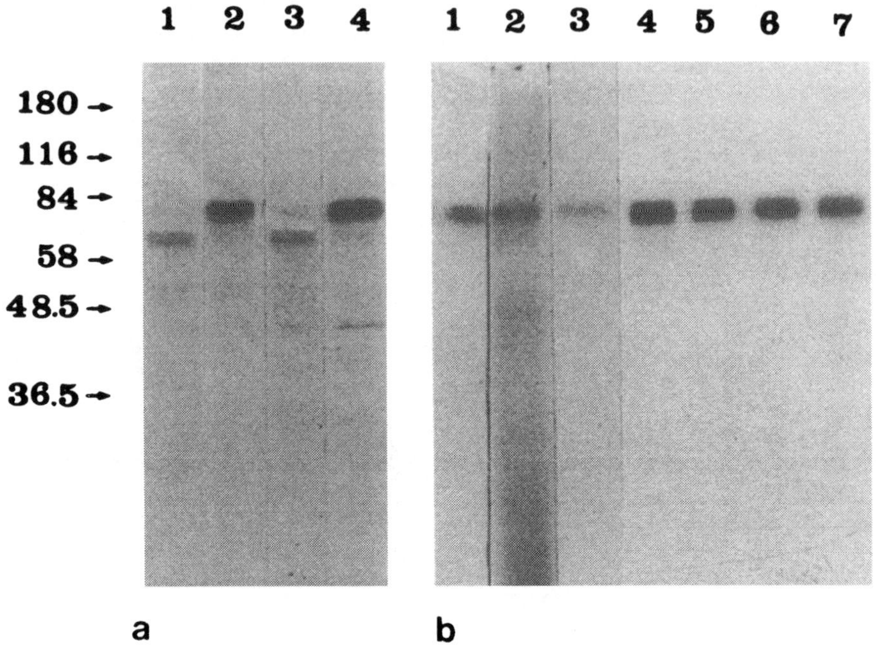

Figure 5. Biochemical analysis of gp80 and its biogenesis. (A) Pulse chase analysis was performed by incubating vegetative AX2-T80 cells (lanes 1 and 2) or aggregation-competent AX2-TC cells (lanes 3 and 4) with $[^{3}H]$methionine for 5 min in 20 mM phosphate buffer, pH 6.4. Cells were washed free of the radiolabel and incubated with 10 mM methionine for 30 min (lanes 2 and 4). Incorporation of label into gp80 was determined by immunoprecipitation, SDS-PAGE, and autoradiography. Molecular weight markers, (in kDa), are indicated on the left. (B) Vegetative cells were incubated with $[^{35}S]SO_4$ (lane 1); $[^{32}P]PO_4$ (lane 2); $[^{3}H]$ethanolamine (lane 3); $[^{3}H]$myristate (lanes 4 and 6); $[^{3}H]$palmitate (lanes 5 and 7). Lanes 6 and 7 show the retention of fatty acid label when gp80 was treated with PI-PLC (from da Silva and Klein, 1990).

recently confirmed by another laboratory (Faiz et al, 1990). The similarity between the protein synthesized by vegetative and starved cells was further demonstrated by examining the cross reactivity with the anti-CRD antibody. Fig 6 shows that during growth, AX2-T80 cells release a portion of the protein into the medium in a soluble form, that membranes incubated under the appropriate conditions retain this releasing activity, and that the released material is recognized by the anti-CRD antibody, as is that released from developing cells. The data indicate that the anchor synthesized by growing cells appears to be the same as that produced by developing cells and provides evidence for the existence of the PI-PLC like enzyme activity also in growing amoebae. That growing cells contain the necessary activities to synthesize and degrade GPI-anchors may implicate these processes in regulating growth phase cell proteins or functions. To date, however, no GPI-anchored proteins has been identified in such cells.

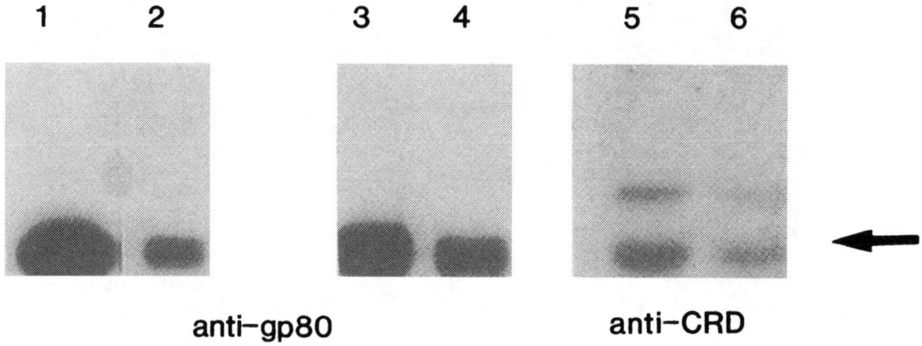

Figure 6. Release of gp80 during growth. AX2-T80 cells (lane 1) and the equivalent amount of growth medium (lane 2) were analyzed by immunoblotting. The release of gp80 from the plasma membrane fraction of aggregation competent and vegetative AX2-T80 cells is shown in lanes 3 and 4, respectively. Lanes 5 and 6 show the material released from plasma membranes of aggregation competent and vegetative AX2-T80 cells that is recognized by anti-CRD antibody. The arrow indicates the position of gp80. Dual immunostaining was performed to verify the identity of the protein recognized by the anti-CRD antibody as gp80 (from da Silva and Klein, 1990).

Since we have been able to express gp80 in growing cells, we can now attempt to dissect the features of the protein important for its biogenesis. Anchoring in other systems appears to occur during, or immediately following the translation of the protein and involves the proteolytic removal of the C-terminal residues and attachment of the anchor to the new terminus (Doeringer et al, 1990). The process appears to require a C-terminal hydrophobic domain located downstream from the cleavage/attachment site and it is predicted by the cDNA sequence of gp80 (Ferguson and Williams, 1988; Kamboj et al, 1988). Currently, we are constructing various fusion proteins to determine what portion of the C-terminus of gp80 is required to render a secreted protein membrane bound and if such an anchored protein is also susceptible to the endogenous releasing enzyme. These, and other projects along this line, will enable us to understand the function and the biogenesis of such anchors in our system.

a

b

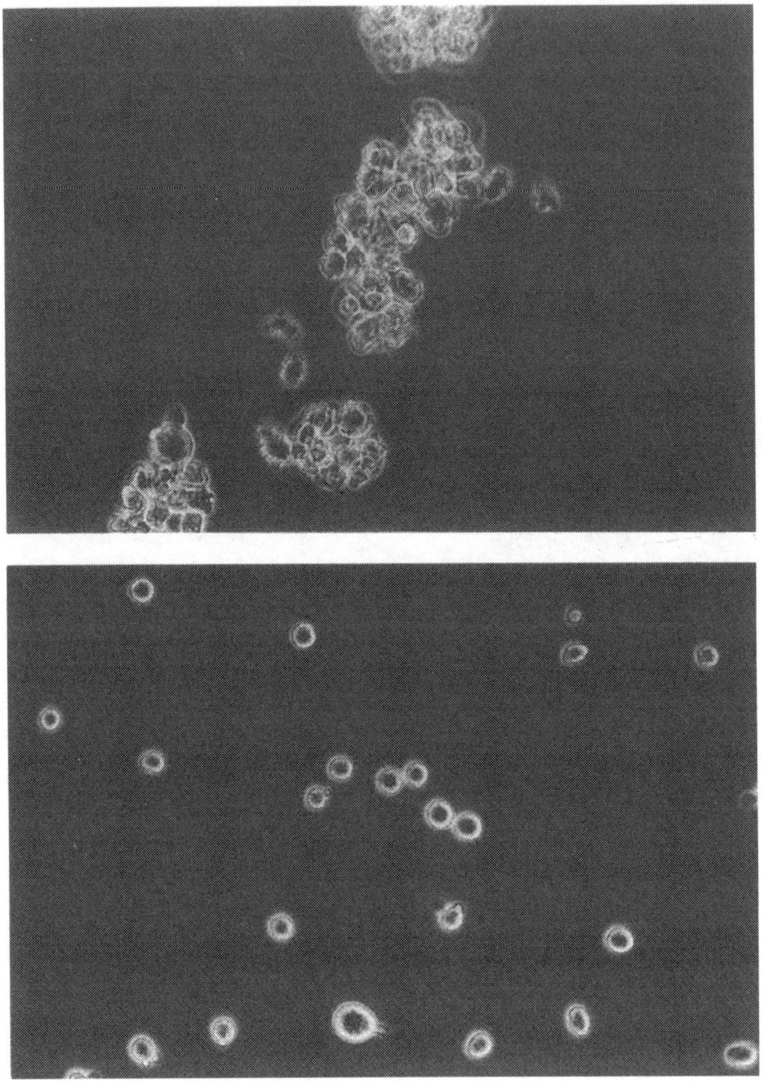

Figure 7. Phase contrast photographs of vegetative AX2-T80 cells (a) and AX2-TC cells (b). Magnification was 50X. Cells were photographed within 5 min of preparing the microscope slide (from da Silva and Klein, 1990).

The gp80 is a putative cell adhesion molecule responsible for the formation of EDTA-resistant cell contacts during aggregation (Muller and Gerisch, 1978). Although the data accumulated until now support this role, definitive proof of it was lacking since specific mutants of gp80 were not as yet reported. As a pointer in this direction, Figure 7 shows that growing AX2-T80 cells display a cohesive phenotype as opposed to growing AX2-TC cells and this provides substantial evidence for gp80s' role as a cohesion molecule. However, the contacts formed by such cells display minimal EDTA-resistance and moreover gene disruption experiments (Harloff et al, 1989) show that gp80 molecule plays an essential adhesion role when cells are subject to shear but it is not essential when the cells are not mechanically disturbed, suggesting that other molecules could be sufficient for their aggregation. Even though there is some controversy between the level of EDTA resistance showed by transformed cells that constitutively express gp80 (da Silva and Klein, 1990; Faix et al, 1990), it is important to consider the existence of other molecules in the determination of EDTA-resistance cell contacts. In addition to their interest as cohesion molecules, such candidates are intriguing as to whether or not they too would be GPI-anchored proteins, providing a commonality of function to a shared protein modification.

ACKNOWLEDGEMENT

We wish to thank Drs N W Andrews (Yale University), M G Low (Columbia University), P T Englund (Johns Hopkins University) and A Noegel (Max Planck Institute) for their gifts of various reagents. We are indebted to Dr J S Ryerse (St Louis University) for taking the photographs presented. Construction of the vector pDNeo67 was done in collaboration with Dr T R Graham and Dr A Kaplan (St Louis University). This work was supported by a NIH research grant GM34561 to CK and by a CNPq (Brazil) post-doctoral fellowship to A.M.S.

REFERENCES

Brodie, C., Klein, C. and Swierkosz, J. (1983). Monoclonal antibodies: use to detect developmentally regulated antigens on *D. Discoideum*. Cell **32:** 1115-1123.

Cardoso de Almeida, M. L. and Turner, M. J. (1983). The membrane form of variant surface glycoproteins of *Trypanosoma brucei*. Nature **302:** 349-352.

Cardoso de Almeida, M. L., Turner, M. J., Stambuk, B. V. and Schenkman, S. (1988). Identification of an acid-lipase in human serum which is capable of solubilising glyco-phosphatidylinositol-anchored proteins. Biochem. Biophys. Res. Comm. **150:** 476-482.

Choi, A. H. C. and Siu, C. H. (1987). Filopodia are enriched in a cell cohesion molecule of Mr 80,000 and participate in cell-cell contact formation in *Dictyostelium discoideum*. J. Cell Biol. **104:** 1375-1387.

Cunningham, B. A., Hemperly, J. J., Murray, B. A., Prediger, E. A., Brackenbury, R. and Edelman, G. M. (1987). Neural cell adhesion molecule: structure, immunoglobulin-like domains, cell surface modulation, and alternative RNA splicing. Science **236:** 799-806.

da Silva, A. M. and Klein, C. (1989). Characterisation of a glycosyl-phosphatidylinositol degrading activity in *Dictyostelium discoideum* membranes. Exp. Cell Res. **185:** 464-472.

da Silva, A. M. and Klein, C. (1990). Cell adhesion in transformed *D.discoideum* cells: Expression of gp80 and its biochemical characterization. Develop. Biol. **140:** 139-148.

Davitz, M. A., Hereld, D., Shak, S., Krakow, J., Englund, P. T. and Nussenzweig, V. (1987). A glycan phosphatidylinositol-specific phospholipase D in human serum. Science **238:** 81-84.

Doering, T. L., Masterson, W. J., Hart, G. W. and Englund, P. T. (1990) Biosynthesis of glycosyl phosphatidylinositol membrane anchors. J. Biol. Chem. **265:** 611-614.

Dustin, M. L., Selvaraj, P., Mattaliano, R. J. and Springer, T. A. (1987) Anchoring mechanisms for LFA-3 cell adhesion glycoprotein at membrane surface. Nature **329:** 846-848.

Faix, J., Gerisch, G. and Noegel, A. A. (1990). Constitutive overexpression of the contact site A glycoprotein enables growth-phase cells of *Dictyostelium discoideum* to aggregate. EMBO J. **9:** 2709-2716.

Ferguson, M. A. J. and Williams, A. F. (1988). Cell-surface anchoring of protein via glycosylphosphatidylinositol structures. Ann. Rev. Biochem. **57:** 285-320.

Gerisch, G. (1980). Univalent antibody fragments as tools for the analysis of cell interactions in *Dictyostelium discoideum*. Curr.Topics Develop. Biol. **14:** 342-270.

Harloff, C., Gerisch, G. and Noegel, A. A. (1989). Selective elimination of the contact site A protein of *Dictyostelium discoideum* by gene disruption. Genes and Development **3:** 2011-2019.

He, H. T., Barbet, J., Chaix, J. C. and Giordis, C.(1986). Phosphatidylinositol is involved in the membrane attachment of NCAM-120, the smallest component of the neural cell adhesion molecule. EMBO J. **5:** 2489-2494.

Hohmann, H. P., Bozzaro, S., Yoshida, M., Merkl, R. and Gerisch, G. (1987). Two-step glycosylation of the contact site A protein of *Dictyostelium discoideum* and transport of an incompletely glycosylated form to the cell surface. J. Biol. Chem. **262:** 16618-16624.

Kamboj, R. K., Wong, L. M., Lam, T. Y. and Siu, C. H. (1988) Mapping of a cell-binding domain in the cell adhesion molecule gp80 of *Dictyostelium discoideum*. J. Biol. Chem. **107:** 1835-1843.

Lederkremer, R. M., Lima, C., Ramirez, M. I. and Casal. O. L. (1990) Structural features of the lipopeptido-phospho-glycan from *Trypanosoma cruzi* common with the glyco-phosphatidylinositol anchors. Eur. J. Biochem. **192:** 337-345.

Low, M. G. (1989). Glycosyl-phosphatidylinositol: a versatile anchor for cell surface proteins. FASEB J. **3:** 1600-1608.

Medof, M. E., Walter, E. I., Roberts, W. L., Haas, R. and Rosenberry, T. L. (1986). Decay accelerating factor of complement is anchored to cells by a C-terminal glycolipid. Biochemistry. **25:** 6740-6747.

Muller, K. and Gerisch, G. (1978). A specific glycoprotein as the target site of adhesion blocking Fab in aggregating *Dictyostelium* cells. Nature **274:** 445-449.

Murray, B. A., Wheeler, S., Jongens, T. and Loomis, W. F. (1984). Mutations affecting a surface glycoprotein, gp80, of *Dictyostelium discoideum*. Mol. Cell Biol. **4:** 514-51.

Roberts, W. H., Myher, J. J., Kuksis, A., Low, M. G. and Rosenberry, T. L. (1988a) Lipid analysis of the glyco-inositol phospholipid membrane anchor of human erythrocyte acetylcholinesterase. Palmitoylation of inositol results in resistance to phosphatidyl-inositol-specific phospholipase C. J. Biol. Chem. **263:** 18766-18775.

Roberts, W. H., Santikarn, S., Reinhold, V. N. and Rosenberry, T. L. (1988b) Structural characterization of the glycoinositol phospholipid membrane anchor of human erythrocyte acetylcholinesterase by fast atom bombardment mass spectrometry. J. Biol. Chem. **263:** 18776-18784.

Sadeghi, H., Williams, K. L. and Klein, C. (1987). Characterization of antigen 117. A developmentally regulated cell surface glycoprotein of *Dictyostelium discoideum*. J. Biol. Chem. **262:** 16294-16299.

Sadeghi, H., da Silva, A. M. and Klein, C. (1988a). Biosynthesis of 117 antigen: a cell cohesion molecule in *Dictyostelium discoideum*. Develop. Genet. **9:** 561-567.

Sadeghi, H., da Silva, A. M. and Klein, C, (1988b). Evidence that a glycolipid tail anchors antigen 117 to the plasma membrane of *Dictyostelium discoideum* cells. Proc. Natl. Acad. Sci. USA. **85:** 5512-5515.

Seed, B. ((1987). An LFA-3 cDNA encodes a phospholipid linked membrane protein homologous to its receptor CD2. Nature **329:** 840-842.

Shak, S., Davitz, M. A., Wolinsky, M. L., Nussenzweig, V., Turner, M. J. and Gurnett, A. (1988). Partial characterization of the cross reacting determinant, a carbohydrate epitope shared by decay accelerating factor and the variant surface glycoprotein of the african *Trypanosoma brucei*. J. Immunol. **140:** 2046-2050.

Siu, C. H., Cho, A. and Choi, A. H. C. (1987). The contact site. A glycoprotein mediates cell-cell adhesion by homophilic binding in *Dictyostelium discoideum*. J. Cell Biol. **105:** 2523-2533.

Stadler, J., Keenan, T. W., Bauer, G. and Gerisch, G. 1989). The contact site A glycoprotein of *Dictyostelium discoideum* carries a phospholipid anchor of a novel type. EMBO J. **8:** 371-377.

Thomas, J. R., Dwek. R. A. and Rademacher, T. W. (1990). Structure, biosynthesis, and function of glycosyl-phosphatidylinositols. Biochemistry **2:** 5413-5422.

Zamze, S. E., Ferguson, M. A., Collins, R., Dwek, R. A. and Rademacher, T. W. (1988). Characterization of the cross-reacting determinant (CRD) of the glycosyl-phosphatidylinositol membrane anchor of *Trypanosoma brucei* variant surface glycoprotein. Eur. J. Biochem. **176:** 527-534.

THE MOLECULAR BIOLOGY OF GPI-ANCHORED BORDER HYDROLASES

***A J Turner, Margaret dos Santos Medeiros and N M Hooper**

Membrane Peptidase Research Group
Department of Biochemistry and Molecular Biology
University of Leeds, Leeds, LS2 9JT, UK.

* Corresponding author.

1. THE FAMILY OF BRUSH-BORDER HYDROLASES

The renal and intestinal brush borders are endowed with a range of hydrolase activities that exist as ectoenzymes. The major group of hydrolases are the peptidases which are especially abundant on renal microvilli where they may play a role in the inactivation of peptide hormones (e.g. luteinizing hormone-releasing hormone and atrial natriuretic peptides (Turner et al, 1987). At other sites the peptidases subserve other functions, for example peptide digestion in the intestine or termination of neuropeptide action in the brain. Glycosidases are principally located in the intestine and are involved in digestive processes, although trehalase and glucosamylase-maltase are also present in the renal brush border where their physiological function is unclear.

Other hydrolases include alkaline phosphatase which exists in several isoenzymic forms (intestinal/liver/placental) and the nucleotide hydrolysing enzymes, 5'-nucleotidase and alkaline phosphodiesterase I. A complete list of brush border hydrolases referred to in this chapter is provided in Table 1.

All these hydrolases share a number of features. They are all anchored to the outer, luminal face of the brush border membrane where they can hydrolyse substrates circulating extracellularly or those released into the luminal environment. They are all glycoproteins and exist as integral membrane proteins, although the mode of anchorage can differ between enzymes and even between the same enzyme in different tissues or species (see below). Three alternative modes of anchorage predominate, typified respectively by endopeptidase-24.11 (Devault et al, 1987; Malfroy et al, 1987), angiotensin converting enzyme (Hooper et al, 1987a; Soubrier et al, 1988) and membrane dipeptidase (Adachi et al, 1990a; Rached et al, 1990). Endopeptidase-24.11 is attached to the cell membrane by a region close to the N-terminus which represents the

Table 1 The family of brush border hydrolases

Enzyme class	Nature of membrane anchor	cDNA sequence available
Peptidases		
Endopeptidase-24.11	N-terminal polypeptide	Devault et al. (1987) Malfroy et al. (1987)
Endopeptidase-2 (meprin)	?	No
Angiotensin converting enzyme	C-terminal polypeptide	Soubrier et al. (1988)
Aminopeptidase N	N-terminal polypeptide	Olsen et al. (1988)
Aminopeptidase A	N-terminal polypeptide	Wu et al. (1990; 1991)
Aminopeptidase P	GPI	No
Aminopeptidase W	?	No
Dipeptidyl peptidase IV	N-terminal polypeptide	Ogata et al. (1989)
Carboxypeptidase M	GPI	Tan et al. (1989)
Carboxypeptidase P	?	No
γ-Glutamyl transpeptidase	N-terminal polypeptide	Coloma & Pitot (1986); Laperche et al. (1986)
Membrane dipeptidase	GPI	Rached et al. (1990); Adachi et al. (1990a)
Glycosidases		
Sucrase-isomaltase	N-terminal polypeptide	Hunziker et al. (1986)
Lactase-glycosyl-ceraminidase	C-terminal polypeptide	Mantei et al. (1988)
Glucosamylase-maltase	?	No
Trehalase	GPI	Ruf et al. (1990)
Other		
Alkaline phosphatase	GPI	Kam et al. (1985); Weiss et al. (1986)
5'-Nucleotidase	GPI	Misumi et al. (1990a; 1990b)
Alkaline phosphodiesterase I	GPI (?)	No

uncleaved signal peptide (Fulcher et al, 1986) preceded by a short hydrophilic, cytoplasmic domain acting as a 'stop-transfer' sequence. As with all the ectoenzymes the bulk of the protein, including the active site, is then located on the extracellular membrane face. Angiotensin converting enzyme on the other hand was shown from partial protein sequencing studies to be anchored via its C-terminus (Hooper et al, 1987a) and this has been confirmed from the cDNA sequence (Soubrier et al, 1988) which predicts a single polypeptide chain of 1306 residues with a 20 amino acid N-terminal signal peptide which is cleaved off to yield the mature protein. There is a putative membrane-spanning hydrophobic sequence of 17 amino acids located 30 residues in from the C-terminus which appears to be retained in the mature protein (Soubrier et al, 1988; Wei et al, 1991). Purified converting enzyme contains no inositol and is not released by phosphatidylinositol-specific phospholipase C (PI-PLC) (Hooper et al, 1987a; Hooper and Turner, 1989a). The other group of hydrolases, such as membrane dipeptidase, exist as glycosyl-phosphatidylinositol (GPI)-anchored membrane proteins. The remainder of this chapter will deal with this latter class of proteins focusing particularly on the peptidases which have been our special interest. The last twelve months have seen the cDNA cloning and sequencing of several of these brush border hydrolases providing important information regarding their biosynthesis, structure and function. Alkaline phosphatase which is dealt with (by Udenfriend) elsewhere in this volume will not be considered further.

2. GPI-ANCHORED PEPTIDASES

2.1 Membrane dipeptidase

Membrane (renal) dipeptidase is a zinc metalloenzyme capable of hydrolysing a range of dipeptides (Armstrong et al, 1974). The enzyme has also been implicated in the renal metabolism of glutathione and its conjugates, especially leukotriene D_4 (Kozak and Tate, 1982). The enzyme was originally shown to be associated with the 'Triton-insoluble' fraction of kidney microvilli, and hence was termed 'microsomal' dipeptidase and an association with the cytoskeleton was suggested. Subsequently we were able to show that the dipeptidase was released from kidney brush border membrane preparations by PI-PLC from several bacterial sources (Hooper et al, 1987b). The relative insolubility of the dipeptidase in Triton X-100 turns out to be a characteristic feature of GPI-anchored proteins and the detergent solubility pattern of a membrane protein can be indicative of the nature of the membrane anchor (Hooper and Turner, 1988a). In addition, differential solubilization and temperature-induced phase separation in Triton X-114 can be used to distinguish between GPI-anchored and polypeptide anchored membrane proteins (Hooper and Bashir, 1991). Subsequent affinity purification and partial sequencing of the dipeptidase allowed us to locate the anchor at the C-terminus of the protein as expected for a glycolipid anchor (Hooper et al, 1990a) and also provided sequence information for the production of suitable oligonucleotide probes for screening a pig kidney cDNA library. However, attempts to isolate cDNA clones by screening with such probes proved unsuccessful. The polymerase chain reaction technique was therefore used to generate and amplify a segment of the dipeptidase cDNA coding for a region of the known N-terminal amino acid sequence. This oligonucleotide probe was then used to screen a pig kidney cortex cDNA library allowing identification of a full length clone

(Rachel et al, 1990). The complete cDNA sequence predicts a protein of 409 amino acids. At the N-terminus there is a 16-residue cleavable signal sequence for translocation into the endoplasmic reticulum and there is also a C-terminal hydrophobic domain (Fig.1).

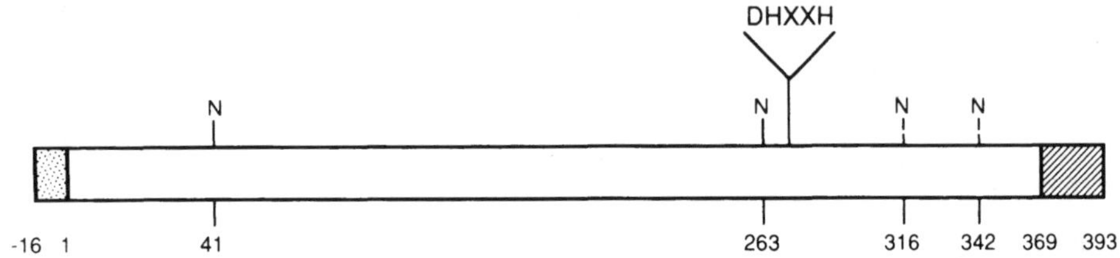

Fig. 1. Schematic representation of pig and human membrane dipeptidase. The sequence shown is that of pig membrane dipeptidase (Rached et al., 1990). The data for the human enzyme is from Adachi et al. (1990a). Residues -16 to -1 represent the cleaved N-terminal signal sequence, and residues 369-393 the cleaved hydrophobic C-terminal region. DHXXH is the proposed zinc binding site. -N, sites of N-linked glycosylation; ---N, additional sites of N-linked glycolsylation in the human enzyme. The two additional amino acid residues in the human enzyme occur in the C-terminal hydrophobic domain.

At about the same time Adachi et al (1990a) succeeded in cloning and sequencing the human membrane dipeptidase. The pig enzyme exists as a disulphide-linked dimer of subunit M_r 47000 whereas the human enzyme is considerably larger with subunits of M_r 59000. We had previously suggested, on the basis of deglycosylation and N-terminal sequence comparison of both human and pig enzymes, that the difference in size between species was solely due to differences in N-linked glycosylation (Hooper et al, 1990a). Comparison of the deduced sequences of human and pig dipeptidase confirmed this prediction (Fig.1). The human enzyme consists of 411 residues. The two additional residues in the human enzyme occur in the C-terminal 25 amino acids which are removed on maturation to the GPI-anchored form. In other respects the human and pig enzymes show extensive similarity except that two additional N-linked glycosylation sites (Fig.1) are predicted in the human, consistent with our previous comparative deglycosylation data.

When dipeptidase mRNA was expressed in *Xenopus laevis* oocytes, the dipeptidase was present as an ectoenzyme on the cell surface which could be released in a hydrophilic form by PI-PLC (Rached et al, 1990). The PI-PLC-solubilized activity was recognised as a polypeptide of M_r 48600 on Western blotting with an antiserum to the pig kidney dipeptidase. Under non-reducing conditions, a single polypeptide of M_r 80700 was observed. Thus, membrane dipeptidase when expressed in *Xenopus* oocytes exists as a disulphide-linked dimer as in pig kidney cortex (Littlewood et al, 1989; Hooper and Turner, 1989b).

2.2 Site of anchor addition on membrane dipeptidase

Recently, Adachi et al (1990b) have identified the membrane anchoring site of the human membrane dipeptidase. The procedure involved trypsin digestion followed by isolation of the C-terminal peptide by anhydro-trypsin column chromatography and reversed-phase h.p.l.c. The amino acid sequence of the peptide corresponded with residues 363-369 of the 395 amino acid protein predicted from the human CDNA sequence. This peptide was also shown to contain 2 mol ethanolamine, 1 mol glucosamine, 5 mol mannose, 1 mol inositol and 1 mol phosphate in addition to its amino acid content. We had previously shown purified pig dipeptidase to contain 1 mol of inositol/mol by G.c.-m.s. and to possess the cross-reacting determinant (CRD; see below) characteristic of the GPI-anchoring domain (Littlewood et al, 1989). Clearly the mature human dipeptidase protein lacks the C-terminal hydrophobic peptide predicted from its CDNA sequence and is anchored at Ser^{369} via GPI to the membrane. Intriguingly, pig dipeptidase contains an alanyl rather than a seryl residue at the equivalent position but the precise site of addition of the anchor to the pig enzyme has yet to be determined. Cassette mutagenesis studies on alkaline phosphatase suggest that both Ser and Ala, among others, can serve as attachment site for the GPI anchor (Berger et al, 1988). The deduced C-terminal sequences of pig and human dipeptidase as well as several other hydrolases discussed in this chapter are listed in Table 2. All have the C-terminal hydrophobic domain characteristic of GPI-anchored proteins.

When CDNA encoding the complete sequence of human dipeptidase was expressed to Cos-1 cells, the enzyme activity was localized to the plasma membrane and was released by treatment of the cells with PI-PLC (Adachi et al, 1990b). In contrast, expression of a cDNA construct lacking the hydrophobic C-terminal sequence 370-395 (see Table 2) led to secretion of the enzyme from the cells. Both recombinant forms were found to be as active as the natural enzyme and the expression system should allow preparation of large amounts of soluble enzyme for further characterization and for crystallisation.

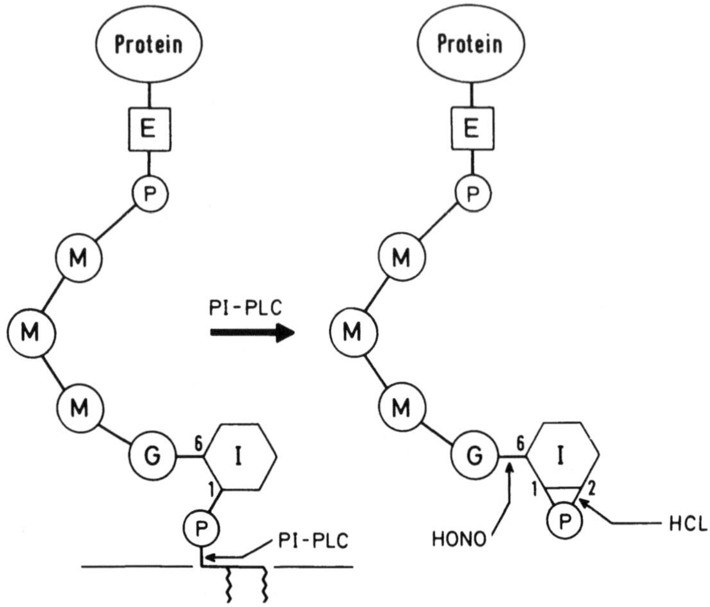

Fig. 2. Schematic representation of the cross-reacting determinant. Cleavage of a GPI membrane anchor with bacterial PI-PLC (or a eukaryotic phospholipase C) leads to the formation of the inositol 1,2-(cyclic)-monophosphate, which appears to be the major epitope involved in the cross-reactivity of mammalian GPI-anchored proteins. The sites of cleavage by nitrous acid and mild acid are also shown. Key: E, ethanolamine; G, glucosamine; I, inositol; M, mannose; P, phosphate.

2.3 Molecular cloning of carboxypeptidase M

Carboxypeptidase M (CP-M) is a widely distributed membrane peptidase with specificity for a C-terminal Lys or Arg residue. The enzyme regulates peptide hormone, particularly bradykinin, activity at the cell surface. It has been purified to homogeneity from human placenta and N-terminal sequence information was used for the construction of suitable oligonucleotide probes for detection of CP-M positive clones (Tan et al, 1989). cDNA sequence analysis predicted a protein of 439 amino acids with hydrophobic regions at both the N- and C-termini. The C-terminal hydrophobic region is flanked by charged lysines (Fig.2) but, at only 15 residues, is relatively short for an α-helical membrane-spanning region. Tan et al (1989) therefore predicted that human placental CP-M may be attached via a GPI anchor, although we had previously failed to show release of pig or human kidney CP-M by *S. aureus* or *B. thuringiensis* 24 PI-PLC (Hooper and Turner, 1988b). Subsequently, Deddish et al (1990) showed that CP-M activity from both Madin-Darby canine kidney (MDCK) cells and from human placenta was released in part by PI-PLC but not by phosphatidylcholine-specific PLC, although the origin of the PI-specific enzyme was not stated. These authors were also able to show the incorporation of [^{3}H]ethanolamine into the enzyme in MDCK cells. The differences between our results (Hooper and Turner, 1988b) and those of Erdos and colleagues (Deddish et al, 1990) may reflect tissue and species differences in the mode of anchorage of CP-M and is consistent with the observation by Northern blotting of two mRNA species for CP-M in human kidney and only one in placenta (Tan et al, 1989). Alternative processing of CP-M mRNA transcripts may therefore occur.

2.4 Purification and characterization of aminopeptidase P

Aminopeptidase P (AP-P), which hydrolyses the N-terminal amino acid from peptides with a penultimate prolyl residue, is located in the brush border of pig and human kidney (Hooper and Turner, 1988b) and has also been characterized in rat intestine (Lasch et al, 1986) and bovine lung (Orawski et al, 1987). It may play a role in the degradation of both collagen and bradykinin. We showed that pig kidney AP-P can be selectively released from pig kidney microvillar membranes by bacterial PI-PLC and has detergent solubility characteristics typical of a GPI-anchored protein (Hooper and Turner, 1988b). We have now been able to purify AP-P to apparent homogeneity following its solubilization from renal membranes by PI-PLC (Hooper et al, 1990b). On SDS/PAGE, purified pig kidney aminopeptidase P migrated with an apparent M_r of 91000 which was not significantly changed under non-reducing conditions. Deglycosylation experiments indicated two distinct populations of N-linked sugars accounting for up to 25% of the weight of the protein. The purified phospholipase-solubilized form of AP-P was recognised by an anti-CRD antiserum confirming the presence of a GPI-anchor on the protein. The availability of pure enzyme should now allow rapid progress on the structure and function of this enzyme. The rat intestinal AP-P, although similar in many respects to the pig kidney enzyme, appears to be resistant to release by PI-PLCs from various bacterial sources (R Koelsch, J Lasch, N M Hooper and A J Turner, unpublished observations). Thus, different modes of anchorage of AP-P may exist in different tissues or species. We are currently clarifying this possibility.

3. OTHER GPI-ANCHORED HYDROLASES

3.1 Molecular cloning of trehalase

Trehalase is a membrane glycosidase found in both the small intestine and the renal brush border. In the intestine it functions in the hydrolysis of ingested trehalose. It is much less abundant than many of the microvillar peptidases comprising less than 0.1% of the total brush border protein of intestine. The enzyme was shown to be released from rabbit kidney and intestinal (Takesue et al, 1986) and pig kidney (Hooper and Turner, 1988a) membranes by PI-PLC indicating a GPI-linkage. It appears to be the only glycosidase anchored in this fashion.

Ruf et al (1990) have purified rabbit intestinal trehalase and cloned and sequenced the enzyme. The cDNA encodes a protein of 578 amino acids which shows no similarity to any other known protein except *E. coli* trehalase (35% identity). The sequence is fully consistent with the mature protein containing a GPI anchor with hydrophobic domains at both N- and C-termini representing the signal peptide and the GPI-anchor signal respectively (Table 2). Four potential N-linked glycosylation sites are present in the protein. Following expression of trehalase cDNA in *Xenopus oocytes*, the enzyme was located to the cell surface and was subsequently releasable by PI-PLC. This result, and the corresponding observations with membrane dipeptidase (Rached et al, 1990 and see above), establish the ability of *Xenopus laevis* oocytes to process expressed proteins correctly to their GPI-anchored forms.

3.2 Molecular cloning of 5'-nucleotidase

5'-nucleotidase is a widely distributed membrane glycoprotein commonly used as a plasma membrane marker. The enzyme is involved in purine nucleotide hydrolysis at the cell surface producing adenosine from 5'-AMP (see Pearson (1987) for review). Data on the release of 5'-nucleotidase by PI-PLC have been conflicting, and incomplete release is commonly obtained. The susceptibility of 5'-nucleotidase to release by PI-PLC is much less than that of alkaline phosphatase (Shukla et al, 1980). For example, a concentration of PI-PLC causing 75% release of rat hepatocyte alkaline phosphatase released no detectable 5'-nucleotidase. At a 100-fold higher concentration of PI-PLC, 85% of the alkaline phosphatase was released but only 3)% of the 5'-nucleotidase. This led Luzio et al (1987) to comment: "... there is good evidence that acetylcholinesterase and alkaline phosphatase are attached to the plasma membrane by a lipid anchor similar to that for variant surface glycoprotein (VSG) and Thy-1. The data for 5'-nucleotidase are less convincing".

Table 2 Deduced C-terminal sequences of some brush border GPI-anchored hydrolases

Enzyme	Species/tissue	C-terminal sequence	Reference
Alkaline phosphatase	Human placental	480 490 500 510 ...A G T T [D] A A H P G R S V V P A L L P L L A G T L L L L E T A T A P	Berger et al. (1988)
Membrane dipeptidase	Human kidney	370 380 390 ...Y G Y S [S] G A S S L H R H W G L L L A S L A P L V L C L S L L	Adachi et al. (1990a)
Membrane dipeptidase	Pig kidney	370 380 390 ...Y G Y S A A P S L H L P P G S L L A S L V P L L L L S L P	Rached et al. (1990)
Trehalase	Rabbit small intestine	550 560 570 ...L D R Y G D R L S S G T Q L A L L E P H C L A A A L L L S F L T R	Ruf et al. (1990)
5'-Nucleotidase	Rat liver	520 530 540 ...R I K F [S] A A S H Y Q G S F P L I L L S F W A V I L V L Y Q	Misumi et al. (1990a)
5'-Nucleotidase	Human placenta	520 530 540 ...R I K F [S] T G S H C H G S F S L I F L S L W A V I F V L Y Q	Misumi et al. (1990b)
Carboxypeptidase M	Human placenta	400 410 420 ...Y R N L P D H S A A T K P S L F L F L V S L L H I F F K	Tan et al. (1989)

The boxed amino acids indicate the amino acid identified as the site of GPI anchor addition. The vertical alignment of sequences where the anchor addition site has not been identified is arbitrary. Note the underlined C-terminal hydrophobic sequences in all cases. Comparison of human and pig kidney membrane dipeptidase reveals that two additional amino acid residues occur within the C-terminal anchoring domain of the human enzyme.

Subsequent work has provided conclusive evidence for a GPI anchor on 5′-nucleotidase. cDNA cloning and sequencing of the rat liver enzyme (Misumi et al, 1990a) revealed a 576 amino acid precursor polypeptide, of predicted M_r 64000, which contained a 29 amino acid N-terminal signal peptide and a C-terminal hydrophobic domain (Table 2). The human placental enzyme has also recently been cloned and sequenced (Misumi et al, 1990b). This predicts a 574 amino acid polypeptide with a predicted M_r of 63000. There is a 26 amino acid N-terminal signal peptide and a C-terminal hydrophobic domain (Table 2). For both rat liver and human placental 5′-nucleotidases the GPI anchor addition site has been identified as Ser-523 (Table 2) (Misumi et al, 1990b; Ogata et al, 1990). For both enzymes the C-terminal peptide was shown to contain 2 mol ethanolamine, 1 mol glucosamine, 3 mol mannose and 1 mol inositol. The only significant difference between the anchor peptide from the two species was the relative amounts of palmitic and stearic acids (rat 0.4 mol and 1.4 mol, respectively; human 0.9 mol and 0.8 mol, respectively). Also, metabolic labelling experiments with rat hepatocytes in primary culture demonstrated incorporation of [^{3}H]ethanolamine, inositol and palmitate and approx 70% of the synthesized enzyme was releasable by PI-PLC with loss of the [^{3}H]palmitate component (Misumi et al, 1990a). Finally, Bailyes et al (1990) have detected 1 mol of *myo*-inositol/mol of 5′-nucleotidase monomer on analysis of immunoaffinity-purified rat liver enzyme by g.c.-m.s. Although 5′-nucleotidase was one of the first proteins shown to be released by PI-PLC (Low and Finean, 1978), it has taken more than a decade to provide unequivocal evidence for a GPI anchor. The variable release of the enzyme reported in different species and tissues may result from acylation or other modifications of the inositol ring in some forms of the enzyme (cf. acetyl-cholinesterase) hindering release by PI-PLC. Alternatively, the existence of transmembrane polypeptide-anchored forms of the enzyme cannot be excluded.

4. ANTIBODIES TO MAMMALIAN GPI-ANCHORS

Polyclonal antisera to the soluble form of the variant surface glycoproteins (VSG) of trypanosomes commonly cross-react with other GPI-anchored proteins. This recognition of GPI-anchored proteins by anti-CRD antiserum is abolished by prior treatment of the protein with mild HCl which decyclizes the inositol 1,2-cyclic phosphate ring epitope formed on PI-PLC cleavage (Zamze et al, 1988) or with HNO_2 which deaminates the glucosamine residue in the GPI-anchor (Ferguson et al, 1985) (see Fig.2). Thus, the inositol 1,2-(cyclic) monophosphate and non-N-acetylated glucosamine residue represent the two CRD epitopes common to all GPI-anchored proteins identified to

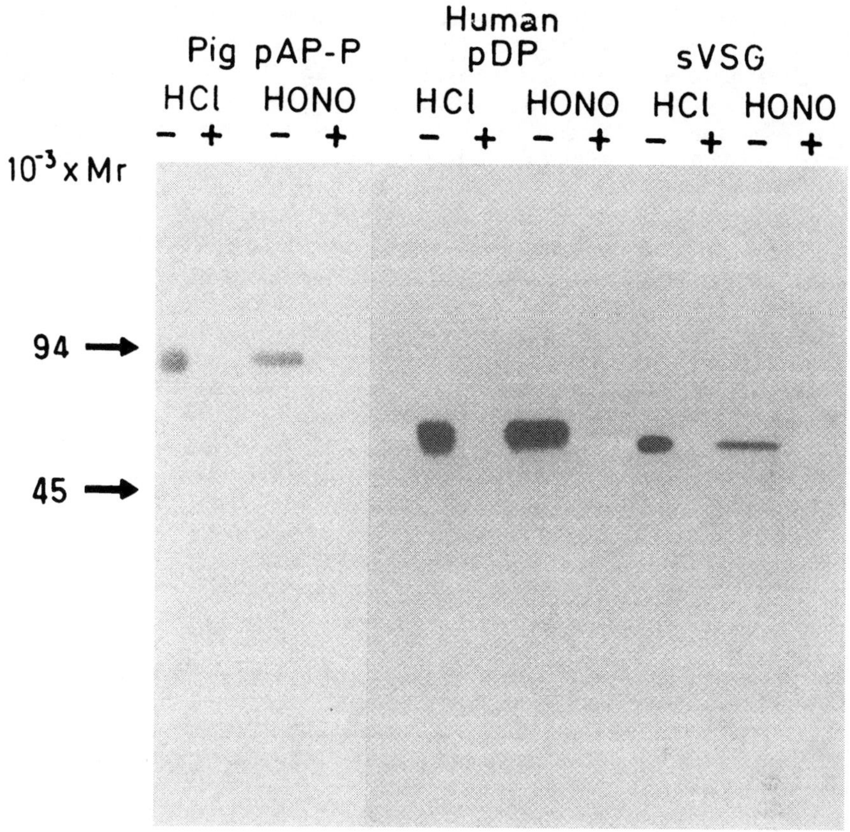

Fig. 3. Western blot of GPI-anchored proteins with pig anti-CRD antiserum. Samples of phospholipase C-solubilized purified proteins (pig pAP-P; pig aminopeptidase P, 5 μg of protein; human membrane dipeptidase, 15 μg; sVSG, trypanosome varient surface glycoprotein, 2 μg) were treated with mild HCl or HNO_2 prior to SDS-PAGE and Western blot analysis. After electrophoretic transfer, the tracks were blotted with a polyclonal antiserum raised in rabbits to purified pig membrane dipeptidase. Bound antiserum was detected with a [^{125}I]-labelled second antibody followed by autoradiography.

date (see Fig.2) (Zamze et al, 1988). A third epitope, the galactose side chain, appears to be specific for VSGs. We recently observed that two polyclonal antisera that we had raised to pig membrane dipeptidase (RP147) and AP-P (RP171) were able to recognize other GPI-anchored proteins from the same and different species including trypanosome VSG (Fig.3). We reasoned that the cross-reactivity must reside in the GPI-anchor and have now characterized the epitopes involved in this cross-reactivity (Hooper et al, 1991). The antibodies recognizing the CRD component were purified by affinity chromatography on a column of the appropriate immobilized peptidase and further examined for specificity. The epitopes involved were characterized by Western blot analysis after milk acid or nitrous acid treatment of the proteins (see Fig.3) and by a competitive ELISA. The primary epitope for both antisera was the inositol 1,2-(cyclic) monophosphate formed on PI-PLC cleavage of the intact GPI anchor (see Fig.2). The epitope was cryptic in the membrane-bound form of the proteins. Inositol 1,2-(cyclic) monophosphate was able to inhibit the binding of both RP147 and RP171 to the immobilized peptidase in the ELISA whereas inositol 1- and 2-monophosphates and inositol had relatively little effect. Phosphoethanolamine partially inhibited the binding of RP147 antibody indicating that this may represent a minor epitope. Other minor epitopes probably involve side-chain modifications to the core anchor structure that appear to be species specific. Anti-CRD antisera have also been generated against *Torpedo* acetylcholinesterase (Jager et al, 1990) and human membrane dipeptidase (S J Broomfield and N M Hooper, unpublished observations). The antibodies against the mammalian membrane peptidases may prove to be more useful than the trypanosome anti-CRD antisera in structural studies on mammalian GPI-anchored proteins, and possibly for characterizing GPI-derived mediators produced in response to the action of insulin and other hormones.

5. PERSPECTIVE

The hydrolases of renal and intestinal brush borders are providing fertile ground for studies of the biosynthesis, intracellular transport and topology of membrane proteins. In addition, the brush border membrane is an ideal model system in which to study the structural and functional differences between GPI- and transmembrane-polypeptide anchors. These membranes provide a relatively abundant source of a number of GPI-anchored proteins, several of which have recently been cloned and sequenced. We are now in a position to examine a number of aspects of mammalian GPI structure and function - its role in trafficking, particularly in directing proteins to the apical rather than basolateral surface (see contribution in this volume by Rodrígues-Boulan and Lisanti); the variations in GPI structures among different species and tissues and its significance; and, finally, the role of phospholipase release in the physiological functions of GPI-anchored proteins.

ABBREVIATIONS

AP-P, aminopeptidase P; CP-M, carboxypeptidase M; CRD, cross-reacting determinant; ELISA. enzyme-linked immunoadsorbent assay; g.c.-m.s., gas chromatography-mass spectroscopy; GPI, glycosyl-phosphatidylinositol; MDCK, Madin-Darby canine kidney; PI-PLC, phosphatidylinositol-specific phosphlipase C; VSG, variant surface glycoprotein.

ACKNOWLEDGEMENT

We thank the Medical Research Council and the Wellcome Trust for financial support of this work. MSM is supported by a fellowship from CNPq, Brazil.

REFERENCES

Adachi, H., Tawaragi, Y., Inuzuka, C., Kubota, I., Tsujimoto, M., Nishihara, T. and Nakazato, H. (1990a). Primary structure of human microsomal dipeptidase deduced from molecular cloning. J. Biol. Chem. **265:** 3992-3995.

Adachi, H., Katayama, T., Inuzuka, C., Oikawa, S., Tsijimoto, M. and Nakazato, H. (1990b). Identification of membrane anchoring site of human renal dipeptidase and construction and expression of a cDNA for its secretory form. J. Biol. Chem. **265:** 15341-15345.

Armstrong, D. J., Mukhopadhyay, S. K. and Campbell, B. J. (1974). Physicochemical characterization of renal dipeptidase. Biochemistry **13:** 1745-1750.

Bailyes, E. M., Ferguson, M. A. J., Colaco, C. A. L. S. and Luzio, J. P. (1990). Inositol is a constituent of detergent-solubilized immunoaffinity-purified rat liver 5′- nucleotidase. Biochem. J. **265:** 907-909.

Berger, J., Howard, A. D., Brink, L. B., Berger, L., Hauber, J., Cullen, B. R. and Udenfriend, S. (1988). COOH- terminal requirements for the correct processing of a phosphatidylinositol-glycan anchored membrane protein. J. Biol. Chem. **263:** 10016-10021.

Coloma, J. and Pitot, H. C. (1986). Characterization and sequence of a cDNA clone of gamma-glutamyltranspeptidase. Nucl. Acids. Res. **14:** 1393-1403.

Deddish, P. A., Skidgel, R. A., Kriho, V. B., Li, X.-Y., Becker, R. P. and Erdos, E. G. (1990). Carboxypeptidase M in Madin-Darby canine kidney cells. Evidence that carboxypeptidase M has a phosphatidyl glycan anchor. J. Biol. Chem. **265:** 15083-15089.

Devault, A., Lazure, C., Nault, C., Le Moual, H., Seidah, N. G., Chretien, M., Kahn, P., Powell, J., Mallet, J., Beaumont, A., Roques, B. P., Crine, P. and Boileau, G. (1987). Amino acid sequence of rabbit kidney neutral endopeptidase-24.11 (enkephalinase) deduced from a complementary DNA. EMBO J. **6:** 1317-1322.

Ferguson, M. A. J., Low, M. G. and Cross, G. A. M. (1985). Glycosyl-sn-1,2-dimyristoylphosphatidylinositol is co-valently linked to *Trypanosoma brucei*. J. Biol. Chem. **260:** 14547-14555.

Fulcher, I. S., Pappin, D. J. C. and Kenny, A. J. (1986). The N-terminal amino sequence of pig kidney endopeptidase-24.11 shows homology with prosucrase - isomaltase. Biochem. J. **240:** 305-308.

Hooper, N. M. and Bashir, A. (1991). Glycosyl-phosphatidylinositol anchored membrane proteins can be distinguished from transmembrane polypeptide anchored proteins by differential solubilization and temperature -induced phase separation in Triton X-114. Unpublished observations.

Hooper, N. M. and Turner, A. J. (1988a). Ectoenzymes of the kidney microvillar membrane: differential solubilization by detergents can predict a glycosyl-phosphatidylinositol membrane anchor. Biochem. J. **250:** 865-869.

Hooper, N. M. and Turner, A. J. (1988b). Ectoenzymes of the kidney microvillar membrane: aminopeptidase P is anchored by a glycosyl-phosphatidylinositol moiety. FEBS Letts. **229:** 340-344.

Hooper, N. M. and Turner, A. J. (1989a). Characterization and isolation of the membrane anchor of angiotensin-converting enzyme. Biochem. Soc. Trans. **17:** 660-661.

Hooper, N. M. and Turner, A. J. (1989b). Ectoenzymes of the kidney microvillar membrane. Isolation and characterization of the amphipathic form of renal dipeptidase and hydrolysis of its glycosyl-phosphatidylinositol anchor by an activity in plasma. Biochem. J. **261:** 811-818.

Hooper, N. M., Keen, J., Pappin, D. J. C. and Turner, A. J. (1987a). Pig kidney angiotensin converting enzyme: isolation of amphipathic and hydrophilic forms of the enzyme establishes C-terminal anchorage to the plasma membrane. Biochem. J. **247:** 85-93.

Hooper, N. M., Low, M. G. and Turner, A. J. (1987b). Renal dipeptidase is one of the membrane proteins released by phosphatidylinositol-specific phospholipase C. Biochem. J. **244:** 465-469.

Hooper, N. M., Keen, J. N. and Turner, A. J. (1990a). Characterization of the glycosyl-phosphatidyl-inositol-anchored human renal dipeptidase reveals that it is more extensively glycosylated than the pig enzyme. Biochem. J. **265:** 429-433.

Hooper, N. M., Hryszko, J. and Turner, A. J. (1990b). Purification and characterization of pig kidney aminopeptidase P. A glycosyl-phosphatidylinositol-anchored ectoenzyme. Biochem. J. **267:** 509-515.

Hooper, N. M., Broomfield, S. J. and Turner, A. J. (1991). Characterization of antibodies to the glycosyl-phosphatidylinositol membrane anchors of mammalian proteins. Biochem. J. **273:** 301-306.

Hunziker, W., Spiess, M., Semenza, G. and Lodish, H. (1986). The sucrase-isomaltase complex: Primary structure, membrane orientation and evolution of a stalked intrinsic brush border protein. Cell **46:** 227-234.

Jager, K., Meyer, P., Stieger, S. and Brodbeck, U. (1990). Production and characterization of antibodies against the cross-reacting determinant of glycosyl-phosphatidylinositol-anchored acetylcholinesterase. Biochim. Biophys. Acta **1039:** 367-373.

Kam, W., Clauser, E., Kim, Y. S., Kan, Y. W. and Rutter, W. J. (1985). Cloning, sequencing and chromosomal localization of human term placental alkaline phosphatase cDNA. Proc. Natl. Acad. Sci. USA **82:** 8715-8719.

Kozak, E. M. and Tate, S. S. (1982). Glutathione-degrading enzymes of microvillus membranes. J. Biol. Chem. **257:** 6322-6327.

Laperche, Y., Bulle, F., Aissani, T., Chobert, M.-N., Aggerbeck, M., Hanoune, J. and Guellaen, G. (1986). Molecular cloning and nucleotide sequence of rat kidney gamma-glutamyl transpeptidase cDNA. Proc. Natl. Acad. Sci. USA. **83:** 937-941.

Lasch, J., Koelsch, R., Ladhoff, A.-M. and Hartrodt, B. (1986). Is the proline-specific aminopeptidase P of the intestinal brush border an integral membrane enzyme? Biomed. Biochem. Acta, **45:** 833-843.

Littlewood, G. M., Hooper, N. M. and Turner, A. J. (1989). Ectoenzymes of the kidney microvillar membrane: affinity purification, characterization and localization of the phospholipase-C solubilized form of renal dipeptidase. Biochem. J. **257:** 361-367.

Low, M. G. and Finean, J. B. (1978). Specific release of plasma membrane enzymes by a phosphatidylinositol-specific phospholipase C. Biochim. Biophys. Acta. **508:** 565-570.

Luzio, J. P., Baron, M. D. and Bailyes, E. M. (1987). Cell Biology. In: Mammalian Ectoenzymes (Ed. Kenny, A. J. and Turner, A. J.) Elsevier, Amsterdam, pp. 111-138.

Malfroy, B., Schofield, P. R., Kuang, K.-J., Seeburg, P. H., Mason, A. J. and Henzel, W. J. (1987). Molecular cloning and amino acid sequence of rat enkephalinase. Biochem. Biophys. Res. Comm. **144:** 59-66.

Mantei, N., Villa, M., Enzler, T., Wacker, H., Boll, W., James, P., Hunziker, W. and Semenza, G. (1988). Complete primary structure of human and rabbit lactase-phlorizin hydrolase: implications for biosynthesis, membrane anchoring and evolution of the enzyme. EMBO J. **7:** 2705-2713.

Misumi, Y., Ogata, S., Hirose, S. and Ikehara, Y. (1990a). Primary structure of rat liver 5'-nucleotidase deduced from the cDNA. J. Biol. Chem. **265:** 2178-2183.

Misumi, Y., Ogata, S., Ohkubo, K., Hirose, S. and Ikehara, Y. (1990b). Primary structure of human placental 5'nucleotidase and identification of the glycolipid anchor in the mature form. Europ. J. Biochem. **191:** 563-569.

Ogata, S., Misumi, Y. and Ikehara, Y. (1989). Primary structure of rat liver dipeptidyl peptidase IV deduced from its cDNA and identification of the NH2-terminal signal sequence as the membrane anchoring domain. J. Biol. Chem. **264:** 3596-3601.

Ogata, S., Hayashi, Y., Misumi, Y. and Ikehara, Y. (1990). Membrane-anchoring domain of rat liver 5'-nucleotidase: identification of the COOH-terminal serine-523 covalently attached with a glycolipid. Biochemistry **29:** 7923-7929.

Olsen, J., Cowell, G. M., Konigshofer, E., Danielson, E. M., Moller, J., Lausten, L., Hanson, O. L., Welinder, K. G., Engberg, J., Hunziker, W., Spiess, M. and Noren, O. (1988). Complete amino acid sequence of human intestinal aminopeptidase N as deduced from cloned cDNA. FEBS Lett. **238:** 307-314.

Orawski, A. T., Susz, J. P. and Simmons, W. H. (1987). Aminopeptidase P from bovine lung: solubilization, properties and potential role in bradykinin degradation. Mol. Cell. Biochem. **75:** 123-132.

Pearson, J. D. (1987). Nucleotide metabolism. In: Mammalian Ectoenzymes (Ed. Kenny, A. J. and Turner, A. J.) Elsevier, Amsterdam. pp. 139-167.

Rached, E., Hooper, N. M., James, P., Semenza, G., Turner, A. J. and Mantei, N. (1990). cDNA cloning and expression in *Xenopus laevis* oocytes of pig renal dipeptidase , a glycosyl-phosphatidylinositol-anchored ectoenzyme. Biochem. J. **271:** 755-760.

Ruf, J., Wacker, H., James, P., Maffia, M., Seiler, P., Galand, G., von Kiekebbusch, A., Semenza, G. and Mantei, N. (1990). Rabbit small intestinal trehalase. Purification, cDNA cloning, expression and verification of glycosyl-phosphatidylinositol anchoring. J. Biol. Chem. **265:** 15034-15039.

Shukla, S. D., Coleman, R., Finean, J. B. and Michell, R. H. (1980). Selective release of plasma membrane enzymes from rat hepatocytes by a phosphatidyl-inositol-specific phospholipase C. Biochem. J. **187:** 277-280.

Soubrier, F., Alhenc-Gelas, F., Hubert, C., Allegrini, J., John, M., Tregear, G. and Corvol, P. (19--). Two putative active centers in human angiotensin I-converting enzyme revealed by molecular cloning. Proc. Natl. Acad. Sci. USA. **85:** 9386-9390.

Takesue, Y., Yokota, K., Nishi, Y., Taguchi, R. and Ikezawa, H. (1986). Solubilization of trehalase from rabbit renal and intestinal brush-border membranes by a phosphatidylinositol-specific phospholipase C. FEBS Lett. **201:** 5-8.

Tan, F., Chan, S. J., Steiner, D. F., Schilling, J. W. and Skidgel, R. A. (1989). Molecular cloning and sequencing of the cDNA for human membrane-bound carboxypeptidase M. J. Biol. Chem. **264:** 13165-13170.

Turner, A. J., Hooper, N. M. and Kenny, A. J. (19--). Metabolism of neuropeptides. In: In: Mammalian Ectoenzymes (Ed. Kenny, A. J. and Turner, A. J.) Elsevier, Amsterdam. pp.211-248.

Wei, L., Alhenc-Gelas, F., Soubrier, F., Michaud, A., Corvol, P. and Clauser, E. (1991). Expression and characterization of recombinant human angiotensin I-converting enzyme. J. Biol. Chem. in press.

Weiss, M. J., Henthorn, P. S., Lafferty, M. A., Slaughter, C., Raducha, M. and Harris, H. (1986). Isolation and characterization of a cDNA encoding a human liver/bone/kidney-type alkaline phosphatase. Proc. Natl. Acad. Sci. USA. **83:** 7182-7186.

Wu, Q., Lahti, J. M., Air, G. M., Burrows, P. D. and Cooper, M. D. (1990). Molecular cloning of the murine BP-1/6C3 antigen: A member of the zinc-dependent metallopeptidase family. Proc. Natl. Acad. Sci. USA. **87:** 993-997.

Wu, Q., Li, L., Cooper, M. D., Pierres, M. and Gorvel, J. P. (1991). Aminopeptidase A activity of the murine b-lymphocyte differentiation antigen BP-1/6C3. Proc. Natl. Acad. Sci. USA. **88:** 676-680.

Zamze, S. E., Ferguson, M. A. J., Collins, R., Dwek, R. A., and Rademacher, T. W. (1988). Characterization of the cross-reacting determinant (CRD) of the glycosyl-phosphatidylinositol membrane anchor of *Trypanosome brucei* variant surface glycoprotein. Eur. J. Biochem. **176:** 527-534.

THE BIOLOGY OF THE GLYCOSYLPHOSPHATIDYLINOSITOL-SPECIFIC PHOSPHOLIPASE C OF *TRYPANOSOMA BRUCEI*.

***Mark Carrington, Dawn Walters and Helena Webb**

Department of Biochemistry, Tennis Court Road, Cambridge, CB2 1QW, UK.

* Corresponding author.

The variant surface glycoprotein (VSG) of the protozoan parasite *Trypanosoma brucei* was amongst the first proteins shown to be attached to the plasma membrane by a glycosylphosphatidylinositol (GPI)-anchor (Ferguson et al, 1985b; Schmitz et al, 1987; Ferguson et al, 1988). The VSG is the key component in the process of antigenic variation that allows a population of *T.brucei* to persist in the bloodstream of the mammalian host. The surface of the trypanosome is covered by a 12-15 nm thick coat (Vickerman, 1969), composed of a dense monolayer of a single polypeptide species, the VSG (Cross, 1975). Antigenic variation occurs when an individual trypanosome stops expressing one VSG gene and starts expressing another which encodes an antigenically distinct VSG (reviewed in Pays and Steinert, 1988; Borst, 1986) thus delaying clearance by the host immune system. Upon ingestion by a tsetse fly vector the trypanosome differentiates to the procyclic form in the insect midgut and the VSG is replaced by procyclin as the predominant surface protein (Roditi et al, 1989; Ziegelbauer et al, 1990) to reappear on the metacyclic trypanosome in the salivary gland of the fly.

The characterisation of VSGs began with a relatively simple method for purification (Cross, 1975) that produced milligram amounts of soluble VSG. Biochemical and sequence analyses of VSGs have shown that the mature polypeptides are about 400 to 500 residues long, and can be divided into a highly variable N-terminal domain of around 350 residues and a C-terminal domain of around 100 residues that shows weak sequence homology between different VSGs (Cross, 1984; Carrington et al, 1991). There is one feature of the primary structure that is conserved; at the C-terminus of all VSGs there is an immunogenic oligosaccharide, the cross-reacting determinant (CRD), that results in immunological cross-reactivity between different purified VSGs (Holder and Cross, 1981). The CRD is not exposed to the host immune system on living trypanosomes.

The discovery of the GPI-anchor on the VSG and subsequently the *T.brucei* GPI-specific phospholipase C

(GPI-PLC) was a result of an observation that the VSG in a total cell lysate (prepared in the presence of detergent at 100°C) had a slightly higher electrophoretic mobility than purified VSG. This led to experiments which demonstrated that during the standard VSG purification an endogenous enzyme converted a hydrophobic membrane form VSG (mfVSG) to a (water) soluble VSG (sVSG) by removing a hydrophobic lipid moiety (Cardoso de Almeida and Turner, 1983). It was subsequently shown that the enzyme released dimyristyl glycerol from both mfVSG and a purified C-terminal glycopeptide, containing the C-terminal residue and the CRD (Ferguson et al, 1985a); and that the dimyristyl glycerol was part of a phosphatidylinositol covalently linked to the carbohydrate CRD (Ferguson et al, 1985b). These results indicated that the mfVSG is anchored in the cell membrane by covalent attachment of the C-terminus to a phosphatidylinositol and that the enzyme that converts mfVSG to sVSG acted like a phospholipase C (PLC). Further support for this came from the observation that the same conversion of mfVSG to sVSG with the release of dimyristyl glycerol could be obtained using *Staphylococcus aureus* phosphatidylinositol-specific phospholipase C (PI-PLC) (Ferguson et al, 1985b). These conclusions led to the development of a standard assay for the *T.brucei* PLC in which the substrate is mfVSG prepared after metabolic labelling with [^{3}H]-myristic acid. After incubation of substrate and enzyme the reaction is extracted to partition the released [^{3}H]-dimyristyl glycerol in an organic phase. The activity of the enzyme determines the amount of ^{3}H in the organic phase.

This assay led to the purification of the enzyme by a range of methods: antibody affinity chromatography, a monoclonal antibody was selected from a set raised against a partially purified enzyme preparation using depletion of enzyme activity as an assay (Bülow and Overath, 1986); a series of protein fractionation methods (Hereld et al, 1986); and affinity chromatography of an enriched fraction on phosphatidylinositol sepharose (Fox et al, 1986). The purified enzyme is 170 times more active against mfVSG than phosphatidylinositol (234 against 1.4 nmol/min/mg protein) at substrate concentration of 175 nM (Bülow and Overath, 1986). In a mixture of trypanosome lipids the enzyme only cleaved the glycosyl-phosphatidylinositol precursor of the GPI-anchor (Hereld et al, 1986). The enzyme had little activity against PI at 1000 fold the concentration of enzyme that fully cleaved the GPI derived from mfVSG (Fox et al, 1986). These results with the purified enzyme confirmed that the conversion of mfVSG to sVSG is catalysed by a GPI-PLC that cleaves mfVSG to produce dimyristyl glycerol and leaves a cyclic phosphate on the inositol - the cyclic phosphate was found when the primary structure of the GPI-anchor was solved (Ferguson et al, 1988). Although the *T.brucei* GPI-PLC has little activity against phosphatidyl-inositols it is active against GPI-anchored proteins other than mfVSG; for example it generates a water soluble form of *Leishmania* sp. surface glycoprotein p63 (Bordier et al, 1986), and acetylcholinesterase from both human erythrocytes and *Torpedo californica* (Steiger et al, 1986).

The GPI-PLC has an estimated molecular weight of 37-39.8 kD (Bülow and Overath, 1986; Hereld et al, 1986; Fox et al, 1986) and occurs as a monomer (Hereld et al, 1986). It acts as an integral membrane protein - the first step in all purification protocols is a crude membrane preparation, to which detergent is then added to solubilise the activity. Also using the purified protein greater than 90% of the enzyme activity partitions into the detergent phase in a Triton X-114 phase separation (Fox et al, 1986).

The topology of the reaction of the enzyme with mfVSG in vitro has been characterised. If the enzyme and substrate are each in detergent micelles then conversion of mfVSG to sVSG occurs readily on mixing. If the two components are reconstituted separately into phospholipid vesicles and then mixed the conversion does not occur; however it can be made to occur on fusing the vesicles either by freeze thawing or the addition of detergent (Bülow and Overath, 1986).

The enzyme has a pH optimum of between 7.0 and 8.5 and is inhibited strongly by some reagents that modify free sulphydryl groups especially Zn^{2+} and Hg^{2+} ions (Bülow and Overath, 1986; Hereld et al, 1986; Fox et al, 1986). The most effective inhibitor is p-chloromercurphenylsulphonate and it is included in the buffer used in the preparation of mfVSG to prevent the conversion to sVSG that occurs on hypotonic lysis of the cells. This implies that a free sulphydryl group is essential for enzyme activity, some further evidence is the observation that 25 nM dithiothreitol added to the assay stimulates the activity six fold (Hereld et al, 1986); although this could be a result of the reduction of the disulphide bonds in the mfVSG making the GPI more accessible to the enzyme.

The ability to purify the enzyme inevitably led to the isolation of a cDNA encoding GPI-PLC using degenerate oligonucleotides derived from amino acid sequence of tryptic peptides (Hereld et al, 1988; Carrington et al, 1989). The *T.brucei* GPI-PLC has weak sequence homology with PI-PLCs from both bacteria and to a lesser extent higher eukaryotes (Mengaud et al, 1991). An alignment of two conserved regions found in the bacterial PI-PLCs and

T.brucei GPI-PLC is shown in Figure 1. Any greater homology than this would possibly not be expected when the different substrate preferences are considered.

```
A

                *  ** *          **   **           **  *******
L m  (1)   GKAYSLNNWNKPIKNSVTTKQWMSALPDTTN--------LAALSIPGTHDTMSY--N
B c  (1)  ASSVNELENWSK----------WMQPIPDSIP--------LARISIPGTHDSGTFKLQ
T b   1      MFGGVKWS-P--QS-----WMSDTRSSIEKKCIGQVYMVGAHNAGTHGIQMFSPF
                   ** *   *     ***    **                ***    *


            *   *  *        *         **             ***  *   *
L m       GDI--TWTLTKPLAQTQTMSLYQQLEAGIRYIDIRAKDNLNIYHGPIFLNASLSGVLETI  130
B c       NPIKQVWGMTQEYDFRYQMDHG-ARIFDIRGRLT-DDNTIVLHHGPLYLYVTLHEFINEA  129
T b       GLDAPEKLRSLPPYVTFLLRFLTVGVSSRWGRCQNLSIRQLLDHGVRYLDLRMNISPDQE  107
          *          *                             *  **  *


B

                     *   *  *    * **** **
L m  166  IYKDYFYTTPRTDTSNK-IPTLKDVRGKILLLSENHTKKPLVINSRKFGMQFGAPNQV  222
B c  165  KKYFVDPIFLKTEG-N--IKLG-DARGKIVLLKRYSGSNEPGGYNNFYWPDNETFTTT  218
T b  292  LKWFYDLNVIK-EGERVTIRRGNNTHGNILLLDFVQEGTCTVKGVDKPMNAVALCVHL  350
           * * *    * **    *  *    * ****
```

Fig.1 Alignment of amino acid sequences of the PI-PLCs from *Listeria monocytogenes(L m)* (Menguad, et al., 1991; Leimeister-Wächter, et al., 1991) and *Bacillus cereus (B c)* (Kuppe, et al., 1989) with the *Trypanosoma brucei* GPI-PLC *(T b)*. These regions show the highest percentage identity between the GPI-PLC and the bacterial enzymes. The * above the sequences indicate the positions of identity between the PI-PLCs and the * below the sequences indicate the location of identities between the GPI-PLC and at least one of the PI-PLCs. The sequence of the *B. thuringensis* PI-PLC (Henner, et al., 1988) is not shown as it has only 8/329 differences when compared with the *B. cereus* sequence.

The primary structure of the enzyme posed one immediate question - GPI-PLC acts as an integral membrane protein yet there is no long run of hydrophobic amino acids that could account for this behaviour. This includes the absence of an N-terminal signal sequence indicating that the mRNA is translated on free cytoplasmic polysomes and is not co-translationally secreted into the endoplasmic reticulum. It is probable that the GPI-PLC itself is post-translationally modified by the covalent attachment of a hydrophobic moiety. A hint as to the nature of this modification comes from the observation that on elution from a monoclonal antibody affinity column the enzyme undergoes an apparent shift in M_r from 39 kD to 42 kD (as measured by SDS-PAGE) (Bülow and Overath, 1986). In the initial fractions eluted from the column with 50 mM sodium phosphate, pH 12.0, only a minor fraction exhibited the shift in M_r, however on storage roughly 50% converted but no more. This is reminiscent of the conditions used to chemically convert mfVSG to sVSG by cleavage of a phosphate ester (Jackson and Voorheis, 1985). The cleavage of the phosphate ester occurred in these mild conditions due to the proximity of a hydroxyl group on the adjacent carbon (the same reason why RNA with a 2'hydroxyl is hydrolysed by mild alkali but DNA is not). However these conditions are also sufficient to hydrolyse O-alkyl esters and S-alkyl esters, such as fatty acyl esters on a serine, threonine or cysteine residue. Since only part of the GPI-PLC converts even with further treatment at pH 12, there must be more than one modified form. The nature of these modifications is yet to be resolved.

What is known about the location and expression of GPI-PLC? The enzyme is on the cytoplasmic face of intracellular membranes, predominantly vesicles of unknown function (Bülow et al, 1989b). The association with the cytoplasmic face of these membranes was demonstrated by immunoelectron microscopy and supported by organelle fractionation and immuno-fluorescence microscopy. These findings are consistent with the translation of GPI-PLC mRNA on free cytoplasmic polysomes due to the absence of an N-terminal signal sequence. This location also immediately poses a topological problem - by the normal mechanisms of membrane fusion the enzyme would never gain access to mfVSG as the potential substrate as the location is equivalent to the cytoplasmic face of the cell membrane. To gain access to mfVSG in the cell the GPI-PLC would have to pass through a membrane. This is not such a daunting problem as it might first appear as it is not without precedent in trypanosomes in which glycolytic enzymes traverse the glycosome membrane, albeit by a not fully characterised mechanism (Fung and Clayton, 1991).

On differentiation of bloodstream trypanosomes to procyclic forms GPI-PLC activity is lost at a rate that indicates that synthesis is repressed and any remaining enzyme is diluted out by cell growth and general protein turnover (Bülow and Overath, 1985). The activity is absent from established procyclic cell lines (Bülow and Overath, 1985) due to a repression of synthesis on differentiation as GPI-PLC mRNA is virtually undetectable in procyclic trypanosomes (Carrington et al, 1989; Mensa-Wilmot et al, 1990). It remains to be determined whether the enzyme is present in the other life cycle stages that are found in the tsetse fly.

So far there has been a discussion of more or less hard facts. It is now time to try and interpret these to see if it is possible to come to any firm conclusions about the role of the GPI-PLC in the life of a trypanosome. This can take the form of answers to four questions:

1. The enzyme was purified on its ability to convert mfVSG to sVSG, but is this a reaction it catalyses in the living cell?

The strongest line of evidence for mfVSG being the in vivo substrate for GPI-PLC is the amount of enzyme present. The enzyme is very specific for GPIs and mfVSG is by far the most abundant GPI-anchored protein in bloodstream trypanosomes, the other GPI-anchored protein characterised thus far is a transferrin binding protein (Schell et al, 1991). There are roughly 1 x 10^7 VSG molecules (Cross, 1975) and 1000 molecules of the transferrin binding protein per cell (Schell et al, 1991) per cell. These numbers can be contrasted with an estimate of 3 x 10^4 molecules of GPI-PLC per cell (Hereld et al, 1986), the estimated turnover number is 100 mfVSG molecules per minute (under assay conditions, Hereld et al, 1986). There is enough enzyme present to release the entire coat in a few minutes and this does indeed happen on hypotonic lysis. So whilst the low abundance GPI-anchored proteins such as the transferrin binding protein may act as substrates for GPI-PLC, the amount of enzyme present indicates that these are not the sole substrates. The strongest argument against the identity of mfVSG as the major substrate is that topologically the GPI-PLC is on the wrong side of a membrane to gain access to mfVSG, or the transferrin binding protein, as substrate by usual mechanisms form a vesicle fusion. Remembering the translocation of glycolytic enzymes into the glycosome this is not such a strong objection as it may first appear.

2. Does the conversion of mfVSG to sVSG only occur in dying cells?

A possibility is that mfVSG is the substrate for GPI-PLC but only in dying (lysing) trypanosomes which then release sVSG in order to confuse or overburden the host immune system by an as yet undefined mechanism. This is presumably of selective advantage to closely related cells and not an altruistic action. This model appears to be supported by the observation that release of sVSG into the peripheral blood of infected mice correlates with the appearance of the non-dividing stumpy form of the parasite and that the release of VSG could be accounted for if it were simply a result of the degeneration of these (Black et al, 1982; Shapiro, 1986). By culturing bloodstream trypanosomes it is possible to follow sVSG release over a longer time scale and to estimate the half-life of VSG on the surface of the cell (Bülow et al, 1989a); in this system release of sVSG into the medium occurred from living trypanosomes and was by far the predominant mode of VSG turnover. The estimate of 32 hours for the half-life of a VSG molecule goes a long way towards explaining the discrepancy between these and earlier results as the longest time period over which VSG release was investigated previously was nine hours (Black et al, 1982). It can be concluded that living bloodstream trypanosomes turn over VSG very slowly by releasing sVSG, - the half life represents several cell generations - and in experiments this effect can be masked by the much more rapid release of sVSG from lysed cells. The release of sVSG from living trypanosomes is almost certainly the result of the action of an endogenous enzyme as the function of the VSG coat is a highly effective barrier that protects the trypanosome cell surface from exogenous proteins.

On induction to differentiate to procyclic forms in vitro the synthesis of VSG is rapidly inhibited (Overath et al, 1983) and in synchronously differentiating trypanosomes (derived from a pleomorphic bloodstream population), after a lag of 4 hours after the signal to differentiate, the half life of mfVSG is reduced to roughly 8 hours, this results in loss of the VSG coat within 36 hours (Ziegelbauer et al, 1990). In an experiment using asynchronously differentiating trypanosomes (derived from a monomorphic bloodstream population), it was shown that roughly one third of the released VSG is found as sVSG and the remaining two thirds as a proteolytic fragment containing the N-terminal domain of the VSG (Bülow et al, 1989a; Shapiro, 1986). These two forms of released VSG appear to be the results of separate events as the released sVSG is not digested to the N-terminal fragment (Bülow et al, 1989a). Since virtually all the cells remain viable over the time course of this experiment at least a third of the VSG is released into the surrounding medium as a result of GPI-PLC action.

Treatment of trypanosomes with the calcium ionophore A-23187 in the presence of external Ca^{2+} causes the release of the sVSG from intact trypanosomes in a process that is inhibited by Zn^{2+} ions, as is GPI-PLC (Voorheis et al, 1982 and Voorheis, P. pers. comm.). The release of sVSG was greater than could be accounted for by lysis of trypanosomes during the experiment.

On balance this evidence argues there is a release of sVSG from living cells during the culturing of bloodstream trypanosomes and during differentiation to procyclic forms and that this turnover can be stimulated by cytoplasmic Ca^{2+}. This turnover can be masked by the rapid hydrolysis of mfVSG by GPI-PLC on lysis of trypanosomes.

3. If mfVSG is not the substrate for GPI-PLC is there an obvious alternative?

There are two rapid processing events in VSG synthesis: first, the pre-VSG is co-translationally secreted into the endoplasmic reticulum with the removal of an N-terminal signal sequence (McConnell et al, 1981). Second, a C-terminal hydrophobic extension (Boothroyd et al, 1981) on pre-VSG is removed and replaced with the GPI-anchor within one minute of polypeptide synthesis (Ferguson et al, 1986; Bangs et al, 1986). The precursor of the GPI-anchor is glycolipid A, which is rapidly turned over, indicating that the GPI-anchor is assembled first then added to the pre-VSG (Krakow et al, 1986; reviewed in Doering et al, 1990). When total lipids extracted from cells metabolically labelled with [^{3}H]-myristate were treated with purified GPI-PLC one lipid - glycolipid A - was clearly hydrolysed (Hereld et al, 1986). Glycolipid A represents a precursor of the GPI-anchor not an alternative substrate for the GPI-PLC. It is also worth noting that because of the rapidity of addition of the GPI-anchor the reaction almost certainly occurs on the lumen face of the endoplasmic reticulum membrane, this is topologically the same as the cell surface so access of GPI-PLC to glycolipid A has the same difficulties as access to mfVSG. It could be considered that the cytoplasmic location of the GPI-PLC is to keep it away from its substrate so that there is no interference with the synthesis of VSG, and that part of the activation of GPI-PLC includes traversing a membrane.

4. What is the role of GPI-PLsC in *T.brucei* and how does it fulfil this function?

To try and speculate on this is particularly challenging at the moment as with the advent of homologous gene targeting in trypanomastid protozoa (Cruz and Beverley 1990; ten Asbroek et al 1990; Lee and Van der Ploeg 1990), it should not be long before the null phenotype of the GPI-PLC gene is known, possibly by the time this article appears.

Under what circumstances is the enzyme active and when and where does it gain access to mfVSG? First, there is some release of sVSG into the medium on growth of bloodstream trypanosomes in culture, however this release is slow and although almost certainly produced as a result of GPI-PLC action it does not account for the amount of GPI-PLC present in each cell. Second, on cell lysis the enzyme very rapidly converts mfVSG to sVSG. The role of neither of these releases of sVSG is clear, the first could simply be necessary protein turnover and the second could be part of a more subtle interaction with the host immune system.

The pH optimum of GPI-PLC (7.0 - 8.5) and the turnover of sVSG by release into the medium suggest that the site of action is probably the cell surface and not a lysosomal compartment. There is some indirect evidence that the GPI-PLC can only gain access to mfVSG on the cell surface. In pulse chase experiments investigating the biosynthesis of VSG it was found that there was a latent period with a half-life of 15 minutes before newly synthesised mfVSG was converted to sVSG on hypotonic lysis (Ferguson et al, 1986; Bangs et al, 1986). This is the same kinetics as for the appearance of newly synthesised VSG on the cell surface. In the presence of detergent there was conversion of the newly synthesised VSG, so the failure to convert results from the GPI-PLC not gaining access to the newly synthesised VSG as substrate. The reason for this could be either selective fusion of the GPI-PLC vesicles with the plasma membrane or, more probably, only the plasma membrane has the factors necessary to activate the GPI-PLC.

On induction to differentiate to procyclic stages the half-life of the VSG is reduced from 32 to 8 hours but most of this increased release can be accounted for by proteolysis of the VSG to release the N-terminal domain into the surrounding medium in what appears to be a separate event to the GPI-PLC mediated release. So it is possible to visualise the change in surface coat during differentiation to procyclic forms resulting from the rapid cessation of VSG synthesis and increased VSG turnover due to the appearance of a protease at the cell surface. The background release of sVSG by GPI-PLC action continues until the enzyme is diluted out by growth and protein turnover. This argues against a specific role in differentiation although it remains unclear what happens to the residual GPI-anchored C-terminal domain after proteolytic release of the N-terminal domain of the VSG.

The one aspect of the life and times of a trypanosome that is now left for consideration are the events that occur within the individual cells that are undergoing antigenic variation, that is, replacing one VSG on the cell surface with another that is antigenically distinct. At one extreme it is conceivable that this replacement occurs by dilution caused by cell growth and turnover. This would take several cell generations. There is an obvious selective advantage in a rapid change over of the surface VSG from the viewpoint of a trypanosome that has undergone the genetic or epigenetic changes necessary for antigenic variation as the host is mounting an immune response. The role of GPI-PLC could be to increase the release of VSG in the individual cells that are switching so as to speed the exchange of surface coats. Since the processes that lead to switching are stochastic the enzyme has to be present in every cell in case an antigen switch occurs. An apparent weakness in this model is again the amount of enzyme, it is unlikely that a trypanosome can shed its old VSG coat and replace it with new VSG in a few minutes. Possibly the enzyme at the cell surface is less active than after lysis of trypanosomes.

It is possible to make some testable predictions to distinguish between the suggested models.

1. If the GPI-PLC is involved in exchanging VSG coats then the phenotype of the null mutation would be a persistence of the old VSG on the surface of the trypanosomes after antigenic switching so that in a switching population the proportion of trypanosomes with two VSGs on the surface would be dramatically increased.

2. On taking a blood meal an infected tsetse fly introduces metacyclic trypanosomes into the mammalian host, these differentiate to bloodstream forms over the next few days. During this time most of the trypanosomes switch from a metacyclic VSG to a bloodstream VSG. This is unlike antigenic variation in bloodstream trypanosomes when the new antigenic type arises from an individual or a very small subset of the original population. In some metacyclic populations undergoing this process at least 2.7% of the cells can be observed to have both VSGs on their surface (Esser and Schoenbechler, 1985). The rate of release of sVSG from a population of these cells would be predicted to be higher than in non-switching cells, assuming that the mechanism of the metacyclic to bloodstream switch is the same as the bloodstream to bloodstream switch. It would also be interesting to know whether these cells are able to specifically release just the metacyclic sVSG and the subcellular location of the GPI-PLC.

3. If the hypothesis that the GPI-PLC is present only to release sVSG from lysing cells is correct then the null mutant would show normal antigenic variation, but would have an altered interaction with the host immune system.

None of these roles are strictly mutually exclusive so perhaps the GPI-PLC performs all these functions, hopefully many of the unknowns concerning the function of GPI-PLC in *T.brucei* will be solved in the near future.

REFERENCES

Bangs, J. D., Andrews, N. W., Hart, G. and Englund, P. T. (1986). Post translational modification and intracellular transport of a trypanosome variant surface glycoprotein. J. Cell Biol. **103:** 255-263.

Black, S. J., Hewett, R. S. and Sendashonga, C. N. (1982). *Trypanosoma brucei* variable surface antigen is released by degenerating parasites but not by actively dividing parasites. Parasite Immunol. **4:** 233-244.

Boothroyd, J. C., Paynter, C. A., Cross, G. A. M., Bernards, A. and Borst, P. (1981). Variant surface glycoproteins of *Trypanosoma brucei* are synthesised with cleavable hydrophobic sequences at the carboxy and amino termini. Nucl. Acids Res. **9:** 4735-4743.

Bordier, C., Etges, R. J., Ward, J., Turner, M. J. and Cardos de Almeida, M. L. (1986). *Leishmania* and *Trypanosoma* surface glycoproteins have a common glycophospholipid membrane anchor. Proc. Natl. Acad. Sci. USA **83:** 5988-5991.

Borst, P. (1986). Discontinuous transcription and antigenic variation in Trypanosomes. Ann. Rev. Biochem. **55:** 701-732.

Bülow, R. and Overath, P. (1985). Synthesis of a hydrolase for the membrane form variant surface glycoprotein is repressed during transformation of *Trypanosoma brucei*. FEBS Lett. **187:** 105-110.

Bülow, R. and Overath, P. (1986). Purification and characterisation of the membrane-form variant surface glycoprotein hydrolase of *Trypanosoma brucei*. J. Biol. Chem. **261:** 11918-11923.

Bülow, R., Nonnengasser, C. and Overath, P. (1989a). The release of the variable surface glycoprotein on the transformation of *Trypanosoma brucei* from bloodstream to procyclic form. Mol. Biochem. Parasitol. **32:** 85-92.

Bülow, R., Griffiths, G., Webster, P., Stierhof, Y.-D., Opperdoes, F. and Overath. P. (1989b). Intracellular location of the glycosyl-phosphatidylinositol-specific phospholipase C of *Trypanosoma brucei*. J. Cell Sci. **93:** 233-240.

Cardosa de Almeida, M. L. and Turner, M. J. (1983). The membrane form of variant surface glycoproteins of *Trypanosoma brucei*. Nature (London) **302:** 349-353.

Carrington, M., Bülow, R., Reinke, H. and Overath, P. (1989). Sequence and expression of the glycosyl-phospatidyl-inositol-specific phospholipase C of *Trypanosoma brucei*. Mol. Biochem. Parasitol. **33:** 289-296.

Carrington, M., Miller, N., Blum, M., Roditi, I., Wiley, D. and Turner, M. (1991). The variant specific glycoprotein of *Trypanosoma brucei* consists of two domains each having an independently conserved pattern of cysteines. J. Mol. Biol. in press.

Cross, G. A. M. (1975). Identification, purification and properties of clone specific glycoprotein antigens constituting the surface coat of *Trypanosoma brucei*. Parasitology, **71:** 393-417.

Cross, G. A. M. (1984). Structure of the variant glycoproteins and surface coat of *Trypanosome brucei*. Phil. Trans. R. Soc. Lond. **B307:** 3-12.

Cruz, A. and Beverley, S. M. (1990). Gene replacement in parasitic protozoa. Nature (Lond) **348:** 171-173.

Doering, T. L., Masterson, W. J., Hart, G. W. and Englund, P. T. (1990). Biosynthesis of glycosylphosphatidylinositol membrane anchors. J. Biol. Chem. **265:** 611-614.

Esser, K. M. and Schoenbechler, M. J. (1985). Expression of two variant surface glycoprotein genes on individual African trypanosomes during antigen switching. Science **229:** 190-192.

Ferguson, M. A. J., Haldar, K. and Cross, G. A. M. (1985a). *Trypanosoma brucei* variant surface glycoprotein has a sn-1,2-dimyristyl glycerol membrane anchor at its COOH terminus. J. Biol. Chem. **260:** 4963-4968.

Ferguson, M. A. J., Low, M. G. and Cross, G. A. M. (1985b). Glycosyl-sn-1,2-dimyristylphosphatidylinositol is covalently linked to *Trypanosoma brucei* variant surface glycoprotein. J. Biol. Chem. **260:** 14547-14555.

Ferguson, M. A. J., Duszenko, M., Lamont, G. S. Overath, P. and Cross, G. A. M. (1986). Biosynthesis of *Trypanosoma brucei* variant glycoprotein. J. Biol. Chem. **261:** 356-362.

Ferguson, M. A. J., Homans, S. W., Dwek, R. A. and Rademacher, T. W. (1988). Glycosylphosphatidylinositol moiety that anchors *Trypanosoma brucei* variant surface glycoprotein to the membrane. Science, **239:** 753-759.

Fox, J. A., Druszenko, M., Ferguson, M. A. J., Low, M. G. and Cross, G. A. M. (1986). Purification and characterisation of a novel glycanphosphatidylinositol specific phoshpolipase C from *Trypanosoma brucei*. J. Biol. Chem. **261:** 15767-15771.

Fung, K. and Clayton, C. (1991). Recognition of a peroxisomal tripeptide signal by the glycosomes of *Trypanosoma brucei*. Mol. Biochem. Parasitol. **45:** 261-264.

Henner, D., Yang, M., Chen, E., Helmiss, R., Rodriguez, H. and Low, M. (1988). Sequence of the *Bacillus thuringensis* phosphatidylinositol-specific phosphalipase C. Nucl. Acids Res. **16:** 10383.

Hereld, D., Krakow, J. L., Bangs, J. D. Hart, G. W. and Englund, P. T. (1986). A phospholipase C from *Trypanosoma brucei* which selectively cleaves the glycolipid on the variant surface glycoprotein. J. Biol. Chem. **261:** 13813-13819.

Hereld, D., Hart, G. W. and Englund, P. T. (1986). cDNA encoding the glycosyl-phosphatidylinositol-specific phospholipase C of *Trypanosoma brucei*. Proc. Natl. Acad. Sci. USA. **85:** 8914-8918.

Holder, A. A. and Cross, G. A. M. (1981). Glycopeptides from variant surface glycoproteins of *Trypanosoma brucei*. C-terminal location of antigenically cross-reacting carbohydrate moieties. Mol. Biochem. Parasitol. **2:** 135-150.

Jackson, D. G. and Voorheis,HH. P. (1985). Release of the variable surface coat glycoprotein from *Trypanosoma brucei* requires the cleavage of a phosphate ester. J. Biol. Chem. **260:** 5179-5183.

Krakow, J., Hereld, D., Bangs, J. D., Hart, G. W. and Englund. P. T. (1986). Identification of a glycolipid precursor of the *Trypanosoma brucei* variant surface glycoprotein. J. Biol. Chem. **261:** 12147-12153.

Kuppe, A., Evans, L., McMillen, D. and Griffith, O. (1989). Phosphatidylinositol-specific phospholipase C of *Bacillus cereus*: cloning,sequencing, and relationship to other phospholipases. J. Bacteriol. **171:** 6077-6083.

Leimeister-Wachter, M., Domann, E. and Chakraborty, T. (1991). Detection of a gene encoding a phosphatidylinositol-specific phospholipase C that is co-ordinately expressed with listeriolysin in *Listeria monocytogenes*. Mol. Microbiol. **5:** 361-366.

Lee, M. G.-S. and Van der Ploeg, L. H. T. (1990). Homologous recombination and stable transfection in the parasitic protozoan *Trypanosoma brucei*. Science **250:** 1583-1587.

McConnell, J., Gurnett, A. M., Cordingley, J. S., Walker, J. E. and Turner, M. J. (1981). Biosynthesis of *Trypanosoma brucei* variant surface glycoprotein. I. Synthesis, size, and processing of an N-terminal signal peptide. Mol. Biochem. Parasitol. **4:** 225-242.

Mengaud, J., Braun-Breton, C. and Cossart, P. (1991). Identification of phosphatidylinositol-specific phospholipase C activity in *Listeria monocytogenes*: a novel type of virulence factor? Mol. Microbiol. **5:** 367-372.

Mensa-Wilmot, K., Hereld, D. and Englund, P. T. (1990). Genomic organisation, chromosomal localisation, and developmentally regulated expression of the glycosyl-phosphatidylinositol-specific phospholipase C of *Trypanosoma brucei*. Mol. Cell. Biol. **10:** 720-726.

Overath, P., Czichos, J., Stock, U. and Nonnengässer, C. (1983). Repression of glycoprotein synthesis and release of surface coat during transformation of *Trypanosoma brucei*. EMBO. J. **2:** 1721-1728.

Pays, E. and Steinert, M. (1988). Control of antigen gene expression in African Trypanosomes. Annu. Rev. Genet. **22:** 107-126.

Roditi, I., Schwarz, H., Pearson, T. W., Beecroft, R. P., Liu, M. K., Richardson, J. P., Bühring, H.-J., Pleiss, J., Bülow, R., Williams, R. O. and Overath, P. (1989). Procyclin gene expression and loss of the variant surface glycoprotein during differentiation of *Trypanosoma brucei*. J. Cell Biol. **108:** 737-746.

Schell, D., Evers, R., Preis, D., Zeigelbauer, K., Keifer, H., Lottspeich, F., Cornelisson, A. and Overath, P. (1991). A transferrin binding protein of Trypanosoma brucei is encoded by one of the genes in the variant surface glycoprotein gene expression site. EMBO. J. **10:** 1061-1066.

Schmitz, B., Klein, R. A., Duncan, I., Egge, H., Gunawan,J., Peter-Katalinic, J., Dabrowski, U. and Dabrowski, J. (1987). MS and NMR analysis of the cross-reacting determinate glycan from *Trypanosoma brucei brucei* MITat 1.6 variant specific glycoprotein. Biochem. Biophys. Res. Comm. **146:** 1055-1063.

Shapiro, S. (1986). *Trypanosoma brucei:* release of variant surface glycoprotein during the parasite life cycle. Exp. Parasitol. **61:** 432-437.

Steiger, A., Cardoso de Almeida,M.L., Blatter, M.-C., Brodbeck, U. and Bordier, C. (1986). The membrane-anchoring system of vertebrate acetylcholine esterase and variant surface glycoproteins of African trypanosomes share a common antigenic determinant. FEBS Lett. **199:** 182-186.

ten Asbroek, A. L. M. A., Ouellette, M. and Borst, P. (1990). Targeted insertion of the neomycin phosphotransferase gene into the tubulin gene cluster of *Trypanosoma brucei*. Nature (London) **348:** 174-176.

Vickerman, K. (1969). On the surface coat and flagellar adhesion in trypanosomes. J. Cell Sci. **5:** 163-193.

Voorheis, H. P., Bowles, D. J. and Smith, G. A. (1982). Characteristics of the release of the surface coat protein from bloodstream forms of *Trypanosoma brucei*. J. Biol. Chem. **257:** 2300-2304.

Zeigelbauer, K., Quinten, M., Schwartz, H., Pearson, T. and Overath, P. (1990). Synchronous differentiation of *Trypanosoma brucei* from bloodstream to procyclic forms in vitro. Eur. J. Biochem. **192:** 373-378.

BACTERIAL PIPLCs-UNIQUE PROPERTIES AND USEFULNESS IN STUDIES ON GPI ANCHORS

*[1]Ikezawa, H.

Faculty of Pharmaceutical Sciences,
Nagoya City University,
3-1 Tanabe-dori, Mizuho-ku, Nagoya,
Aichi 467, Japan.

* Corresponding author.

INTRODUCTION

Recent advances in the studies on biomembranes have revealed the presence of GPI-anchored proteins on the cell surface of eukaryotes from protozoa to mammals. In the studies of GPI anchors, bacterial phosphatidylinositol-specific phospholipases C (PIPLCs) have been used as valuable, analytical tools. Especially, the discovery of GPI-anchors originated from the action of bacterial PIPLCs toward the plasma membranes of mammalian cells. In 1962, Slein and Logan showed that culture filtrates of certain *Bacillus* sp. contained a phosphatasemia factor (PF) which markedly increased blood alkaline phosphatase when injected intravenously into rabbit ears (Slein and Logan, 1962). At the same time, they also showed that this factor induced the release of alkaline phosphatase from slices of rabbit epiphyseal bone or kidney cortex, or suspensions of leukocytes. Afterwards, they further showed that the phosphatasemia factor present in culture filtrates of *Bacillus cereus* eluted with phosphatidylinositol hydrodyzing phospholipase C when chromatographed in DEAE-cellulose columns (Slein & Logan, 1965). In spite of these unique results, their work has been overlooked for a decade, in that they used a crude enzyme preparation with a crude phospholipid mixture, so-called "soybean lecithin", as the substrate. In 1976, however, PIPLC was purified from the culture supernatant of *B. cereus* to a homogeneous state as indicated by polyacrylamide gel electrophoresis, and the purified PIPLC actually released alkaline phosphatase from rat liver slices, in proportion to the amount of PIPLC or slices and to the time after PIPLC addition (Ikezawa et al., 1976). Indeed, the phenomenon of alkaline phosphatase release by PIPLC seemed to be analogous to the product formation observed in the conventional enzyme-substrate reaction. Thereafter, erythrocytic acetyl-cholinesterase (Low and Finean,

1977a), 5'-nucleotidase (Taguchi and Ikezawa, 1978; Low and Finean, 1978), *Torpedo* acetylcholinesterase (Futerman et al, 1983), alkaline phosphodiesterase I (Nakabayashi and Ikezawa, 1984) have been shown to be released from the cell surface by the action of bacterial PIPLCs. As the result of discovery of a novel glycolipid anchor (GPI anchor) binding VSG to the cell surface of *Trypanosoma brucei* (Ferguson et al., 1985), bacterial PIPLCs have become the most essential reagents for the study of GPI-anchored proteins, most of which are liberated from the plasma membranes of eucaryotic cells by this enzyme.

PROPERTIES OF BACTERIAL PIPLCS

Both PIPLCs from *B.cereus* and *Staphylococcus aureus* were found almost at the same time (Slein and Logan, 1965; Doery et al., 1965), then purified and well characterized (Ikezawa et al., 1976; Ikezawa and Taguchi, 1981; Low and Finean, 1977b; Low, 1981). Bacterial PIPLC degrades phosphatidylinositol (PI) as follows (Ikezawa et al., 1976):

Phosphatidylinositol + H_2O ->

1,2-diacylglycerol + D-myoinositol 1,2-cyclic phosphate or D-myoinositol 1-phosphate

At first, both myoinositol cyclic and 1-phosphates were detected as the products by the action of *B. cereus* PIPLC, suggesting that the hydrolytic process involves cyclization of phosphate ester followed by the cleavage of cyclic diester bond (Ikezawa et al., 1976). However, there was no evidence at that time. Recently, Volwerk et al., (1990) demonstrated the sequence of two processes such as phosphatidylinositol -> D-myoinositol 1,2-cyclic phosphate (fast step) and D-myoinositol 1,2-cyclic phosphate -> D-myoinositol 1-phosphate (slow step), using ^{31}P-NMR. This suggests that the reaction mechanism of PIPLC is similar to the two-step mechanism of pancreatic ribonuclease (RNase A). On the one hand, phosphatidylnucleosides such as phosphatidylcytidine, phosphatidyluridine and phosphatidyladenosine which contain CMP, UMP and AMP in their respective polar heads are not hydrolyzed by PIPLC of *Bacillus thuringiensis* (Kume et al., unpublished data). Since the conformation of the ribofuranose ring in phosphatidylnucleosides is completely different from that of the myoinositol ring in PI, the action of PIPLCs must be strictly confined to the inositol ring with high stereospecificity. Due to this rigorous specificity, it is conceivable that bacterial PIPLCs are able to hydrolyze PI and lyso PI (Sundler et al., 1978; Taguchi et al., 1980) as well as PI-glycans and GPI-anchored proteins, and that those enzymes cannot act on other phospholipids such as phosphatidylcholine, phosphatidylethanolamine, phosphatidylserine, phosphatidylglycerol and sphingomyelin under the conditions examined (Ikezawa et al., 1976; Low and Finean, 1976; Low and Finean, 1977b; Taguchi et al., 1980).

Besides PIPLC, *B. cereus* produces two phospholipases C, i.e. phosphatidylcholine-hydrolyzing phospholipase C (PCPLC) and sphingomyelinase C (SMPLC), secreting them into the culture broth. On the other hand, *Bacillus thuringiensis* was shown to produce a great deal of phosphatasemia factor (PF) whereas the production of lecithinase seemed not so high in this organism (Slein and Logan, 1962). In view of this observation, four strains of *B.thuringiensis*, and all the strains tested proved to produce the enzyme in the yield much higher than *B. cereus* IAM 1208 (Taguchi et al., 1980). Furthermore, the activities of PCPLC and SMPLC were hardly detected in the cultures of some strains. Thus PIPLC was purified homogeneously in sufficient quantity from the culture supernatant of *B.thuringiensis*, and studied for its properties (Taguchi et al., 1980; Ikezawa and Taguchi, 1981; Ikezawa et al., 1983). Now, this enzyme is the most widely used PIPLC in the studies of GPI-anchored proteins and PI-glycans. In addition to PIPLCs from *S. aureus* and *Bacillus* sp., the enzyme was also shown to be produced from a strain of *Clostridium novyi* Type A and proved to release alkaline phosphatase and 5'-nucleotidase from rat liver slices (Taguchi and Ikezawa, 1978), like other PIPLCs. Recently, sequencing of cDNA revealed the amino acid sequences of PIPLCs from *B.cereus* (Kuppe et al., 1989) and *B.thuringiensis* (Henner et al., 1988; Lechner et al., 1989), differing only in eight amino acid residues. Both of the mature PIPLCs are composed of 298 amino acid residues, and does not contain cysteine, in spite of inhibition by iodoacetate, p-chloromercuribenzoate, Hg^{2+} and Cu^{2+} as reported previously (Low and Finean, 1977b; Ikezawa and Taguchi, 1981).

The properties of bacterial PIPLCs reported are shown in Table I. (see next page).

Unlike mammalian enzymes, bacterial PIPLCs are not metalloenzymes, since they are not inhibited by chelating agents such as EDTA and 1,10-phenanthroline (Ikezawa et al., 1976; Low and Finean, 1976; Taguchi and Ikezawa, 1978; Sundler et al., 1978; Taguchi et al, 1980). Divalent metal ions such as Ca^{2+}, Mg^{2+}, Zn^{2+} and Mn^{2+} are inhibitory for PIPLC activities (Low and Finean, 1976; Taguchi and Ikezawa, 1978; Sundler et al., 1978; Taguchi et al., 1980). Thus bacterial PIPLCs are different from other bacterial phospholipases C such as PCPLC and SMPL, since PCPLCs of *Clostridium perfringens* and *B. cereus* are zinc metalloenzymes susceptible to inhibition by 1,10-phenanthroline (Ispolatovskaya, 1970; Sato and Murata, 1973; Ottolenghi, 1965; Little and Otnass, 1975) and SMPLCs of *S. aureus* and *B. cereus* require Mg^{2+}, Mn^{2+} and Co^{2+} for their activities (Doery et al., 1965; Tomita et al., 1982; Ikezawa et al., 1986).

Table 1 Properties of Bacterial PIPLCs

Source	optimal pH	molecular weight	pI	substrate specificity	other properties
Bacillus cereus					
	7.2-7.5[1)]	34,466[2)*]	5.4[3)]	PI[1)], lyso PI[4)]	activated by Triton X-100[4)], inhibited by Ca^{2+},Mg^{2+},Zn^{2+}, and Mn^{2+}
Bacillus thuringiensis					
	7.5[5)]	34,586[6)*]	5.4[5)] 4.9[7)] 5.15[8)]	PI, lyso PI[5)] myoinositol cyclic phosphate[9)]	activated by Triton X-100 or SDC[5)], inhibited by Ca^{2+},Mg^{2+}, and Zn_{2+}[5)]
Staphylococcus aureus					
	4.6[10)]	20,000[11)**]		PI[10)]	activated by SDC[12)], inhibited by Triton X-100[12)] and Ca^{2+}[13)]
Clostridium novyi					
		30,000[14)]	5.6,6.0[14)]	PI[14)]	activated by SDC[14)], inhibited by Ca^{2+},Mg^{2+} and Zn^{2+}

* estimated by cDNA sequence, ** calculated from gel filtlation

References: 1)Ikezawa et al., 1976. 2)Kuppe et al., 1989. 3)Ikezawa et al., 1978. 4)Sundler et al., 1978. 5)Taguchi et al., 1980. 6)Lechner et al., 1989. 7)Ikezawa et al., 1983. 8)Kupke et al., 1989. 9)Volwerk et al., 1990. 10)Doery et al., 1965. 11)Low and Finean, 1977b. 12)Low et al., 1988. 13)Low and Finean, 1976. 14)Taguchi and Ikezawa, 1978.

The activities of bacterial PIPLCs are also inhibited by more than 0.1 M NaCl or KCl (Taguchi et al., 1980; Low and Finean, 1976; Sundler et al., 1978). A striking difference exists between PIPLCs of *Bacillus sp.* and *S. aureus*, as shown in Table I. The enzyme activities of *B. cereus* and *B. thuringiensis* are stimulated by Triton X-100 especially at lower concentrations (Sundler et al., 1978; Taguchi et al., 1980) whereas the activity of staphylococcal PIPLC is inhibited by this detergent (Low et al., 1988). However, all of the four bacterial PIPLCs are stimulated by sodium deoxycholate (SDC) at lower concentrations (Ikezawa and Taguchi, 1981; Taguchi et al., 1980; Low et al., 1988; Taguchi and Ikezawa, 1978). Usually, PIPLC activities of *B. cereus* and *B. thuringiensis* are determined at 2 mM PI in the presence of 0.1% Triton X-100 or 0.16% SDC, where the activities are maximally stimulated and measurable with high sensitivity (Ikezawa and Taguchi, 1981).

As substrates for PIPLC assay, preparations of PI from soybean (Slein and Logan, 1965; Low and Finean, 1977b), yeast (Ikezawa et al., 1976) and rat liver (Low and Finean, 1976) have been used. However, the preparations from these natural sources are the mixtures of various molecular species of PI, containing several unsaturated fatty acids such as arachidonic acid. Therefore, natural PIs are not stable substrates, being susceptible to oxidation on long exposure in air. Recently, synthetic PIs such as 1,2-dilauroyl-, 1,2-dimyristoyl-, 1,2-dipalmitoyl- and 1,2-dioleoyl PIs were tested as the substrates for PIPLC assay, and gave good results (Kume et al., in preparation). In particular 1,2-dilauroyl- and 1,2-dimyristoyl PIs are much more sensitive to *B. thuringiensis* PIPLC than yeast PI, enhancing the sensitivity of enzyme assay. These PIs will probably become available as substrates in the near future.

APPLICATION OF BACTERIAL PIPLCS TO STUDIES ON GPI ANCHORS

In the early stage of our studies, bacterial PIPLCs were directly applied to several vertebrate tissues (Ikezawa et al., 1976; Low and Finean, 1977c, 1978; Taguchi and Ikezawa, 1978; Futerman et al., 1983; Nakabayashi and Ikezawa, 1984; Taguchi et al., 1985), cultured cells (Ikezawa et al., 1985), microsomes (Low and Finean, 1978) and blood cells such as erythrocytes, lymphocytes and platelets (Low and Finean, 1977a, 1977c, 1978; Taguchi et al., 1983, 1984; Majumdar and Balasubramanian, 1985). From the plasma membranes of these materials, the ectoenzymes such as alkaline phosphatase, 5′-nucleotidase, acetylcholinesterase and alkaline phosphodiesterase I were liberated in the active forms keeping their molecular size and several enzymological properties, which were consistent with those reported previously.

In these experiments, PIPLC preparations used were proved to be protease-free, so far as the conventional substrates were utilized in the protease assays. Thus, liberation of these enzymes proved not to be due to the contamination of any protease with known specificity. Furthermore, the fact that *B. cereus* PIPLC is responsible for the liberation of alkaline phosphatase from rat kidney tissues was demonstrated in the end of 1970s, since nearly equivalent amounts of anti-PIPLC IgG purified from rabbit antiserum completely neutralized both PIPLC and phosphatase-releasing activities (Ohyabu et al., 1978). Also, alkaline phosphatase was shown to be released in vivo into the blood stream of rat by administration of purified *B.*

cereus PIPLC into tail veins, and this in vivo release of alkaline phosphatase could be neutralized by rabbit anti-PIPLC IgG (Ohyabu et al., 1978). Prior to the discovery of GPI anchors (Ferguson et al., 1985), therefore, bacterial PIPLCs had been shown to release certain ectoenzymes such as alkaline phosphatase both in vitro and in vivo from the eukaryotic plasma membranes.

As the tools for investigation of GPI protein-anchors and PI-glycans, PIPLCs from *B. thuringiensis* and *B. cereus* have now become easily available, since the strains of these bacteria can be aerobically cultivated on a mass scale (Taguchi et al., 1980; Ikezawa et al., 1976) thereby allowing PIPLCs to be made commercially available. Furthermore, bacterial PIPLCs solubilize GPI-anchored proteins in hydrophilic forms favourable for the subsequent purification whereas these proteins are solubilized by butanol and several detergents in the amphiphilic, somewhat aggregated forms (Low and Zilversmit, 1980; Taguchi et al., 1984, 1985; Nakabayashi and Ikezawa, 1984, 1986; Futerman et al., 1985; Takesue et al., 1986; Taguchi and Ikezawa, 1987). For instance, butanol-solubilized acetylcholinesterase of bovine erythrocytes (Taguchi et al., 1984) were converted by the action of *B. thuringiensis* PIPLC into the low molecular-weight forms, as indicated by gel filtration. From rabbit renal and intestinal brush-border membranes, trehalase can be solubilized either with Triton X-100 or with PIPLC. When the preparations of solubilized trehalase were applied to a phenyl-Sepharose column, almost all the activity of Triton X-100-solubilized trehalase was adsorbed onto the column and then eluted by the buffer containing the detergent, whereas virtually no PIPLC-solubilized trehalase was adsorbed on phenyl-Sepharose (Takesue et al., 1986). This indicates that Triton X-100- and PIPLC-solubilized trehalases are amphiphilic and hydrophilic, respectively. Also, when Lubrol- or PIPLC-solubilized acetylcholinesterase of bovine erythrocytes was applied to a butyl-Toyopearl column, the major activity of Lubrol-solubilized enzyme was eluted at the lowest concentration of the decreasing NaCl gradient, while most of the PIPLC-solubilized enzyme appeared in the breakthrough fractions without absorption on the column, as shown in Fig. 1 (Taguchi and Ikezawa 1987). The results show that Lubrol-solubilized acetylcholinesterase still retain amphiphilic characteristics whereas the PIPLC-solubilized enzyme is hydrophilic.

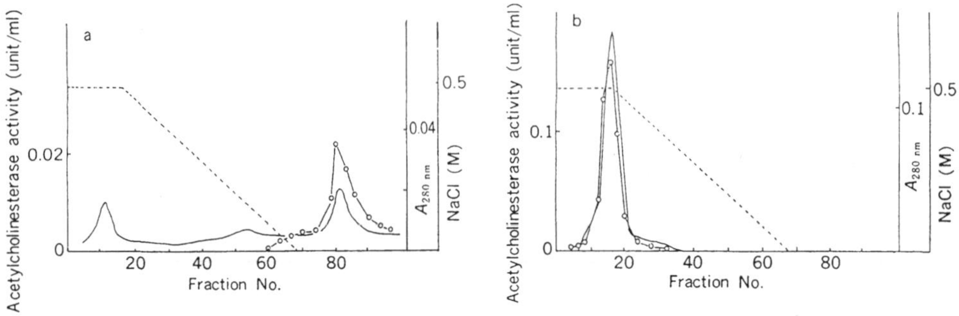

Fig 1. Butyl-Toyopearl column chromatography of acetylcholin-esterase of bovine erythrocytes. (a) Lubrol-solubilized acetylcholinesteras in 10mM Tris-HCl (pH 7.5) containing 0.5M NaCl was applied on to a Butyl-Toyopearl column (1.3 x 11.5cm), flushed with 20ml of starting buffer, and then eluted with 50ml of a decreasing linear gradient of 0.5-0M NaCl in 10mM Tris-HCl (pH 7.5) at a flow rate of 1ml/min. (b) Acetylcholinesterase solubilized with PI-specific phospholipase C was applied on to the same column. The conditions for elution were the same a described in (a). —o— , acetylcholinesterase activity; —— , A_{280nm}; --- , NaCl concentration. (Taguchi and Ikezawa, 1987).

Furthermore, butanol-extracted alkaline phosphatase of porcine kidney (Low and Zilversmit, 1980), detergent-solubilized acetylcholinesterases of *Torpedo* electric organ (Futerman et al., 1985) and bovine erythrocytes (Taguchi and Ikezewa, 1985) can combine with several phospholipids and their mixtures but PIPLC-treatment impairs the phospholipid-binding ability of these enzymes. These results suggest that the GPI-anchored proteins extracted with butanol or detergents possess the ability to associate with other lipids and the extracting solvents through the diacyl- or diradylglycerol moiety of GPI anchors whereas PIPLC-treatment deprives the protein anchors of their hydrophobic moiety, rendering the proteins hydrophilic.

Recently, Bordier developed a new method for phase separation of membrane proteins in a Triton X-114 solution (Bordier, 1981), which is homogeneous at 0°C, which then partitioned into the two phases at 30° according to their hydrophobicity. This method is now widely used in the studies of GPI-anchored proteins, since it is rapid and convenient. Most of GPI-anchored proteins are usually partitioned in the detergent-rich phase, and migrate into aqueous phase after PIPLC treatment. In particular the method of Triton X-114 phase separation is feasible for estimating sensitivity of GPI-anchored proteins to the action of PIPLC.

For screening and confirmation of the presence of GPI anchors in the membrane proteins, bacterial PIPLCs have been used as powerful indicators, since most of GPI-anchored proteins are liberated from the membrane by PIPLCs. However, some of GPI-anchored proteins are resistant to bacterial PIPLCs. By the action of PIPLC from *S. aureus* or *B. thuringiensis*, only 5-10 % of the acetylcholinesterase was released from human erythrocyte membrane whereas the enzyme was almost completely liberated from bovine erythrocyte membrane (Low and Finean, 1977a; Taguchi and Ikezawa, 1984; Futerman et al., 1985). Recently, Rosenberry and his colleagues succeeded in resolving this problem (Roberts et al., 1988). Through the structural analysis for the GPI anchor of acetylcholinesterase from human erythrocytes, they demonstrated the presence of a palmitoyl group acylating an inositol hydroxyl group. Since the reaction mechanism of PIPLC inevitably involves the formation of cyclic phosphate diester between two hydroxyl groups (1 and 2) of the inositol ring in the molecule of GPI anchor or PI, palmitoylation of inositol must result in prevention of anchor cleavage by PIPLC. Indeed, it was shown that the removal of this palmitoyl group by treatment with hydroxylamine at pH 10.7 allowed PIPLC to degrade the anchor, generating a substantial amount of hydrophilic acetylcholinesterase. Certainly, the presence of esterified fatty acid(s) in the inositol ring may be responsible for PIPLC resistance of GPI anchors in many cases. On the one hand, the accessibility of PIPLC to the respective protein anchors in the membrane must be also important. For instance, PIPLC causes the release of 5′-nucleotidase as well as alkaline phosphatase from rat kidney tissues. However, release of 5′-nucleotidase occurs after a delay and then proceeds slowly whereas that of alkaline phosphatase proceeds much faster without any delay, as shown in Fig.2 (Taguchi et al., 1985).

In fact, most of 5′-nucleotidase activity is released from rat hepatocytes and from microsomal fractions of rat liver and heart, although many more units of PIPLC are needed for release of 5′-nucleotidase than for that of alkaline phosphatase (Low and Finean, 1978; Shukla et al., 1980; Prasad and Low, 1987; Low, 1990). From these observations, it seems that 5′-nucleotidase is located in the membrane at the locus where the accessibility to PIPLC is not so well defined.

Bacterial PIPLCs are useful tools in the purification and analysis of GPI-anchored proteins. For instance, acetylcholinesterase was effectively isolated from intact bovine erythrocytes with a high recovery rate (31 %), after exhaustive solubilization by *B. thuringiensis* PIPLC (Taguchi et al., 1984). Also in the analysis of COOH-terminal GPI anchors in the membrane proteins, PIPLCs are now extensively used in the separation of glycans from diacyl- or diradylglycerols, as shown originally in the pioneering analysis of VSG anchor from *Trypanosoma brucei* (Ferguson et al., 1985).

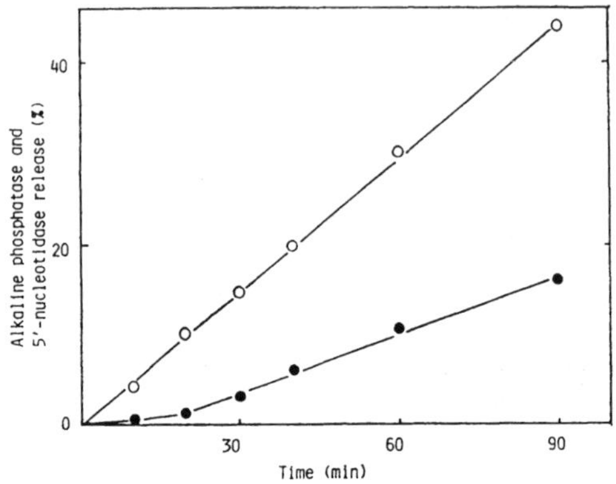

Fig 2. Time course curves of alkaline phosphatase and 5'-nucleotidase release from rat kidney slices by PI-specific phospholipase C. The reaction mixtures (4 ml) containing 60 munits of PI-specific phospholipase C and 0.5 g of kidney slices were incubated at 37°C. At the indicated times, the released alkaline phosphatase (○) and 5' nucleotidase (●) activities were estimated. (Taguchi and Ikezawa, 1985).

Since bacterial PIPLCs are not highly toxic, they can be administered into animals or added to the culture medium of well-known cell lines without causing cell lysis, especially in studies on the biological role of GPI-anchored proteins. However, it must be noted that the sources of these enzymes are more or less virulent; *S. aureus* and *C. novyi* have been known as the causative bacteria for food poisoning and infectious diseases of mammals, and *B. cereus* and *B. thuringiensis* are well-known insect pathogens. In the infection of PIPLC-producing bacteria against the host eukaryotes, PIPLC may play some important but unknown role as one of the exotoxins. In fact PIPLCs from *B. cereus* and *B. thuringiensis* caused growth inhibition of transplanted Sarcoma 180 ascites tumor cells in vivo with deformation of the cells (Ikezawa et al., 1982-1983) and *B. thuringiensis* PIPLC proved to inhibit the growth of KB III (Ikezawa et al., 1985) and TN-368 (Ikezawa et al., 1989) cells when added to the culture media. *B. thuringiensis* PIPLC also caused the release of GPI-anchored ectoenzymes such as alkaline phosphatase, 5'-nucleotidase and B-glucosidase from the surface of these cells.

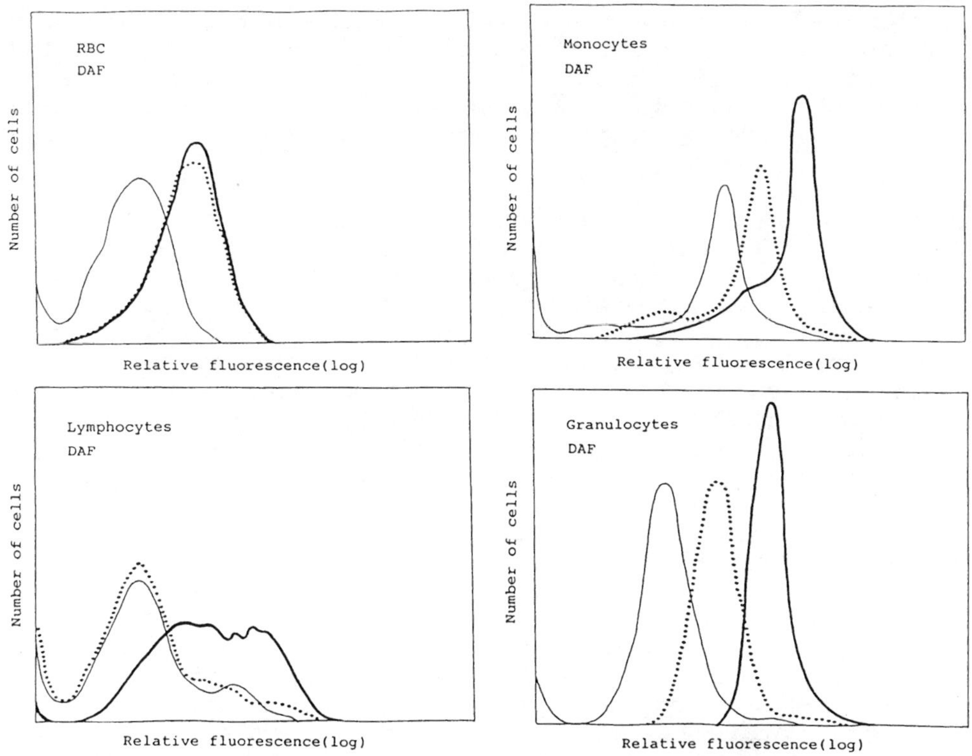

Fig 3. FACS analysis of DAF expression on normal blood cells. PIPLC-treated and -untreated cells were reacted with monoclonal mouse anti-DAF IgG(1), then stained with FITC-conjugated rabbit anti-mouse IgG. Control cells were directly stained with FITC-conjugated nonimmune mouse IgG(1) or FITC-conjugated rabbit anti-mouse IgG. —— , Control; --- , PIPLC-treated; —— , PIPLC-untreated.

Recently, many surface proteins such as VSG (Ferguson et al., 1985), Thy-1 (Low and Kincade, 1985), DAF (Davitz et al., 1986; Medof et al., 1986) and NCAM-120 (He et al., 1986, Hemperly et al., 1986), *Leishmania* surface protease gp63 (Bordier et al., 1986) and *Paramecium* SAgs (Capdeville et al., 1986) proved to be GPI-anchored proteins. PIPLCs are also available for detection of GPI-anchors in these antigens by the common epitope. PIPLC treatment of VSG exposes a cross-reacting determinant (CRD) containing oligosaccharide chains and myoinositol phosphate. Anti-CRD antibody has been shown to cross-react with *Leishmania* gp63 (Bordier et al., 1986), *Paramecium* SAgs (Capdeville et al., 1986) and acetylcholinesterase from human erythrocytes and *Torpedo* (Stieger et al., 1986), which were solubilized by PIPLC. Therefore, this cross reaction is a convenient hallmark for the presence of GPI anchors in the membrane proteins. Furthermore, the recent development of FACS analysis provides the means of sorting various cells and cell surface antigens for the immunological studies. Thus the expression of GPI-anchored surface antigens such as Thy-1 and several cell-differentiation antigens such as DAF (CD55) can now be quantitatively estimated in thymocytes as well as in blood cells.

Fig. 3 shows the example of FACS analysis of DAF on the surface of human erythrocytes (RBS), monocytes, lymphocytes and granulocytes, using *B. thuringiensis* PIPLC. As reported previously by Davitz et al. (1986) and Medof et al. (1986), DAF on human erythrocytes was resistant to PIPLC, while a significant amount of DAF was removed by PIPLC treatment from the surface of granulocytes (32 %), monocytes (48 %) and lymphocytes (75 %) (Taguchi et al., 1990). By such analysis, the sensitivity of a specific cell-surface antigen to PIPLC can be easily estimated for every blood cell and the difference in the sensitivity can be compared among various cells. Also in each cell species, the sensitivity of the surface antigens to PIPLC will be different by the stage and the extent of cell differentiation. In addition, it is now recognized that in the blood cells from the patients of PNH, paroxysmal nocturnal hemoglobinuria there is partial or complete deficiency of one or more GPI-anchored proteins (Davitz et al., 1986). Thus FACS analysis with PIPLC treatment will become as regularly used in clinical analysis as it is in immunological studies.

REFERENCES

Bordier, C. (1981). Phase separation of integral membrane proteins in Triton X-114 solution. J. Biol. Chem. **256:** 1604-1607.

Bordier, C., Etges, R. J., Eard, J., Turner, M. J. and Cardoso de Almeida, M. L. (1986). *Leishmania* and *Trypanosoma* surface glycoproteins have a common glycophospholipid membrane anchor. Proc. Natl. Acad. Sci. USA **83:** 5988-5991.

Capdeville, Y., Baltz, T., Deregnaucourt, C. and Keller, A. M. (1986). Immunological evidence of a common structure between *Paramecium* surface antigens and *Trypanosoma* variant surface glycoproteins. Exp. Cell Res. **167:** 75-86.

Davitz, M. A., Low, M. G. and Nussenweig, V. (1986). Release of decay-accelerating factor (DAF) from the cell membrane by phosphatidylinositol-specific phospholipase (PIPLC). Selective modification of complement regulatory protein. J. Exp. Med. **163:** 1150-1161.

Doery, H. M., Magnusson, B. J., Gulasekharam, J. and Pearson, J. E. (1965). The properties of phospholipase enzymes in staphylococcal toxins. J. Gen. Microbiol. **40:** 283-296.

Ferguson, M. A. J., Low, M. G. and Cross, G. A. M. (1985). Glycosyl-sn-1,2-dimyristoylphosphatidylinositol is covalently linked to *Trypanosoma brucei* variant surface glycoprotein. J. Biol. Chem. **260:** 14547-14555.

Futerman, A. H., Low, M. G. and Silman, I. (1983). A hydrophobic dimer of acetylcholinesterase from *Torpedo californica* electric organ is solubilised by phosphatidyl-specific phospholipase C. Neuroscience Letts. **40:** 85-89.

Futerman, A. H., Fiorini, R. M., Roth, E., Low, M.G and Silman, I. (1985). Physicochemical behaviour and structural characteristics of membrane-bound acetylcholinesterase from *Torpedo* electric organ. Biochem. J. **226:** 369-377.

He, H.-T., Barbet, J., Chaix, J.-C. and Goridis, C. (1986). Phosphatidylinositol is involved in the membrane attachment of NCAM-120, the smallest component of the neural cell adhesion molecule. EMBO J. **5:** 2489-2494.

Hemperly, J. J., Edelman, G. M. and Cunningham, B. A. (1986). cDNA clones of the neural cell adhesion molecule (N-CAM) lacking a membrane-spanning region consistent with evidence for membrane attachment via a phosphatidylinositol intermediate. Proc. Natl. Acad. Sci. USA **83:** 9822-9826.

Henner, D. J., Yang, M., Chen, E., Hellmiss, R., Rodriguez, H. and Low, M. G. (1988). Sequence of the *Bacillus thuringiensis* phosphatidylinositol specific phospholipase C. Nucl. Acids Res. **16:** 10383.

Ikezawa, H. and Taguchi, R. (1981). Phosphatidylinositol-specific phospholipase C from *Bacillus cereus* and *Bacillus thuringiensis*. Meths. Enzymol. **71:** 731-741.

Ikezawa, H., Yamanegi, M., Taguchi, R., Miyashita, T. and Ohyabu, T. (1976). Studies on phosphatidylinositol phosphodiesterase (phospholipase C type) of *Bacillus cereus*. Biochim. Biophys. Acta. **450:** 154-164.

Ikezawa, H., Taguchi, R., Asahi, Y. and Tomita, M. (1982-1983). The physiological actions of bacterial phospholipases C on eukaryotic cells and their membranes. J. Toxicology-Toxin Revs. **1:** 233-255.

Ikezawa, H., Nakabayashi, T., Suzuki, K., Nakajima, M., Taguchi, T. and Taguchi, R. (1983). Complete purification of a phosphatidylinositol-specific phospholipase C from a strain of *Bacillus thuringiensis*. J. Biochem.(Tokyo). **93:** 1717-1719.

Ikezawa, H., Kato, T., Ohta, S., Nakabayashi, T., Iwata, Y. and Ono, K. (1985). Physiological actions of phosphatidylinositol-specific phospholipase C from *Bacillus thuringiensis* on KB III cells: Alkaline phosphatase release and growth inhibition. Toxicon. **23:** 145-155.

Ikezawa, H., Hashimoto, A., Taguchi, R., Nakabayashi, T. and Himeno, M. (1989). Release of PI-anchoring enzymes and other effects of phosphatidylinositol-specific phospholipase C from *Bacillus thuringiensis* on TN-368 cells from a moth ovary. Toxicon. **27: 637-645.**

Ispolatovskaya, M. V. (1970). Role of metals in the activity of lecithinase of *Cl. perfringens*. Biokhimiya **35:** 434-439.

Kume, T., Taguchi, R., Ikezawa, H., Tokuyama, S., Nakachi, O. and Hirano, H. unpublished observations.

Kuppe, A., Evans, L. M., McMillen, D. A. and Griffith, O. H. (1989). Phosphatidylinositol-specific phospholipase C of *Bacillus cereus:* Cloning, sequencing, and relationships to other phospholipases. J. Bacteriol. **171:** 6077-6083.

Lechner, M., Kupke, T., Stefanovic, S. and Gotz, F. (1989). Molecular characterization and sequence of phosphatidylinositol-specific phospholipase C of *Bacillus thuringiensis*. Mol. Microbiol. **3:** 621-626.

Little, C. and Otnass, A.-B. (1975). The metal ion dependence of phospholipase C from *Bacillus cereus*. Biochim. Biophys. Acta. **391:** 326-333.

Low, M. G. (1981). Phosphatidylinositol-specific phospholipase C from *Staphylococcus aureus*. Methods in Enzymology, **71:** 741-746.

Low, M. G. (1990). Degradation of glycosyl-phosphatidylinositol anchors by phospholipases. Mol. Cell Biol. Membrane Proteins. Glycolipid Anchors of Cell-surface Proteins, 35-63.

Low, M. G. and Finean, J. B. (1976). The action of phosphatidylinositol-specific phospholipase C on membranes. Biochem. J. **154:** 203-208.

Low, M. G. and Finean, J. B. (1977a). Non-lytic release of acetylcholinesterase from erythrocytes by a phosphatidylinositol-specific phospholipase C. FEBS Letts. **82:** 143-146.

Low, M. G. and Finean, J. B. (1977b). Modification of erythrocyte membranes by a purified phosphatidylinositol-specific phospholipase C (*Staphylococcus aureus)*. Biochem. J. **162:** 235-240.

Low, M. G. and Finean, J. B. (1977c). Release of alkaline phosphatase from membranes by a phosphatidylinositol-specific phospholipase C. Biochem. J. **167:** 281-284.

Low, M. G. and Finean, J. B. (1978). Specific release of plasma membrane enzymes by a phosphatidylinositol-specific phospholipase C. Biochim. Biophys. Acta **508:** 565-570.

Low, M. G. and Kincade, P. W. (1985). Phosphatidylinositol is the membrane-anchoring domain of the Thy-1 glycoprotein. Nature **318:** 62-64.

Low, M. G. and Zilversmit, D. B. (1980). Role of phosphatidylinositol in attachment of alkaline phosphatase to membranes. Biochemistry **19:** 3913-3918.

Low, M. G., Stiernberg, J., Waneck, G. L., Flavell, R. A. and Kincade, P. W. (1988). Cell-specific heterogeneity in sensitivity of phosphatidylinositol-anchored membrane antigens to release by phospholipase C. J. Immunol. Methods. **113:** 101-111.

Majumdar, R. and Balasubramanian, A. s. (1985). The solubilisation of platelet membrane-bound acetylcholinesterase and aryl acylaminase by exogenous and endogenous phosphatidylinositol-specific phospholipase C. Biochem. Pharmacol. **34:** 4109-4115.

Medof, M. E., Walter, E. I., Roberts, W. L., Haas, R. and Rosenberry, T. L. (1986). Decay-accelerating factor of complement is anchored to cells by a C-terminal glycolipid. Biochemistry **25:** 6740-6747.

Nakabayashi, T. and Ikezawa, H. (1984). Release of alkaline phosphodiesterase I from rat kidney plasma membrane produced by the phosphatidylinositol-specific phospholipase C of *Bacillus thuringiensis*. Cell Struct. Func. **9:** 247-263.

Nakabayashi, T. and Ikezawa, H. (1986). Alkaline phosphodiesterase I release from eukaryotic plasma membranes by phosphatidylinositol-specific phospholipase C. II. The release from brush border membranes of porcine intestine. Toxicon. **24:** 975-984.

Ohyabu, T., Taguchi, R. and Ikezawa, H. (1978). Studies on phosphatidylinositol phosphodiesterase (phospholipase C type) of *Bacillus cereus*. II. In vivo and immunochemical studies of phosphatase-releasing activity. Arch. Biochem. Biophys. **190:** 1-7.

Ottolenghi, A. C. (1965). Phospholipase C from *Bacillus cereus*. A zinc-requiring metalloenzyme. Biochim. Biophys. Acta **106:** 510-518.

Prasad, A. R. S. and Low, M. G. (1987). Release of 5'nucleotidase from rat liver by bacterial phosphatidylinositol-specific phospholipase C. Fed. Proc. **46:** 2111.

Robert, W. L., Santikarn, S., Reinhold, V. N. and Rosenberry, T.L. (1988). Structural characterization of the glycoinositol phospholipid membrane anchor of human erythrocyte acetylcholinesterase by fast atom bombardment spectrometry. J. Biol. Chem. **263:** 18776-18784.

Sato, H. and Murata, R. (1973). Role of zinc in the production of *Clostridium perfringens* alpha toxin. Infection and Immunity **8:** 360-369.

Shukla, S. D., Coleman, R., Finean, J. B. and Michell, R. H. (1980). Selective release of plasma membrane enzymes from rat liver hepatocytes by a phosphatidylinositol-specific phospholipase C. Biochem. J. **187:** 277-280.

Slein, M. W. and Logan, G. F. Jr. (1962). Mechanism of action of the toxin of *Bacillus anthracis*. Alkaline phosphatasemia produced by culture filtrates of various bacilli. J. Bact. **83:** 359-369.

Slein, M. W. and Logan, G. F. Jr. (1965). Characterization of the phospholipases of *Bacillus cereus* and their effects on erythrocytes, bone and kidney cells. J. Bact. **90:** 69-81.

Stieger, S., Cardoso de Almeida, M. L., Blatter, M.-C., Brodbeck, U. and Bordier, C. (1986). The membrane-anchoring systems of vertebrate acetylcholinesterase and variant surface glycoproteins of African trypanosomes share a common antigenic determinant. FEBS LETTS. **199:** 182-186.

Sundler, R., Alberts, A. W. and Vagelos, P. R. (1978). Enzymatic properties of phosphatidyl-inositol inositol-phosphohydrorase from *Bacillus cereus*. Substrate dilution in detergent-phospholipid micelles and bilayer vesicles. J. Biol. Chem. **253:** 4175-4179.

Taguchi, R. and Ikezawa, H. (1978). Phosphatidylinositol-specific phospholipase C from *Clostridium novyi* Type A. Archives Biochem. Biophys. **186:** 196-201.

Taguchi, R. and Ikezawa, H. (1987). Properties of bovine erythrocyte acetylcholinesterase solubilized by phosphatidylinositol-specific phospholipase C. J. Biochem.(Tokyo), **102:** 803-811.

Taguchi, R., Asahi, Y. and Ikezawa, H. (1980). Purification and properties of phosphatidylinositol-specific phospholipase C of *Bacillus thuringiensis*. Biochim. Biophys. Acta **619: 48-57.**

Taguchi, R., Mizuno, M., Inoue, M. and Ikezawa, H. (1983). Increase in osmotic fragility of bovine erythrocytes induced by bacterial phospholipases C. J. Biochem. (Tokyo). **93:** 403-412.

Taguchi, R., Suzuki, K., Nakabayashi, T. and Ikezawa, H. (1984). Acetylcholinesterase release from mammalian erythrocytes by phosphatidylinositol-specific phospholipase C of *Bacillus thuringiensis* and characterization of the released enzyme. J. Biochem. (Tokyo). **96:** 437-446.

Taguchi, R., Asahi, Y. and Ikezawa, H. (1985). Ectoenzyme release from rat liver and kidney by phosphatidylinositol-specific phospholipase C. J. Biochem. (Tokyo). **97:** 911-921.

Taguchi, R., Funahashi, Y., Ikezawa, H. and Nakashima, I. (1990). Analysis of PI (phosphatidylinositol)-anchoring antigens in a patient of paroxysmal nocturnal hemoglobinuria (PNH) reveals deficiency of 1F5 antigen (CD 59), a new complement-regulatory factor. FEBS Letts. **261:** 142-146.

Takesue, Y., Yokota, K., Nishi, Y., Taguchi, R. and Ikezawa, H. (1986). Solubilization of trehalase from rabbit renal and intestinal brush-border membranes by a phosphatidylinositol-specific phospholipase C. FEBS Letts. **201:** 5-8.

Tomita, M., Taguchi, R. and Ikezawa, H. (1982). Molecular properties and kinetic studies on sphingomyelinase of *Bacillus cereus*. Biochim. Biophys. Acta. **704:** 90-99.

Volwerk, J. J., Shashidar, M. S., Kuppe, A. and Griffith, O. H. (1990). Phosphatidylinositol-specific phospholipase C from *Bacillus cereus* combines intrinsic phosphotransferase and cyclic phosphodiesterase activities: A ^{31}P NMR study. Biochemistry **29:** 8059-8062.

A CHEMICAL MODIFICATION THAT MAKES GLYCOINOSITOL PHOSPHOLIPIDS RESISTANT TO PHOSPHOLIPASE C CLEAVAGE: FATTY ACID ACYLATION OF INOSITOL

***Terrone L Rosenberry**

Department of Pharmacology, Case Western Reserve University, Cleveland, Ohio 44120, USA.

* Corresponding author.

INTRODUCTION

Integral membrane proteins maintain intimate contact with the hydrophobic phase of phospholipid bilayer membranes, require detergents for solubilization, and bind detergents in extracts. Such proteins generally interact with the bilayer through hydrophobic amino acid sequences in one or more transmembrane segments (Sabatini et al, 1982), although membrane-association of several intracellular proteins is mediated at least in part by acylation of protein residues with the fatty acids myristate or palmitate (Sefton and Buss, 1987). Recently a new class of more than 50 extracellular membrane proteins has emerged that are anchored, not by a transmembrane peptide segment, but solely by a glycoinositol phospholipid (GPL) linked covalently to the protein C-terminus. In this chapter the term "anchor" will refer exclusively to such GPLs. Many features of proteins with GPL anchors are covered in excellent review articles by Low (1987), Ferguson and Williams (1988), Cross (1990), and elsewhere in this volume. This article will address the susceptibility of protein anchors to cleavage by certain phospholipases, and it will focus on an unusual chemical modification of GPLs in which the acylation of an inositol hydroxyl group by a fatty acid conveys resistance to phospholipase C hydrolysis.

Susceptibility of many proteins with GPL anchors to cleavage by phospholipases.

The discovery of proteins with GPL anchors is intimately linked to the susceptibility of these anchors to cleavage by purified bacterial phosphatidylinositol-specific phospholipase C (PIPLC). Slein and Logan (1963) first suggested that a bacterial phospholipase C could solubilize alkaline phosphatase from tissue slices. Later workers demonstrated that highly purified PIPLC from *B. cereus* (Ikezawa et al, 1976), *Staphylococcus aureus*

(Low and Finean, 1977a), or *B. thuringiensis* (Taguchi et al, 1980) released alkaline phosphatase and other enzymes from cell surfaces, although the mechanistic basis of the release was unclear. The criterion of release by PIPLC has become the most widely applied for the identification of proteins with GPL anchors. Selective release from intact cells can be measured by flow cytometry or by assay of cell supernatants, and release from membrane fractions also has been described (see Rosenberry et al, 1989). Highly purified bacterial PIPLC preparations have excellent phospholipase C specificity, are free of protease contamination, and have yet to cleave a protein that has failed to contain a GPL anchor on further analysis.

Despite their phospholipase C specificity, a disadvantage in the use of bacterial PIPLCs to characterize GPL anchors and to identify free cellular glycolipids that correspond to GPLs is the unknown specificity of these PIPLCs for the characteristic GPL core structure described below. Preliminary experiments in our laboratory indicate that *B. thuringiensis* PIPLC has a specific activity with GPL anchors that is roughly comparable to that with phosphatidyl [^{3}H]inositol, although comparison is difficult because the specific activities can vary more than 1000-fold in various mixed nonionic detergent-phospholipid micelles. Other phospholipases have been identified which cleave GPL anchors with greater selectivity. An endogenous membrane-bound trypanosome phospholipase C has been purified to homogeneity and found to cleave dimyristoylglycerol from intact VSG at 170 times the rate that it cleaves phosphatidylinositol (Bülow and Overath, 1986; Hereld et al, 1986; Fox et al, 1986). Purification of a phospholipase C with an even greater specificity for the VSG anchor from rat liver plasma membranes has been reported (Fox et al, 1987). Like bacterial PIPLC, these anchor-specific phospholipases C do not require Ca ions and are fully active in buffers containing EGTA. An enzyme in mammalian serum cleaves between phosphate and inositol in a number of GPL-anchored proteins and thus corresponds to a phospholipase D (Davitz et al, 1987; Low and Prasad, 1988). This enzyme appears highly specific for GPLs and shows no activity toward phosphatidylinositol or other nonglycosylated phospholipids.

Use of these more GPL-specific phospholipases C and D has been limited by observations that these enzymes appear to be much more efficient in detergent extracts or reconstituted liposomes than on intact cells. Bülow and Overath (1966) noted that the trypanosome phospholipase C showed virtually no activity toward GPL-anchored VSGs when the purified proteins were reconstituted into separate liposomes before mixing but that hydrolysis rapidly occurred on fusing the liposomes or adding Triton X-100. Both for this reason and for their greater availability, bacterial PIPLCs have found much wider use in the identification of GPL anchors.

Explanations for the failure of certain GPL-anchored proteins to be released from cells by bacterial PIPLCs.

Several caveats must accompany the interpretation of experiments that measure the release of a protein from cells or membranes by a bacterial PIPLC, particularly when only partial release is achieved. First, a GPL-anchored protein may be incompletely accessible to PIPLC at the cell surface. This conclusion is appropriate when the protein is shown to be completely cleaved by PIPLC in detergent extracts by techniques described below. Very few examples of cell surface inaccessibility have been documented. The most striking is that of the trypanosome variant surface glycoproteins (vsgS), which apparently fail to be cleaved by PIPLC because of their unusually tight packing on the trypanosome surface (Low et al, 1986). Inaccessibility also might underlie the curious observation that GPL-anchored proteins in the R1.1 murine T cell line are released by *B. thuringiensis* PIPLC but not by *S. aureus* PIPLC (Low et al, 1988). No other systematic comparisons of these two PIPLCs have been reported, although this study showed that the *B. thuringiensis* enzyme is active over a wider range of solvent conditions in nonionic detergent solutions. In other cases of partial release of proteins with PIPLC, technical limitations like the failure to use saturating concentrations of PIPLC or the localization of a significant percentage of the GPL-anchored protein in intracellular transport organelles can prevent clear conclusions about cell surface accessibility.

A second explanation for partial release with PIPLC is the expression of a membrane protein in both GPL-anchored and non-GPL forms. Polymorphic forms of acetylcholinesterase are known to arise from alternative mRNA splicing of a single gene. Two alternative third exons encode either a C-terminus that becomes linked to a GPL anchor or a hydrophilic C-terminus that can attach to distinct membrane-binding subunits or simply be secreted (Sikorav et al, 1988; Rachinsky et al, 1990; Roberts et al, 1991). Alternative mRNA splicing also occurs with the neural cell adhesion molecule N-CAM, where 180 kDa and 140 kDa forms with membrane-spanning domains and a 120 kDa species with a GPL anchor have been identified. The C-terminus of the GPL-anchored chicken N-CAM form is encoded by exon 15, while exon 14 is spliced directly to exon 16 in the 180 kDa and 140 kDa forms (Owens et al, 1987). All three N-CAM forms are expressed concurrently in C6 glioma cells and cultured cerebellar cells, and the 120 kDa form appears to be released selectively to the culture medium (He et al, 1987). However, release may reflect expression of yet another alternatively spliced mRNA isoform, as illustrated with the Qa-2 antigen in splenocytes. Mitogenic stimulation of these cells with concanavalin A results in the appearance of a soluble, secreted form of the Qa-2 antigen that partially replaces a GPL-anchored cell surface Qa-2 of approximately the same size. The C-terminus of GPL-anchored Qa-2 is encoded by exon 5, while in the secreted Qa-2 form exon 4 is spliced to exon 6 to give a hydrophilic C-terminus (Ulker et al, 1990). The distinction between expression of soluble Qa-2 from an alternatively spliced mRNA and release of GPL-anchored Qa-2 by a putative endogenous PIPLC or protease can be difficult to make (Robinson 1987).

A third concern is that cell surface associations can either prevent the PIPLC-induced release of a GPL-anchored protein that is tightly associated with a membrane protein without a GPL anchor or promote release of a protein without a GPL anchor that interacts strongly with only a GPL-anchored protein. No conclusive evidence for interactions of these types have yet been reported, but lipoprotein lipase is a possible example. This enzyme is reported to have a GPL anchor (Chan et al, 1988), although no indication of an exon encoding an appropriate C-terminal sequence for GPL addition is apparent from the lipoprotein lipase gene structure (Deeb and Peng, 1989). Chan et al point out that a tight association of lipoprotein lipase with another GPL-anchored protein cannot be excluded and note that lipoprotein lipase interacts strongly with heparan sulfate proteoglycans, one form of which has recently been shown to contain a GPL anchor (David et al, 1990).

A fourth explanation for the failure of bacterial PIPLC to release a GPL-anchored protein from the cell surface is a chemical modification of the GPL discovered by detailed analysis of the GPL anchor on human erythrocyte acetylcholinesterase (AChE). Erythrocyte AChEs represent a subclass of dimeric (G_2) AChEs that also includes AChE from *Drosophila* and one AChE form from *Torpedo* electric organ. It was initially perplexing that bacterial PIPLC did not release AChE from human erythrocytes even though it did release other cellular G_2 AChEs, including erythrocyte AChE of other mammalian species such as the cow (Low and Finean, 1977b). This difference in GPL susceptibility between human and bovine AChEs remained when the purified AChEs were treated with PIPLC in detergent solution and was correlated with a significant difference in the polarity of the inositol phospholipids released from both AChEs by treatment with nitrous acid

(Roberts et al, 1987). The structural characterization outlined in the following section identifies the chemical modification on the human erythrocyte AChE GPL anchor as an acylation of an inositol hydroxyl group with palmitic acid and demonstrates that this modification is sufficient to block the cleavage of the GPL anchor by PIPLC.

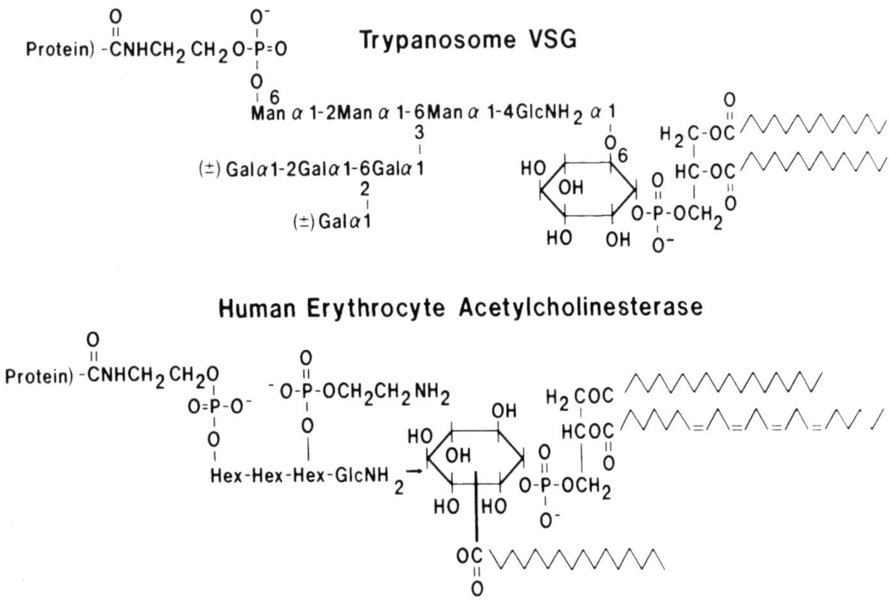

Fig. 1. GPL anchor structures. GPL anchors were released from the purified proteins by digestion with proteases and isolated by high-pressure liquid chromatography (HPLC). The structure of the anchor of VSG from *Trypanosoma brucei* was obtained by a combination of nuclear magnetic resonance (NMR) spectroscopy, chemical modification, gas chromatography-mass spectrometry (GC-MS), and exoglycosidase digestions (Ferguson et al., 1988). The structure of the human erythrocyte AChE anchor was deduced by lipid composition analyses and fast atom bombardment-mass spectrometry (FAB-MS) (Roberts et al., 1988a, b). The VSG anchor lipid is entirely composed of dimyristoylphosphatidylinositol, while the AChE anchor includes an alkylglycerol, 83% of which contains an 18:0 or 18:1 1-alkyl group and a 22:4, 22:5 or 22:6 2-acyl group. The AChE anchor also includes a fatty acid that directly acylates an inositol hydroxyl group at a position yet to be determined. This fatty acid is primarily palmitate.

Fatty acid acylation of inositol in the human erythrocyte AChE GPL anchor is responsible for resistance to PIPLC cleavage.

Rigorous identification of fatty acid acylation of an inositol hydroxyl group in the GPL requires direct determination of the chemical structure. Complete GPL structural determination involves a variety of analytical approaches to deduce the core oligosaccharide and inositol sequence and linkage position, the lipid composition and structure, and the location and nature of all branching groups. Further compounding the analysis is the heterogeneity apparent with several GPL anchors. The first complete structure was obtained for a trypanosome VSG anchor (Ferguson et al, 1988), and important elements are conserved in partial structures reported for the anchors of human erythrocyte AChE (Roberts et al, 1988b) and the rat Thy-1 (Homans et al, 1988). As noted in Fig. 1, the VSG GPL anchor contains a core linear sequence of 6-O-(ethanolamine-PO_4)-α-Man*p*-(1 -> 2)-α-Man*p*-(1 -> 6)-α-Man*p*-(1 -> 4)-α-GlcNH_2*p*-(1 -> 6)-*myo*-inositol-1-PO_4. This core glycan sequence is completely conserved in the Thy-1 GLP, and the less detailed information currently available from the AChE GPL is entirely consistent with it. Variation in the glycerolipid groups constitutes one source of heterogeneity in GPL structures, as noted in Fig. 1 (also see Roberts et al, 1988a). Additional contributors to heterogeneity are groups which branch from the linear core sequence. Many mammalian GPL anchors have a phosphorylethanolamine moiety with a free amino group, and in the Thy-1 GPL this group is in phosphodiester linkage to the 2-hydroxyl on the Man residue adjacent to GlcNH_2. A similar substitution is found in the AChE anchor (Fig. 1). The VSG anchor lacks this substituent but contains a branched galactose oligosaccharide of variable size linked to the 3-hydroxyl of this Man residue. The rat brain Thy-1 GPL anchor has a single GalNAc linked ß-(1 - 4) to this Man residue and a single Man linked α-(1 - 2) to the Man residue at the nonreducing terminus of the core sequence. These branching substituents may be relatively specific to mammalian brain, as they are also found in the GPL anchor of human brain scrapie prion protein (Baldwin et al, 1990) but nearly absent from the rat thymocyte Thy-1 GPL anchor. The only GPL structures yet known to deviate from the core glycan sequence noted above are found in *Leishmania,* where both the GPL anchor of a lipophosphoglycan (Turco et al, 1989) and free GPL glycolipids (McConville et al, 1990) retain only the Man-GlcNH_2-inositol-PO_4 portion of the core but replace other residues toward the nonreducing terminus of the core with sugars that include galactose.

Despite considerable heterogeneity, all the GPL structures described in the preceding paragraph can be cleaved by bacterial PIPLC except for that of human erythrocyte AChE. The structural modification responsible for this PIPLC resistance was first identified after cleavage of the AChE GPL anchor with nitrous acid, a procedure which breaks the glycosidic bond between $GlcNH_2$ and inositol in the core glycan sequence. This cleavage releases dimyristoylphosphatidylinositol from the VSG GPL anchor, but FAB-MS analysis revealed that the inositol phospholipid released from the AChE GPL anchor contains a palmitate group that directly acylates an inositol hydroxyl (Fig. 1). To document that the palmitoylation of inositol was directly responsible for the resistance of this inositol phospholipid to bacterial PIPLC, the experiment outlined in Fig. 2 was conducted. This figure shows clearly that alkylacylglycerol could be released from the inositol phospholipid by PIPLC only after the fatty acyl group was removed from inositol by NH_3 treatment.

The conversion from a PIPLC-resistant to a PIPLC-susceptible form following base treatment has become an important criterion for the identification of inositol acylation in a GPL, as illustrated by examples in following sections. However, while few other chemical bonds in a GPL should be as sensitive to base cleavage as this fatty acid ester, it is possible that other structural modifications could give a similar conversion. For example, esters of carboxylic acids other than fatty acids would also undergo conversion. Thus definitive identification of the basis for GPL resistance to PIPLC should still involve complete structural determination of the GPL. One report that suggests a different mechanism for GPL anchor resistance to PIPLC has appeared. The anchor of contact site A glycoprotein of *Dictyostelium discoideum* resembles a GPL in which the glycerolipid is replaced by ceramide, and it is not cleaved by PIPLC (Stadler et al, 1989). It is unclear, however, that the resistance arises solely from the ceramide component as the GPL anchor of the lipopeptidophosphoglycan from *Trypanosoma cruzi* also has a ceramide substitution for glycerolipid and is susceptible to bacterial PIPLC cleavage (Previato et al, 1990; Lederkremer et al, 1990).

Indications that other GPLs are resistant to PIPLC cleavage because of inositol acylation.

To determine whether inositol acylation is the basis of PIPLC resistance of other GPL-anchored proteins and of free GPLs, a procedural strategy derived from that for human erythrocyte AChE in Fig. 2 often has ben followed. PIPLC cleavage is conveniently monitored either by radiolabelling the glycerolipid with an exogenous reagent like $[^{125}I]$TID (Fig. 2) or through biosynthetic labelling with tritiated fatty acids or by observing a change in

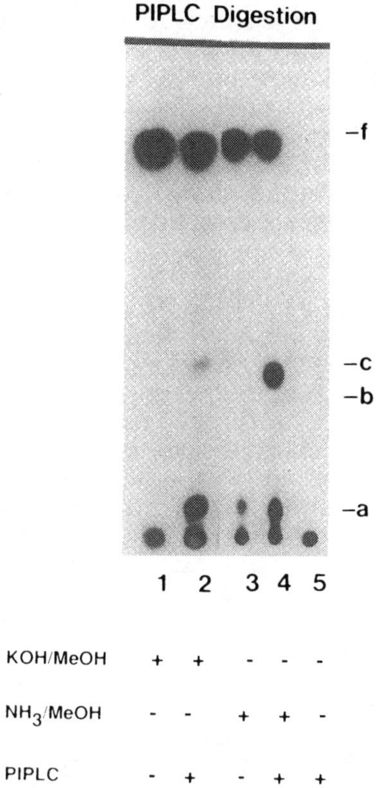

Fig. 2. Palmitoylation of an inositol hydroxyl is responsible for the resistance of the human erythrocyte GPL anchor to bacterial PIPLC (Roberts et al., 1988b). To detect lipid components, intact human erythrocyte AChE was radiolabelled with 3-(trifluoromethyl)-3-(*m*-[^{125}I]iodophenyl)diazirine ([^{125}I]TID), a reagent that is about 90% specific for lipid groups in the GPL anchor (Roberts et al., 1987). The GPL inositol phospholipid was then released by treatment with nitrous acid and purified by HPLC. Lipid fragments produced by the following procedures were extracted with chloroform, analysed by TLC on Silica Gel GF with hexane/diethyl ether/acetic acid (60:30:1), and detected by autoradiography. Treatment only with purified bacterial PIPLC failed to cleave the inositol phospholipid, which remained at the origin in this TLC system (lane 5). Methanolic KOH (lane 1) or NH_3-saturated methanol (lane 3) released label which moved predominantly at the position of a fatty acid methyl ester standard (f). If the KOH-treated sample was then digested with PIPLC, released label was also detected (lane 2) at the position of an alkylglycerol standard (a). However, after the milder NH_3 treatment of the sample, digestion with PIPLC released label at the positions of both alkylglycerol and alkylacylglycerol (b,c) standards (lane 4).

property of the protein or radiolabelled glycan on cleavage. The PIPLC resistance of human erythrocyte decay accelerating factor (DAF), for example, was investigated after radiolabelling the purified protein with [^{125}]TID. Nitrous acid released the same radiolabelled inositol phospholipid from DAF as from AChE, and the fatty acid compositions of the DAF and AChE were similar enough to suggest that DAF and AChE received anchors from the same pool of acylated GPL precursors in human erythrocytes (Walter et al, 1990).

While isolation and analysis of the inositol phospholipids released from GPLs by nitrous acid is a route to rigorous identification of inositol acylation, less demanding tests were needed that did not require protein purification and could be applied to cell extracts. One test introduced by Toutant et al (1989) to examine the effect of deacylation on the susceptibility of GPL-anchored proteins to PIPLC took advantage of the fact that in all GPL-anchored proteins that have been studied, the inositol phospholipid is the only nonionic detergent-binding domain. In GPL anchors susceptible to PIPLC, removal of the hydrophobic diacyl- or alkylacylglycerol generates a hydrophilic protein that retains the glycan portion of the anchor. Techniques that can demonstrate this loss of detergent-binding capacity include phase partitioning in Triton X-114 (Etges et al, 1986), phenyl-Sepharose chromatography (Fatemi et al, 1987), and polyacrylamide gel electrophoresis in nonionic detergents (Toutant et al, 1989). Since associated detergent micelles contribute substantially to the Stokes radii of intact anchored proteins, loss of detergent binding can generate a several-fold increase in the electrophoretic mobility of the cleaved proteins. This effect of cleavage by PIPLC was shown for bovine erythrocyte AChE (Toutant et al, 1989), whose GPL anchor is completely susceptible to PIPLC, and it is illustrated for DAF purified from HeLa cells in Fig. 3B, where nearly complete conversion to a higher mobility form was produced by PIPLC treatment (compare lanes 2 and 2). In contrast, DAF purified from human erythrocytes is acylated on inositol and shows no conversion with PIPLC (Fig. 3A, lanes 1 and 2). The inositol acyl group was removed from the lipid in Fig. 2 with methanolic base, but similar treatment of proteins would result in denaturation and loss of native protein mobility on gel electrophoresis. Therefore, a milder deacylation protocol involving hydroxylamine at pH 10.5-11 was introduced to prevent protein denaturation and to take advantage of the fact that, in most mammalian GPL anchors examined to date, the inositol phospholipid contains alkylacylglycerol. The alkylglycerol cannot be cleaved by hydroxylamine at pH 10.5-11, and GPL-anchored proteins retain their interactions with nonionic detergent micelles. Subsequent treatment with PIPLC resulted in partial cleavage of human erythrocyte AChE (Toutant et al, 1989), but complete conversion of AChE from a PIPLC-resistant to a PIPLC-susceptible form under these conditions was difficult to achieve without denaturing most of the protein. DAF, on the other hand, was more resistant to denaturation during incubation at high Ph with hydroxylamine. This treatment alone gave little conversion to the hydrophilic form (Fig. 3, lanes 3), but subsequent treatment of human erythrocyte DAF with PIPLC gave complete cleavage (Fig. 3A, lane 4).

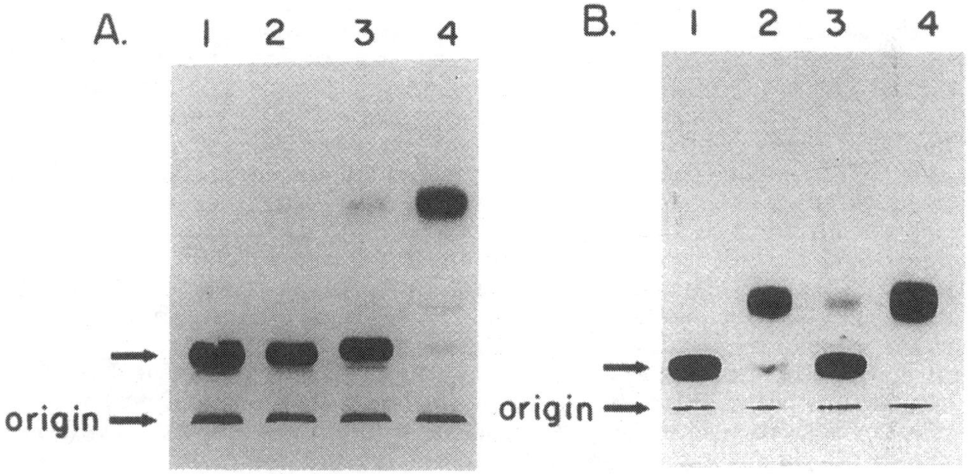

Fig. 3. Conversion of human erythrocyte DAF from a PIPLC-resistant to a PIPLC-susceptible form by pretreatment at high pH with hydroxylamine (Walter et al., 1990). DAF from human erythrocytes (panel A) was purified by immunoaffinity chromatography and radiomethylated by reductive methylation, and DAF was HeLa cells (panel B) was labelled biosynthetically with [^{35}S]cysteine and similarly purified. Samples were treated overnight at 0° C at pH 7.4 (lanes 1 and 2) or at pH 10.7 with 1 M hydroxylamine (lanes 3 and 4). Samples were then incubated for 2 hr at 37° with (lanes 2 and 4) or without (lanes 1 and 3) purified bacterial PIPLC and subjected to electrophoresis in 7.5% polyacrylamide gels containing 50 mM Tris/glycine (pH 8.9) and 0.5% Triton X-100. The gel position where sample was loaded is marked as the origin, and the detergent-DAF complex is indicated by the upper arrow in each panel. DAF cleaved by PIPLC does not interact with detergent and corresponds to the bands farthest from the origin in each panel.

The nondenaturing gel electrophoresis procedure in Fig. 3 can be applied to GPL-anchored proteins in other cell types to estimate the extent of inositol acylation. Analysis of AChE in two sublines of the human erythroleukemia K562 was conducted on crude cell extracts without AChE purification, because the enzyme activity of AChE is retained in nondenaturing gels and AChE migration can be assessed with great sensitivity by a histochemical stain. Subline K562-243 (ATCC) expressed AChE that was mostly susceptible to PIPLC without treatment by hydroxylamine at Ph 10.7, whereas AChE in subline K562-48 required basic hydroxylamine treatment before PIPLC cleavage could occur (Toutant et al, 1990). The susceptibility of DAF to PIPLC cleavage in the two K562 sublines paralleled that of AChE, with most DAF in K562-243 (ATCC) directly susceptible to PIPLC and DAF in K562-48 requiring prior treatment with basis hydroxylamine for PIPLC cleavage (Walter et al, 1991). These observations suggested that the extent of inositol acylation is regulated in a cell-specific manner and is similar for all GPL-anchored proteins in a given cell. To test this hypothesis, Walter et al. (1991) transfected placental alkaline phosphatase into each of the K562 sublines. More than 90% of the alkaline phosphatase isolated from placenta was directly susceptible to PIPLC, but virtually all of the alkaline phosphatase expressed in K5652-48 was cleaved by PIPLC only after prior alkaline hydroxylamine treatment. This observation provided strong support for the hypothesis. Part of the alkaline phosphatase expressed in K562-243 (ATCC) was directly susceptible to PIPLC, but a larger percentage than observed for the endogenous DAF required prior treatment with alkaline hydroxylamine for PIPLC cleavage. It is unclear whether this discrepancy involves preferential expression of alkaline phosphatase in a small subset of phenotypically distinct transfected cells or a quantitative difference between GPL anchors of alkaline phosphatase and DAF in the same cells. Walter et al (1991) note that the K562-48 subline expresses more hemoglobin and glycophorin than K562-243 (ATCC) and suggest that the appearance of acylated GPL anchors is correlated with differentiation along the erythroid pathway.

One might conclude from the preceding observations that inositol acylation is an idiosyncrasy of human erythroid cells. While it is true that human erythrocytes show a much higher percentage of acylated GPL anchors than other mammalian cells that have been examined, 10-15% of the DAF in HeLa cells, polymorphonuclear cells, lymphocytes, and stimulated T cells does appear to have an acylated GPL anchor (Walter et al, 1990). These observations suggest that small amounts of GPL-anchored proteins may be acylated in many tissues. However, it has recently become apparent that palmitoylation of inositol occurs to a major extent in GPLs from the procyclic stage (insect midgut form) of *Trypanosoma brucei* (Field et al, 1991). Trypanosomes in this stage do not produce VSGs but do express a major cell surface protein PARP (procyclic acidic repetitive protein). Biosynthetic labelling of PARP with [^{3}H]ethanolamine and [^{3}H]myristic acid suggested a GPL anchor, but PARP was completely resistant to PIPLC cleavage (Clayton and Mowatt, 1989; Field et al, 1991). Nitrous acid cleavage of [^{3}H]myristate-labelled PARP released the radiolabel in an inositol phospholipid which was shown to consist primarily of 1-stearyl-2-lysophosphatidyl-palmitylinositol (Field et al, 1991).

The structure of the palmitylated inositol phospholipid in trypanosome PARP was obtained by detailed comparison to a highly similar free GPL in procyclic trypanosomes designated PP1 (Field et al, 1991). The core glycan sequences of PP1 and of two free GPLs from the bloodstream form of trypanosomes denoted P2 and P3 appear identical to that indicated for the GPL anchor of trypanosome VSG in Fig. 1 (Mayor et al, 1990a). Because the inositol phospholipid in P2 is also dimyristoylphosphatidyl-inositol, P2 is thought to be the glycolipid precursor of the VSG GPL anchor (Mayor et al, 1990b). A similar precursor-product relationship is indicated for PP1 and PARP (Field et al, 1991). The P2 and P3 GPLs were first identified in [^{3}H]myristate-labelled trypanosome extracts as lipids referred to as A and C, respectively, that were sensitive to nitrous acid cleavage (Krakow et al, 1989; Mayor et al, 1990b). Meanwhile, in vitro trypanosome membrane preparations were shown to synthesize radiolabelled lipids resembling GPLs from UDP-[^{3}H]GlcNAc or GDP-[^{3}H]mannose (Masterson et al, 1989), and a progressive core glycan biosynthesis extending from an inositol phospholipid precursor was deduced for trypanosome GPLs from these in vitro studies (Doering et al, 1990). Detailed structural studies demonstrated that these in vitro products included P2 and P3 species indistinguishable from the corresponding in vivo GPLs (Menon et al, 1990). Recent observations indicate that an array of acylated GPLs may be present in mammalian cells. A nitrous acid-sensitive glycolipid that could be biosynthetically labelled with tritiated inositol, glucosamine or ethanolamine was identified in T cell hybridoma cells (DeGasperi, et al, 1990). This glycolipid was not cleaved by bacterial PIPLC, and it was absent in mutant hybridoma cells that were deficient in GPL anchor biosynthesis. A set of [^{3}H]mannose-labelled glycolipids detected in murine lymphoma cells also was missing in lymphoma mutants that do not express GPL-anchored proteins (Lemansky et al, 1991). These glycolipids were not cleaved by PIPLC, but base treatment of the most polar species converted it to a form susceptible to PIPLC. In vitro radiolabelling of microsomal preparations from these lymphoma mutants with UDP-[^{3}H]GlcNac or GDP-[^{3}H]mannose revealed that mutants A, C and H produce no glycolipids consistent with GPL precursors. In contrast, mutants E and I make mannolipids not observed in the wild-type lymphoma cells that become susceptible to PIPLC after base treatment (Hirose et al, 1991a and b). Further work is required to establish that these intriguing mammalian glycolipids correspond to acylated GPLs.

Possible involvement of acylation of GPL inositol in cell function.

The conservation of inositol palmitylation in GPL-anchored proteins of cells that range from trypanosomes to human erythrocytes suggests that this chemical modification of GPLs plays an important cellular role. Further observations that VSG proteins selectively link to PIPLC-sensitive P2 and not PIPLC-resistant acylated P3 in bloodstream trypanosomes while PARP does link to PIPLC-resistant acylated PP1 in procyclic trypanosomes has led to additional speculation about the function of inositol acylation. Mayor et al (1990b) postulate two alternative models for the involvement of inositol acylation in GPL anchor biosynthesis. In the first, all GPLs are derived from precursors with acylated inositol, and the activity of a putative inositol deacylating enzyme determines the steady-state level of nonacylated GPLs. Potential roles of acylation in this model include the protection of newly synthesized GPLs from degradation by endogenous phospholipases, the facilitation of putative obligatory transmembrane flip-flop during biosynthesis of GPLs, or the direct stereochemical acceleration of biosynthetic catalytic enzymes. Enthusiasm for this model is dampened by an inability as yet to detect any deacylase activity in cell extracts and by the lack of any observable precursor-product relationship between trypanosome P2 and P3 (Krakow et al, 1986; Mayor et al, 1990b). The second model assumes two substrate-specific protein-GPL transferases. These transferases are proposed as recognizing distinct protein signals and/or GPLs. In trypanosomes, for example, one transferase could be selective for the transfer of P2 to VSGs while the second is specific for the transfer of PP1 to PARP, and differential expression of the transferases could be controlled by the trypanosome life cycle. The assumption of transferase specificity for protein in this model seems incompatible with observations in a previous section on mammalian GPL-anchored proteins. These proteins appear to be acylated on inositol in a cell-specific rather than a protein-specific fashion, as shown in particular for transfected alkaline phosphatase. Transferase specificity for acylated versus nonacylated GPLs at first seems incompatible with the observed in vitro transfer of both P2 and P3 to VSGs (Mayor et al, 1991). However, successful transfer in these studies required solubilization by small amounts of nonionic detergents, a treatment that could abolish any putative transferase selectivity among GPLs. Therefore, current data seem least incompatible with the hypothesis that 1) steady-state levels of acylated GPLs are regulated by the level of acyl transferase activity, not by deacylase activity; and 2) at least two protein-GPL transferases can be expressed which selectively recognize either acylated or nonacylated GPLs. It is noteworthy that the second tenet of this hypothesis is based primarily on the selective transfer of P2 to VSGs under conditions in which both P2 and P3 are present, and other examples of selective transfer of only acylated or nonacylated GPLs are needed to establish this point.

More information about the function of GPLs themselves will probably be required before a clear role can be assigned to the acylation of inositol in GPLs. The identification of phospholipases that are highly specific for GPLs together with the fact that inositol acylation blocks phospholipase C cleavage of GPLs suggests that inositol palmitoylation may be a general mechanism for regulating the susceptibility of anchored proteins and free GPLs to endogenous phospholipase C. This proposal is difficult to examine, however, because so few examples of GPL cleavage by endogenous phospholipases in intact cells have been found. Activation of an anchored latent protease in malarial parasites by apparent endogenous phospholipase cleavage has been reported (Braun-Breton et al, 1988), and insulin interaction with insulin receptors has been proposed to lead to activation of endogenous phospholipases that cleave free GPLs to generate insulin second messengers (see Low and Saltiel, 1988). However, compelling structural analyses are still required to establish that GPLs or their glycan products are involved in insulin action. GPL-specific phospholipase C in intact bloodstream trypanosomes may play a role in the degradation of VSGs, but other conditions under which this phospholipase C cleaves the anchor of VSGs have yet to be identified (Bülow et al, 1989).

An alternative functional role for GPL acylation can be envisioned in which phospholipase cleavage is only of secondary importance. This role may relate to the possible importance of GPLs as ligands in cell signal transduction during antibody-mediated T cell activation. The attachment of a GPL anchor to a T cell membrane protein enables antibody to that protein to activate the T cell (Robinson et al, 1989). The mechanism that involves the GPL anchor in this process is unclear, but it presumably involves GPL interaction with a separate transmembrane protein that can transduce an activation signal to the cell interior. Such GPL-binding proteins can be viewed as receptors for membrane-bound GPL ligands, and distinct receptors for acylated and nonacylated GPLs can be contemplated. Thus a rationale for the regulation of GPL acylation may arise from signal transduction processes that distinguish these two GPL forms.

REFERENCES

Baldwin, M. A., Stahl, N., Reinders, L. G., Gibson, B. W., Prusiner, S. B. and Burlingame, A. L. (1990). Analyt. Biochem. In press.

Braun-Breton, C., Rosenberry, T. L. and Pereira da Silva, L. (1988). Nature **332:** 457-459.

Bülow, R., and Overath, P. (1986). J. Biol. Chem. **261:** 11918-11923.

Bülow, R., Nonnengasser, C. and Overath, P. (1989). Molec. Biochem. Parasitol. **32:** 85-92.

Chan, B. L., Lisanti, M. P., Rodríguez-Boulan, E. and Saltiel, A. R. (1988). Science **241:** 1670-1672.

Clayton, C. E. and Mowatt, M. R. (1989). J. Biol. Chem. **264:** 15088-15093.

Cross, G. A. M. (1990). Ann. Rev. Cell Biol. **6:** 1-39.

David, G., Lories, V., Decock, B., Marynen, P., Cassiman, J-J. and Van den Berghe, H. (1990). J. Cell Biol. **111:** 3165-3176.

Davitz, M. A., Hereld, D., Shak, S., Krakow, J., Englund, P. T. and Nussenzweig, V. (1987). Science **238:** 81-84.

Deeb, S. S. and Peng, R. (1989). Biochemistry **28:** 4131-4135.

DeGasperi, R., Thomas, L. J., Sugiyama, E., Chang, H. M., Beck, P. J., Orlean, P., Albright, C., Waneck, G., Sambrook, J. F., Warren, C. D. and Yeh, E. T. H. (1990). Science **250:** 988-991.

Doering, T. L., Masterson, W. J., Hart, G. W. and Englund, P. T. (1990). J. Biol. Chem. **265:** 611-614.

Etges, R., Bouvier, J. and Bordier, C. (1986). EMBO J. **5:** 597- 601.

Fatemi, S. H., Haas, R., Jentoft, N., Rosenberry, T. L. and Tartakoff, A. M. (1987). J. Biol. Chem. **262:** 4728-4732.

Ferguson, M. A. J. and Williams, A. F.(1988). Ann. Rev. Biochem. **57:** 285-320.

Ferguson, M. A. J., Homans, S. W., Dwek. R. A. and Rademacher, T. W. (1988). Science **239:** 753-759.

Field, M. C., Menon, A. K. and Cross, G. A. M. (1991). J. Biol. Chem. **266:** in press.

Fox, J. A., Duszenko, M., Ferguson, M. A. J., Low, M. G. and Cross, G. A. M. (1986). J. Biol. Chem. **261:** 15757-15771.

Fox, J. A., Soliz, N. M. and Saltiel, A. R. (1987). Proc. Natl. Acad. Sci. USA. **84:** 2663-2667.
He, H-T., Finne, J. and Goridis, C. (1987). J. Cell Biol. **105:** 2489-2500.
Hereld, D., Krakow, J. L., Bangs, J. D., Hart, G. W. and Englund, P. T. (1986). J. Biol. Chem. **261:** 13813-13819.
Hirose, S., Ravi, L., Tartakoff, A. and Medof, M. E. (1991a). FASEB J. **5:** A 793.
Hirose, S., Ravi, L., Tartakoff, A. and Medof, M. E. (1991b). Personal communication.
Homans, S. W., Fergusion, M. A. J., Dwek, R. A., Rademacher, T. W., Anand, R. and Williams, A. F. (1988). Nature **333:** 269-272.
Ikezawa, H., Yamanegi, M., Taguchi, R., Miyashita, T. and Ohyabu, T. (1976). Biochim. Biophys. Acta **450:** 154-164.
Krakow, J. L., Hereld, D., Bangs, J. D., Hart, G. W. and Englund, P. T. (1986). J. Biol. Chem. **261:** 12147-12153.
Krakow, J. L., Doering, T. L., Masterson, W. J.,Hart, G. W. and Englund. P. T. (1989). Molec. Biochem. Parasitol. **36:** 263-270.
Lederkremer, R. M., Lima, C., Ramirez, M. I. and Casal, O. L. (1990). Eur. J. Biochem. **192:** 337-345.
Lemansky, P., Gupta, D. K., Meyale, S., Tucker, G. and Tartakoff, A. M. (1991). Mol. Cell Biol. in press.
Low, M. G. (1987). Biochem. J. **244:** 1-13.
Low, M. G. and Finean, J. B. (1977a). Biochem. J. **167:** 281-284.
Low, M. G. and Finean, J. B. (1977b). FEBS Lett. **82:** 143-146.
Low, M. G. and Prasad, A. R. S. (1988). Proc. Natl. Acad. Sci. USA **85:** 980-984.
Low, M. G. and Saltiel, A. R. (1988). Science **239:** 268-275.
Low, M. G., Ferguson, M. A. J., Futerman, A. H. and Silman, I. (1986). Trends Biochem. Sci. **11:** 212-215.
Low, M. G., Stiernberg, J., Waneck, G. L., Flavell, R. A. and Kincade, P. W. (1988). J. Immunol. Meths. **113:** 101-111.
Masterson, W. J., Doering, T. L., Hart, G. W. and Englund, P. T. (1989). Cell **56:** 793-800.
Mayor, S., Menon, A. K., Cross, G. A. M., Ferguson, M. A. F., Dwek, R. A. and Rademacher, T. W. (1990a). J. Biol. Chem. **265:** 6164-6173.
Mayor, S., Menon, A. K. and Cross, G. A. M. (1990b). J. Biol. Chem. **265:** 6174-6181.
Mayor, S., Menon, A. K. and Cross, G. A. M. (1991). J. Cell Biol. in press.

McConville, M. J., Homans, S. W., Thomas-Oates, J. E., Dell, A. and Bacic, A. (1990). J. Biol. Chem. **265:** 7 385-7394.

Menon, A. K., Schwartz, R. T., Mayor, S. and Cross, G. A. M. (1990). J. Biol. Chem. **265:** 9033-9042.

Owens, G. C., Edelman, G. M. and Cunningham, B. A. (1987). Proc. Natl. Acad. Sci. USA **84:** 294-298.

Previato, J. O., Gorin, P. A. J., Mazurek, M., Xavier, M. T., Fournet, B., Wieruszesk, J. M. and Mendonca-Previato, L. (1990). J. Biol. Chem. **265:** 2518-2526.

Rachinsky, T. L., Camp, S., Li, Y., Ekstrom, T. J., Newton, M. and Taylor, P. (1990). Neuron **5:** 317-327.

Roberts, W. L., Kim, B. H. and Rosenberry, T. L. (1987). Proc. Natl. Acad. Sci. USA **84:** 7817-7821.

Roberts, W. L., Myher, J. J., Kuksis, A., Low, M. G. and Rosenberry, T. L. (1988a). J. Biol. Chem. **263:** 18766-18775.

Roberts, W. L., Santikarn, S., Reinhold, V. N. and Rosenberry, T. L. (1988b). J. Biol. Chem. **263:** 18776-18784.

Roberts, W. L., Doctor, B. P., Foster, J. D. and Rosenberry, T. L. (1991). J. Biol. Chem. **266:** 7481-7487.

Robinson, P. J. (1987). Proc. Natl. Acad. Sci. USA **84:** 527- 531.

Robinson, P. J., Millrain, M., Antoniou, J., Simpson, E. and Mellor, A. L. (1989). Nature **342:** 85-87.

Rosenberry, T. L., Toutant, J.-P., Haas, R. and Roberts, W. L. (1989). Meths. Cell Biol. **32:** 231-255.

Sabatini, D. D., Kreibich, G., Morimoto, T. and Adesnik, M. S. (1982). J. Cell Biol. **92:** 1-22.

Sefton, B. M. and Buss, J. E. (1987). J. Cell Biol. **104:** 1449-1453.

Sikorav, J-L., Duval, N., Anselmet, A., Bon, S., Krejci, E., Legay, C., Osterlund, M. Reimund, B. and Massoulié, J. (1988). EMBO J. **7:** 2983-2993.

Slein. M. W. and Logan, G. F. (1963). J. Bacteriol. **85:** 369-381.

Stadler, J., Keenan, T. W., Bauer, G. and Gerisch, G. (1989). EMBO J. **8:** 371-377.

Taguchi, R., Asahi, Y. and Ikezawa, H. (1980). Biochim. Biophys. Acta. **619:** 48-57.

Toutant, J-P., Roberts, W. L., Murray, N. R. and Rosenberry, T. L. (1989). Eur. J. Biochem. **180:** 503-508.

Toutant, J-P., Richards, M. K., Krall, J. A. and Rosenberry, T. L. (1990). Eur. J. Biochem. **187:** 31-38.

Turco, S. J., Orlandi, Jr. P. A., Homans, S. W., Ferguson, M. A. J., Dwek, R. A. and Rademacher, T. W. (1989). J. Biol. Chem. **264:** 711-6715.

Ulker, N., Lewis, K. D., Hood, L. E. and Stroynowski, I. (1990). EMBO J. **9:** 3839-3847.

Walter, E. I., Roberts, W. L., Rosenberry, T. L., Ratnoff, W. D. and Medof, M. E. (1990). J. Immunol. **144:** 1030-1036.

Walter, E. I., Lapurga, J. P., Millan, J. L., Ikehara, Y. and Medof, M. E. (1991). Personal communication.

GLYCOSYLPHOSPHATIDYLINOSITOL ANCHORED RECOGNITION MOLECULES THAT FUNCTION IN AXONAL FASCICULATION, GROWTH AND GUIDANCE IN THE NERVOUS SYSTEM

***F S Walsh and P Doherty**

Department of Experimental Pathology, UMDS, Guy's Hospital, London Bridge, London, SE1 9RT.

* Corresponding author.

SUMMARY

A large number of glycoproteins in the central nervous system are attached to the cell membrane via covalent linkage to glycosylphosphatidylinositol (GPI). Many of them, including the drosophila fasciclin 1 as well as the mammalian glycoproteins Thy-1, TAG1, N-CAM and F11,F3, contactin are members of the immunoglobulin gene superfamily. These and other GPI-linked molecules have been implicated in key developmental events including selective axonal fasciculation and highly specific growth to and innervation of target tissues. In model systems fasciclin 1, TAG1 and N-CAM have been shown to be capable of mediating cell-cell adhesion via a homophilic binding mechanism confirming their operational classification as cell adhesion molecules (CAMs). However, of these molecules, only N-CAM has been shown to mediate a complex response (neurite outgrowth) via a homophilic binding mechanism. Whether the other molecules in this family mediate biological responses by binding to themselves and/or other molecules remains to be determined.

Studies on N-CAM provide an ideal model system for understanding the function of GPI anchors since alternative splicing of the NCAM gene generates both lipid-linked and transmembrane N-CAM isoforms. Recent studies have shown that neurons can recognise and respond (by increased neurite outgrowth) to both lipid-linked and transmembrane N-CAM isoforms expressed on the surface of non-neuronal cells following transfection with appropriate cDNAs. The major determinant of neuronal responsiveness was the level of N-CAM expression rather than the isoform type. Neurite outgrowth in response to transfected N-CAM is mediated by transmembrane N-CAM isoforms expressed by neurons and this involves the activation of classical second messenger pathways in the neurons. One possibility is that GPI anchors are utilised when a cell has simply to provide recognition or positional information to a second cell whereas transmembrane molecules might be required for cells that actively respond to such information. The hypothesis is compatible with all the known information on N-CAM expression and function and may be extended to other adhesive events.

INTRODUCTION

In the last 6 years, a large number of proteins have been found to be attached to cell membranes via covalent linkage to a glycosylphosphatidylinositol (GPI) anchor. It is of some considerable interest that one group of proteins namely recognition molecules in the central nervous system (CNS) some of which are members of the immunoglobulin (Ig) gene superfamily appear to frequently utilise this mode of attachment to the cell membrane. In this review we will discuss the evidence for the presence of GPI anchors in these proteins and various ideas as to why this class of molecule might be anchored to the membrane in this manner. Table 1 lists the recognition molecules in the CNS for which there is strong evidence of membrane attachment via a GPI anchor. With the exception of the Thy-1 antigen all have been found to be involved in either axonal growth, guidance or fasciculation. In some cases these molecules have been shown to interact via a homophilic binding mechanism.

Table 1

GPI-anchored recognition molecules implicated in axonal fasciculation, growth and guidance.

NAME / SOURCE	APPARENT MOLECULAR WEIGHT (kD)
Fasciclin 1	72
F3 / F11 / contactin*	130-135
TAG1	135
THY-1	25
NCAM	120-125
Tectal Membranes	33
T-cadherin	90

RECOGNITION MOLECULES INVOLVED IN AXONAL GROWTH AND/OR FASCICULATION

Fasciclin 1 is a homophilic adhesion molecule that is expressed in the nervous system of drosophila and grasshoppers (Bastiani et al, 1987; Zinn et al, 1988; Elkins et al, 1990). Its specific expression on axon fascicles has raised the possibility that it may be involved in specific cell-cell interactions at this location although it is also found outside the nervous system. In terms of structure it is an Ig superfamily member with four Ig domains and a short hydrophobic region at the carboxy terminal (Zinn et al, 1988). This suggested a possible attachment to the cell membrane by a GPI anchor as many proteins with short hydrophobic sequences at their carboxy termini have been found to contain a GPI anchor (Low, 1989). Use was made of the drosophila Schneider 1 cell line to show that the endogenous fasciclin 1 could be released by the enzyme phosphatidylinositol specific phospholipase C (PIPLC) (Hortsch and Goodman, 1990). However, there was a fraction of fasciclin 1 that was resistant to release suggesting heterogeneity in the GPI anchor structure. Further supporting evidence was obtained using a gene transfer approach where Schneider 2 cells that do not express fasciclin 1 were transfected with a full length cDNA construct under the control of a heat shock promoter. Here a percentage of the transfected fasciclin 1 was released by PIPLC, and metabolic labelling studies with ^{14}[C] ethanolamine provided strong evidence for the presence of an endogenous GPI anchor in fasciclin 1 rather than release through binding to a second GPI anchored protein. A soluble form of fasciclin 1 has also been observed but this is likely to be due to release of the GPI anchored form from the cell membrane although whether this cleavage is specific or non-specific is not clear (Hortsch and Goodman, 1990). The possibility that this release reflects a direct secretion from the cells perhaps due to an overloading of the machinery for lipid linkage cannot however be excluded.

Three groups have independently cloned cDNAs for a membrane glycoprotein that appears to be involved in axonal fasciculation. It has been named F11 (Brummendorf et al, 1988), F3 (Gennarini et al, 1989) and contactin (Rancht, 1988). Each group used expression cloning with antibodies to isolate appropriate cDNA clones. The deduced protein sequence of the protein predicts a structure that is a member of the Ig Superfamily with six Ig domains. F11 and F3 have a hydrophobic amino acid at their carboxy terminus again compatible with membrane attachment via a GPI anchor. In contrast, contactin appears to be a transmembrane protein. Whether contactin is an alternatively spliced version of F11 and F3 is not known but it appears a strong possibility. F11 can be released from brain membranes by PIPLC and like fasciclin 1 can be released from F11 transfected cells with PIPLC (Brummendorf et al, 1989). Whether the F11, F3 contactin protein operates via a homophilic or heterophilic binding mechanism is not known but as F11 can be isolated as a complex of several polypeptides, a heterophilic mechanism seems more likely. Antibodies to F11 inhibit neurite outgrowth over other neurites by - 20%; however, this may reflect the ability of the same antibodies to perturb neurite fasciculation (Chang et al, 1987).

TAG1 is a GPI anchored protein that appears to be involved in axonal growth in the CNS (Dodd et al, 1988; Furley et al, 1990). This glycoprotein was shown to be transiently expressed on commissural neurons as they projected towards the ventral midline of the spinal cord, with a down regulation of expression after the axons crossed the floor-plate region. Full length cDNAs for TAG1 were recently isolated by expression cloning (Furley et al, 1990). The protein again has Ig domains showing it is a member of the Ig superfamily. In the case of TAG1 there are six Ig domains and a short hydrophobic region at the carboxy terminus (Furley et al, 1990). PIPLC removes TAG1 from brain membranes and also a neuronal somatic cell hybrid that expressed TAG1 following transfection providing strong evidence that TAG1 is attached to the cell surface via a GPI anchor. Direct binding studies with transfected cells suggest that TAG1 can bind to itself in a homophilic manner. Similarly purified TAG1 can support neurite outgrowth from TAG1 positive neurons when coated to a culture dish. However this is unlikely to be mediated by a homophilic binding mechanism as PIPLC, which removes TAG1 from neurons, does not inhibit the response. A percentage of TAG1 also appears to be released from cells spontaneously (Furley et al, 1990; Dodd, 1991). As there is no evidence of alternative splicing in the TAG1 gene associated with membrane attachment, one possibility is that the soluble form is simply released from the cell membrane by a protease or phospholipase C rather than by any specific cellular mechanism.

THE THY-1 ANTIGEN

The Thy-1 antigen was originally identified as a differentiation antigen on T-lymphocytes and more recently has been clearly found on neurons (Morris, 1985). Although it is extremely abundant in brain, and from its amino acid sequence is a member of the Ig superfamily, its biological function in brain or thymus has not been established (Campbell et al, 1981; Morris, 1985). It is however clear that Thy-1 is attached to the cell membrane by a GPI anchor since PIPLC will remove most Thy-1 from the cell membrane. The GPI anchor from Thy-1 has also been isolated and its structure established (Hortsch et al, 1988). Recent modelling studies have shown that the Thy-1 protein appears to sit on the membrane and that the GPI anchor is buried in the core of the protein (Rademacher et al, 1991). It is also possible that the glycan moiety in the GPI anchor interacts with a carbohydrate binding site on the Ig domain of Thy-1. One of the interesting aspects of Thy-1 is that antibodies to the protein are mitogenic for T-cells. The mechanism whereby lipid-linked molecules in lymphocytes transduce recognition signals remains to be established. However, in many instances the requirement for accessory transmembrane spanning components has been suggested and in the case of the Thy-1 antigen has been demonstrated experimentally (for review, see Robinson, 1991 and in this volume).

THE NEURAL CELL ADHESION MOLECULE (NCAM)

Next to Thy-1 the best characterised recognition molecule in the CNS that has a GPI anchor is the neural cell adhesion molecule (NCAM). NCAM is a sialoglycoprotein that is clearly involved in specific cell-cell interactions in the CNS and muscle and operates via a homophilic binding mechanism (Walsh and Doherty, 1991). Studies with the NCAM gene have proceeded beyond most of the other CNS recognition molecules and an essentially complete picture of the NCAM intron-exon gene structure is now available. This has allowed the observed protein isoforms of NCAM to be related to differential patterns of splicing of the NCAM gene. Figure 1 shows schematically the three major classes of NCAM protein isoforms that have been identified. In outline these are either transmembrane, GPI anchored or soluble. It is clear from studies that combine the protein structure with the NCAM gene structure (Walsh and Doherty, 1991) that, in general, the homophilic binding region which is believed to be contained in the five Ig domains is constitutively expressed in the different NCAM isoforms. Thus the different mechanisms of membrane attachment are unlikely to directly affect NCAM-NCAM recognition per se. A complete picture of the exons in the NCAM gene and their location can be found in a recent review (Walsh and Doherty, 1991). The exons that mediate the different means of membrane attachment have been identified. The exon that dictates a transmembrane attachment binding mechanism is exon 16 (Owens et al, 1987). Inclusion of this exon and constitutive exons downstream of this generate a 140 kD NCAM isoform. An exon downstream of exon 16 is used specifically in neurons and this is associated with a NCAM protein isoform of 180 kD. In terms of mRNAs the 140 kD isoform results from the translation of a 6.7 Kb mRNA while the 180 kD isoform results from the expression of the 7.2 Kb mRNA. The

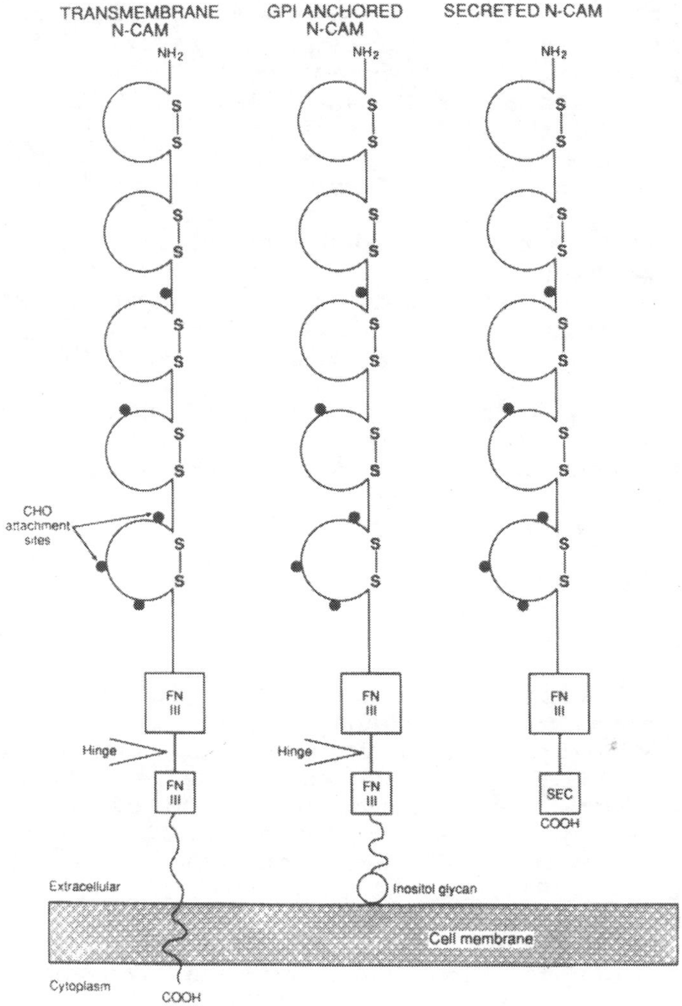

Figure 1. GPI anchored NCAM is one of three main classes of the protein. The diagram shows these in schematic form.

NCAM is found as either transmembrane, GPI anchored or secreted forms. All three share the five immunoglobulin domains (numbered 1-5 from the amino terminus) which are believed to be the sites of homophilic binding. There are five sites of possible N-linked carbohydrate (CHO) attachment although which sites are actually occupied is not known at present. The transmembrane and GPI anchored forms have two type III fibronectin repeats while the soluble NCAM has only one form. The site hinge region is between the two fibronectin repeats. Two main forms of transmembrane NCAM exist (180 and 140 kD) and these have either a large or short cytoplasmic tail.

splicing pathway for GPI membrane attachment is also known. Exon 15 encodes the sequence information for GPI anchoring and encodes 25 amino acids in human and chick NCAM and 24 amino acids in mouse NCAM (Cunningham et al, 1987; Barbas et al, 1988; Barton et al, 1988). Hydropathy analysis of the exon 15 encoded sequence of human NCAM shows a sequence that is relatively hydrophobic (Barton et al, 1988). Figure 2 shows a comparison of the amino acid sequence of the exon 15 encoded sequences for human, chick and mouse NCAM. There is strong conservation between the sequences with an overall 68% match, if the sequences are compared one to one and this can increase to 84% for human to mouse NCAM. In line with other known GPI anchored proteins it is likely that there is a cleavage during protein processing but the residues at which this occurs is not known for NCAM. In terms of mRNAs the GPI anchored NCAM is encoded by the 2.9 Kb mRNA and generates protein isoforms of 120-125 kD. The secreted NCAM forms of about 115 kD are generated by a novel splicing pathway that involves the introduction of an exon called SEC in the extracellular region of the protein upstream of exon 13. A stop codon in the SEC exon terminates the reading frame and generates a soluble protein (Grower et al, 1988).

The expression of GPI anchored NCAM is developmentally regulated and is also found in specific cell types. Perhaps the best studied example of the regulation of GPI anchored NCAM is in skeletal muscle. Mononucleate myoblasts express a transmembrane NCAM of about 140 kD while the myotubes that result from the fusion of the myoblasts express a GPI anchored NCAM of about 125 kD (Moore et al, 1987). At the mRNA level this is associated with a shift from a 6.7 Kb mRNA to a predominantly 2.9 Kb mRNA. Direct evidence for the expression of a GPI anchored NCAM in myotubes was found by treating muscle cell cultures with PIPLC (Moore et al, 1987). The enzyme completely removed 125 kD NCAM from myotubes, but had no effect on NCAM immunoreactivity on myoblasts. Glial cells also express a GPI linked NCAM that can be released by PIPLC (He et al, 1980) but neurons appear to synthesise transmembrane NCAMs only. The glial cell line C6 also appears to shed the GPI anchored NCAM at a high rate from the cell membrane (He et al, 1986). Again as with fasciclin 1 and TAG1, whether this is a specific or non specific cleavage remains to be determined.

We have attempted to assess the role of GPI anchored NCAM in two model systems namely myoblast fusion and neurite outgrowth. During myoblast fusion there is a shift from a transmembrane to GPI anchored NCAM. Studies with PIPLC show that fusion competent myoblasts are inhibited from fusing by addition of the enzyme suggesting that NCAM-NCAM

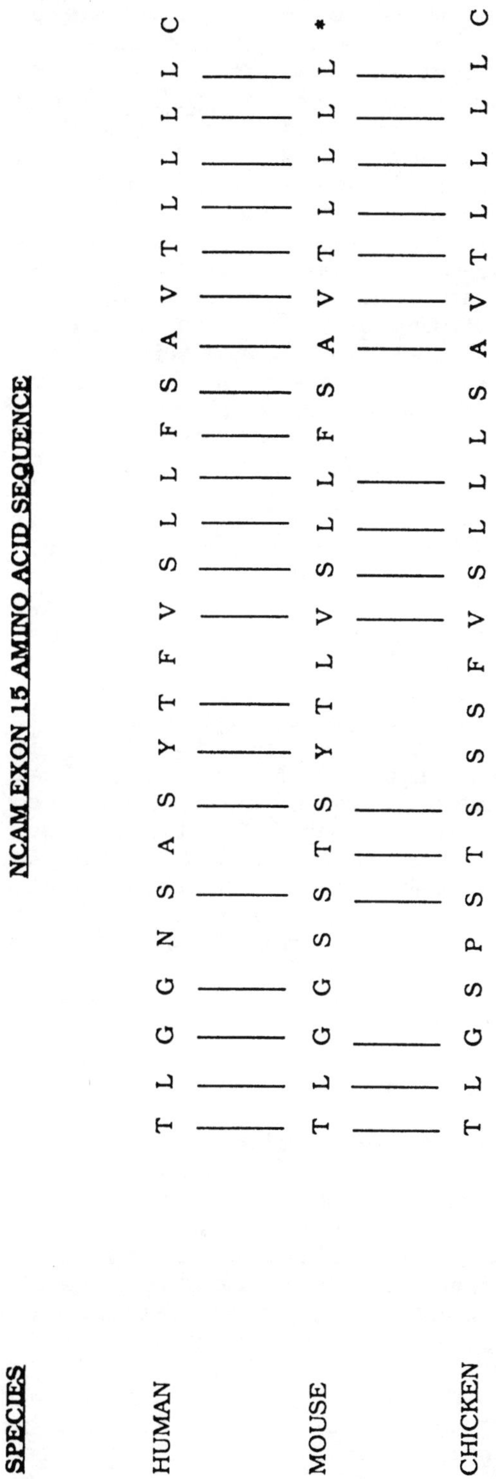

Figure 2. A comparison of the amino acid sequence encoded by exon 15 of the NCAM gene from human, mouse and chicken. This sequence contains the signals for GPI anchoring. Between the three sequences there is 68% similarity. However between human and mouse it is 84%, human and chicken 76%, mouse and chicken 72%. A line between sequences indicates sequence similarity across species. The mouse NCAM sequence is one amino acid shorter than human or chicken (indicated by *) such that the carboxy terminal amino acid in human and chick is cysteine while in mouse it is leucine.

interactions may be an obligatory part of the fusion process (Knudsen et al, 1989). Full length cDNA clones for GPI anchored NCAM have been produced (Barton et al, 1988) and placed in a plasmid based expression vector. This muscle cell line C2 has been transfected with a cDNA construct that allows expression of the 125 kD GPI anchored NCAM isoform (Dickson et al, 1990). In these experiments there is an elevation of the total NCAM expressed by the myoblasts since there is an endogenous NCAM pool in these muscle cells. The results show that the NCAM transfected cells have an enhanced fusion capacity and that there is a relationship between the level of transfected NCAM in the myoblasts and the fusion index (Dickson et al, 1990). It is likely that the high level expression of GPI linked NCAM is mimicking the natural increase in this isoform that occurs during fusion (Moore et al, 1987). The experiments also show that the GPI anchored NCAM is functionally active and that a transmembrane linkage is not an essential requirement for the effect to be observed. However, in the case of GPI-linked NCAM induced fusion, increased adhesion per se may be sufficient to mediate the response. It is not known whether the GPI-linked NCAM directly modulates biochemical events within the muscle cells.

The question of the role of the different NCAM isoforms in recognition events has been dealt with more rigorously in a study on neurite outgrowth. Figure 3 shows a flowsheet of the main aspects of the assay system used. Monolayer cell cultures of 3T3 fibroblasts or 3T3 fibroblasts transfected with different NCAM isoforms are grown to confluence on Titertek eight chamber sides. At that point, populations of dissociated neurons are plated on these different monolayers and allowed to attach and extend neurites. After 24 hours the cultures are fixed with paraformaldehyde and the neurons marked by staining with a specific antibody. The cultures are then viewed under a fluorescence microscope and the lengths of neurites measured following digitising of the fluorescence image. In this way, it is possible to assess the role of transfected NCAM on neuronal growth. The main isoform classes tested to date include the 140 kD transmembrane isoform, the 120 kD GPI anchored isoform and finally the secreted NCAM (see Fig. 1). Neurons have been grown on a large panel of cells that express different isoforms at varying levels. Figure 4 shows that it is possible to readily quantitate expression of either transmembrane or GPI anchored transfected NCAM at the surface of 3T3 cells by enzyme linked immunoassays. PIPLC will remove the transfected NCAM from the 3T3 cells synthesising GPI anchored NCAM but not the transmembrane NCAM. Associated with the loss of NCAM at the cell surface, there is also a loss of responsiveness in terms of enhanced neurite

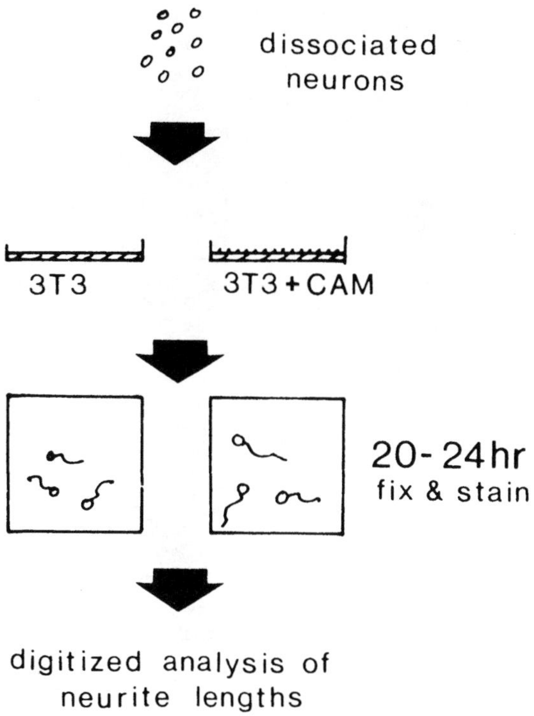

Figure 3. Schematic diagram showing the procedure we use to measure the neurite outgrowth promoting activity of CAMS introduced into 3T3 fibroblasts via gene transfer. Different populations of neurons are added to establish monolayers of parental and transfected cells and allowed to attach and extend neurites over 24 hours. At this point the culture is fixed and the neurons identified by a specific antibody using immunofuorescence. The neuronal morphology is measured by an image analysis system following digitising of the fluorescence image.

outgrowth in the GPI-NCAM synthesising cells but not the transmembrane ones (Fig. 4). This important experiment shows that both transmembrane and GPI-anchored NCAM can support neurite outgrowth. However, PIPLC can specifically inhibit the latter response and this is associated with removal of GPI-linked NCAM from the transfected cells. In a larger study comparing the properties of different cell lines, we confirmed that the transmembrane and GPI anchored NCAMs but not soluble NCAM were able to induce an enhanced pattern of neurite outgrowth. However, the major determinant of growth was the level of NCAM expression rather than the mechanism of membrane association (Doherty et al, 1990). There was a highly cooperative relationship between the level of NCAM expression and neuronal responsiveness with clear evidence of a threshold effect. Above the threshold, relatively small increases in the level of NCAM results in substantial increases in neurite outgrowth (Doherty et al, 1990). The reasons for the highly cooperative nature of the NCAM response are not known but it is possible that membrane mobility is a major

NCAM DEPENDENT NEURITE OUTGROWTH FROM RAT CEREBELLAR NEURONS : PI-PLC SPECIFICALLY BLOCKS THE FUNCTION OF GPI-LINKED NCAM

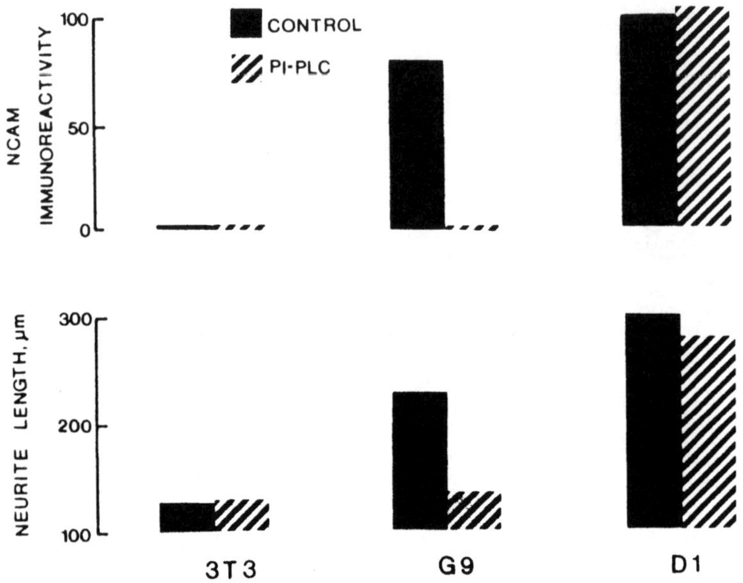

Figure 4. Effect of PIPLC on transfected NCAM immunoreactivity and neurite outgrowth. 3T3 fibroblasts, transfected G9 cells that express GPI linked NCAM and D1 cells that express transmembrane NCAM were used as a substrate for cerebellar neurons. PIPLC was added to each cell type and the effects on NCAM surface expression and neurite outgrowth assessed. PIPLC removes NCAM from the surface of G9 cells with a corresponding decrease in NCAM dependent neurite outgrowth. D1 cells are unaffected by the treatment as are 3T3 cells that do not express NCAM.

determinant. Both the GPI anchored and transmembrane NCAM are freely mobile in the plasma membrane and may be able to form small aggregates of NCAM that allow a cooperative response. With respect to GPI anchored NCAM, it is clear that neurons can recognise and respond to this form and that it is equally as functional as the transmembrane NCAM isoform in providing a recognition signal. However these experiments do not address the question of the ability of different NCAM isoforms in transducing a recognition signal to a cellular response.

OTHER GPI ANCHORED RECOGNITION MOLECULES

In addition to the recognition molecules described so far, there are preliminary reports of two additional proteins in the CNS that are likely to have important functions and are also GPI anchored. These are an inhibitory axonal guidance protein in the chick visual system (Stahl et al, 1990). The protein is of 33 kD and is released from the optic tectum by PIPLC. The second protein is T-cadherin which may be involved in neural crest migration (Ranscht and Bonner-Fraser, 1991). Cadherins are generally believed to be transmembrane proteins (Takeichi, 1991) but T-cadherin appears to be truncated at its carboxy terminus and is probably attached to the cell membrane by a GPI anchor. Further studies will be required to assess the structure and function of these two recognition molecules.

FUNCTION OF GPI ANCHORS ON RECOGNITION MOLECULES

Although the attachment of cell membrane proteins via a GPI anchor is a highly specific and regulated process the functional significance of this system has remained unclear. Studies on recognition molecules in the nervous system may however provide model systems with which to approach this issue. In the CNS most of the recognition molecules that have a GPI anchor are members of the Ig superfamily. They are also involved in cell-cell adhesion mediated either via a homophilic or heterophilic mechanism. One possibility is that the rapid removal of GPI anchored CAMs from the cell membrane would lead to the rapid breakage of established cell-cell bonds. This has been suggested to be important for NCAM interactions but has yet to be demonstrated experimentally (He et al, 1986). In fact, the observation that transmembrane NCAM i neurons mediates neurite outgrowth equally well over monolayers expressing lipid-linked or transmembrane NCAM suggests that this possibility is unlikely. Furthermore, there is also the question of specificity since any membrane protein that is GPI anchored would be partially or completely susceptible to removal. A PIPLC has recently been found in mouse brain but whether it acts intra- or extracellularly is not clear (Fouchier et al, 1990). A second possibility is that GPI anchors are involved in cell signalling events and that cleavage of the extracellular glycoprotein may result in the activation of second messenger pathways. One consequence of PIPLC action on cell membranes would be the release of diacylglycerol and phosphatidic acid. However as these would be generated on the wrong side of the lipid bilayer, they are unlikely to be active in cell signalling. An additional possibility is that GPI anchored proteins may be internalised via an endocytotic mechanism and that the anchor is removed from the protein at that point (Robinson, 1991). The consequences would be a possible increase in intracellular diacylglycerol and phosphatic acid and potential release of the cleared protein by exocytosis. At present this remains a model with no experimental support.

Direct evidence exists to support the notion that NCAM in neurons does not simply mediate cell adhesion per se, and that NCAM can directly induce biochemical changes in these cells. For example, studies on the PC12 neuronal cell line have shown that the binding of both polyvalent antibodies and monovalent Fab fragments to NCAM (as well as the L1 glycoprotein) can induce a pertussis toxin sensitive increase in intracellular calcium (Schuch et al, 1990). Furthermore, NCAM and N-cadherin dependent morphological differentiation of PC12 cells can also be inhibited by pertussis toxin and partially inhibited by calcium channel antagonists (Doherty and Walsh, unpublished observations). The CAMs on neurons (NCAM, N-cadherin and L1) as well as the integrin receptors for extracellular matrix components that, in combination, mediate essentially all of the neurite outgrowth over a wide variety of cell types, including astrocytes, Schwann cells and muscle exist only as transmembrane spanning molecules with cytoplasmic domains (for review see Doherty and Walsh, 1989). This suggests that they all mediate neurite outgrowth by either directly binding to the cytoskeleton and/or by stimulating intracellular second messenger pathways.

It is also clear that cells that act as a substratum for axonal growth and/or regeneration, or target cells to which neurons respond with extensive branching and synapse formation, often express lipid-linked recognition molecules. For example astrocytes and Schwann cells over which neurons grow but do not synapse predominantly express GPI-linked NCAM. Similarly, skeletal muscle exhibits a highly controlled pattern of alternative splicing characterised by the expression of 140 kD transmembrane isoform in myoblasts with a switch to the 125 kD GPI-linked isoform in myotubes (see above). Motor neurons grow over and synapse with the myotubes. In addition, tectal membranes which provide the positional information to growing retinal ganglion cells necessary for the establishment of the retino-tectal map express a 33 kD glycoprotein on their cell surface that is inhibitory for the growth of temporal but not nasal axons. This molecule is expressed in a graded manner across the tectum and ca be released from membrane in an active form by PIPLC treatment (Stahl et al, 1990). Similarly, T-cadherin which may provide positional information for the segmental nature of neural crest cell migration and motor axon segregation in early development also appears to be linked to the cell membrane by a GPI anchor (Ranscht and Bronner-Fraser 1991).

Based on our own work with the differing isoforms of NCAM, and considering the above observations on the relative patterns of expression of lipid-linked and transmembrane recognition molecules, we would postulate that a GPI anchor may be utilised when a cell has simply to provide recognition or positional information to a second cell whereas transmembrane molecules might be required by cells that actively respond to such information by migration and/or axon extension.

Advantages of lipid-linkage would include increased lateral mobility in the cell membrane which may facilitate highly co-operative relationships between expression and neuronal responsiveness. Potential roles of lipid-linkage versus transmembrane insertion in changing the half-life of glycoproteins in the cell membrane should also be actively investigated.

REFERENCES

Barbas, J. A., Chaix, J-C., Steinmetz, M. and Goridis, C. (1988). Differential splicing and alternative poly-adenylation generates distinct NCAM transcripts and proteins in the mouse. EMBO J. **7:** 625-632.

Barton, C. H., Dickson, G., Gower, H. J., Rowett, L. H., Putt, W., Elsom, V., Moore, S. E., Goridis, C. and Walsh, F. S. (1988). Complete sequence and in vitro expression of a tissue specific phosphatidylinositol-linked NCAM isoform from skeletal muscle. Development **104:** 165-173.

Bastiani, M. J., Harrelson, A. L., Snow, P. M. and Goodman, C. S. (1987). Expression of fasciclin I and II glycoproteins on subsets of axon pathways during neuronal development in the grasshopper. Cell **48:** 745-755.

Brummendorf, T., Wolff, J. M., Frank, R. and Rathjen, F. (1989). Neural cell recognition molecule F11: homology with fibronectin type III and immunoglobulin type C domains. Neuron **2:** 1351-1361.

Campbell, D. G., Gagnon, J., Reid, K. B. M. and Williams, A. F. (1981). Rat brain Thy-1 glycoprotein: the amino acid sequence, disulphide bonds and an unusual hydrophobic region. Biochem. J. **195:** 15-30.

Chang, S. F., Rathjen, F. G. and Raper, J. A. (1987). Extension of neurites on axons is impaired by antibodies against specific cell surface glycoproteins. J. Cell Biol. **104:** 355-362.

Cunningham, B. A., Hemperly, J. J., Murray, B. A., Prediger, E. A., Brackenbury, R. and Edelman, G. M. (1987). Neural cell adhesion molecule: structure, immunoglobulin-like domains, cell surface modulation, and alternative RNA splicing. Science **236:** 799-806.

Dickson, G.,Peck, D., Moore, S. E., Barton, C. H. and Walsh, F. S. (1990). Enhanced myogenesis in NCAM transfected mouse myoblasts. Nature **344:** 348-351.

Doherty, P. and Walsh, F. S. (1989). Neurite guidance molecules. Current Opinions in Cell Biology. **1:** 1102-1106.

Doherty, P., Fruns, M., Seaton, P., Dickson, G., Barton, C. H., Sears, T. A. and Walsh, F. S. (1990). A threshold effect of the major isoforms of NCAM on neurite outgrowth. Nature **343:** 464-466.

Dodd, J., Morton, S. B., Karagogeos, D., Yamamoto, M. and Jessel, T. M. (1988). Spatial regulation of axonal glycoprotein expression on subsets of embryonic spinal neurons. Neuron **1:** 105-115.

Dodd, J. (1991). Cell surface and diffusible molecules that guide developing axons. Discussions in Neurosci. **7:** 53-58.

Elkins, T., Hortsch, M., Bieber, A. J., Snow, P. M. and Goodman, C. S. (1990). Drosophila fasciclin I is a novel homophilic adhesion molecule that along with fasciclin III can mediate cell sorting. J. Cell Biol. **110:** 1825-1832.

Fouchier, F., Baltz, T. and Rougon, G. (1990). Identification of glycosylphosphatidylinositol-specific phospholipase C in mouse brain membranes. Biochem. J. **269:** 321-327.

Furley, A. J., Morton, S. B., Manolo, D., Karagogeos, D., Dodd, J. and Jessell, J. M. (1990). The axonal glycoprotein TAG1 is an immunoglobulin superfamily member with neurite outgrowth promoting activity. Cell **61:** 157-170.

Gennarini, G., Cibelli, G., Rougon, G., Mattei, M-G. and Goridis, C. (1989). The mouse neuronal cell surface protein F3: a phosphatidylinositol-anchored member of the immunoglobulin superfamily related to chicken contactin. J. Cell Biol. **109:** 775-788.

Gower, H. J., Barton, C. H., Elsom, V. L., Thompson, J., Moore, S. E., Dickson, G. and Walsh, F. S. (1988). Alternative splicing generates a secreted form of NCAM in muscle and brain. Cell **55:** 955-964.

He, H-T., Barbet, J., Chaix, J-C. and Goridis, C. (1986). Phosphatidylinositol is involved in the membrane attachment of NCAM-120, the smallest component of the neural cell adhesion molecule. EMBO J. **5:** 2489-2494.

Homans, S. W., Ferguson, M. A. J., Dwek, R. A., Rademacher, T. W., Anand, R. and Williams, A. F. (1988). Complete structure of the glycosyl-phosphatidylinositol membrane anchor of rat brain Thy-1 glycoprotein. Nature **333:** 269-272.

Hortsch, M. and Goodman, C. S. (1990). Drosophila fasciclin I, a neural cell adhesion molecule, has a phosphatidyl inositol lipid anchor that is developmentally regulated. J.Biol. Chem. **265:** 15104-15109.

Knudsen, K. A., Smith, L. and McElwee, S. (1989). Involvement of cell surface phosphatidylinositol-anchored glycoproteins in cell-cell adhesion of chick embryo myoblasts. J. Cell Biol. **109:** 1779-1786.

Low, M. G. and Kincade, P. W. (1985). Phosphatidylinositol is the membrane-anchoring domain of the Thy-1 glycoprotein. Nature **318:** 62-64.

Low, M. G. (1989). The glycosylphosphatidylinositol anchor of membrane proteins. Biochim. Biophys. Acta. **988:** 427-454.

Moore, S. E., Thompson, J., Kirkness, V., Dickson, J. G. and Walsh, F. S. (1987). Skeletal muscle neural cell adhesion molecule (NCAM): changes in protein and mRNA species during myogenesis of muscle cell lines. J. Cell Biol.**105:** 1377-1386.

Morris, R. (1985). Thy-1 in developing nervous tissue. Dev. Neurosci. **7:** 133-160.

Owens, G. C., Edelman, G. M. and Cunningham, B. A. (1987). Organisation of the neural cell adhesion molecule gene: alternative exon usage as the basis for different membrane associated domains. Proc. Natl. Acad. Sci. USA **84:** 294-298.

Rademacher, T. W., Edge, C. J. and Dwek, R. (1991). Dropping anchor with the lipophosphoglycans. Current Biology. **1:** 41-42.

Ranscht, B. (1988). Sequence of contactin, a 130kD glycoprotein concentrated in areas of interneuronal contact, defines a new member of the immunoglobulin supergene family in the nervous system. J. Cell Biol. **107:** 1561-1573.

Ranscht, B. and Bonner-Fraser, M. (1991). T-cadherin expression alternates with migrating neural crest cells in the trunk of the avian embryo. Development **111:** 15-22.

Robinson, P. J. (1991). Phosphatidylinositol membrane anchors and T-cell activation. Immunology Today. **12:** 35-41.

Schuch, V., Lohse, M. J. and Schachner, M. (1989). Neural cell adhesion molecules influence second messenger systems. Neuron **3:** 13-200.

Stahl, B., Muller, B., von Buxberg, Y., Cox, E. C. and Bonhoeffer, F. (1990). Biochemical characterisation of a putative axonal guidance molecule of the chick visual system. Neuron **5:** 735-743.

Takeichi, M. (1991). Cadherin cell adhesion receptors as a morphogenetic regulator. Science **251:** 1451-1455.

Walsh, F. S. and Doherty, P. (1991). Structure and function of the gene for neural cell adhesion molecule. Seminars in Neuroscience-in press.

Zinn, K., McAllister, L. and Gooman, C. S. (1988). Sequence analysis and neuronal expression of fasciclin I in grasshopper and drosophila. Cell **53:** 577-587.

Made in the USA
Monee, IL
13 September 2026

62977000R10177